Jamal Chaouki, Rahmat Sotudeh-Gharabagh (Eds.)

Scale-Up Processes

Also of Interest

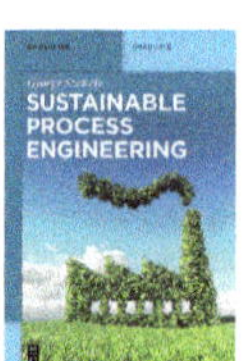

Sustainable Process Engineering
Gyorgy Szekely, 2021
ISBN 978-3-11-071712-9, e-ISBN 978-3-11-071713-6

Process Intensification.
Breakthrough in Design, Industrial Innovation
Practices, and Education
Jan Harmsen and Maarten Verkerk, 2020
ISBN 978-3-11-065734-0, e-ISBN 978-3-11-065735-7

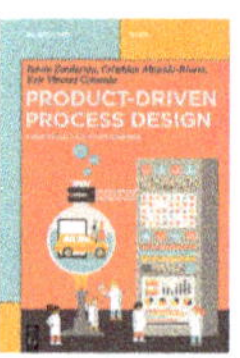

Product-Driven Process Design.
From Molecule to Enterprise
Edwin Zondervan, Cristhian Almeida-Rivera,
Kyle Vincent Camarda, 2020
ISBN 978-3-11-057011-3, e-ISBN 978-3-11-057013-7

Process Engineering.
Addressing the Gap between Study and Chemical Industry
2nd Edition
Michael Kleiber, 2020
ISBN 978-3-11-065764-7, e-ISBN 978-3-11-065768-5

Basic Process Engineering Control
2nd Edition
Paul Serban Agachi, Mircea Vasile Cristea,
Emmanuel Pax Makhura, 2020
ISBN 978-3-11-064789-1, e-ISBN 978-3-11-064793-8

Scale-Up Processes

Iterative Methods for the Chemical, Mineral
and Biological Industries

Edited by
Jamal Chaouki and Rahmat Sotudeh-Gharebagh

DE GRUYTER

Editors
Prof. Jamal Chaouki
Department of Chemical Engineering
Polytechnique Montréal
C.P. 6079, Succ Centre Ville
Montréal, Québec
Canada
jamal.chaouki@polymtl.ca

Prof. Rahmat Sotudeh-Gharebagh
Department of Chemical Engineering
University of Tehran
P.O. Box 11155-4563, Tehran
Iran
sotudeh@ut.ac.ir

ISBN 978-3-11-071393-0
e-ISBN (PDF) 978-3-11-071398-5
e-ISBN (EPUB) 978-3-11-071413-5

Library of Congress Control Number: 2021938130

Bibliographic information published by the Deutsche Nationalbibliothek
The Deutsche Nationalbibliothek lists this publication in the Deutsche Nationalbibliografie; detailed bibliographic data are available on the Internet at http://dnb.dnb.de.

Cover image: Eric Middelkoop/Chemical Installation/Shutterstock 15487942
Typesetting: Integra Software Services Pvt. Ltd.
Printing and binding: CPI books GmbH, Leck

www.degruyter.com

When a human being dies, nothing lives on after him/her except three things: a continuing charity, knowledge he/she imparted that can benefit others, or a pious child, who prays for him/her.

To our honored teachers, whose pens have been devoted to the welfare of others

To our families, for their unconditional support, inspiration, and understanding

Preface

Most industrial processes fail at the early stages of operation, yet limited information is available to design engineers regarding the failure modes of industrial processes due to scale-up. Current scale-up practices, namely, the conventional scale-up method, begin with laboratory-scale experiments and then continue with piloting and commercialization. The degree of failure, however, is not well reported in the literature. Piloting is time-consuming and expensive, but it is widely believed that an integrated or partial pilot plant is essential for process extrapolation during scale-up. With a pilot plant, it is possible to evaluate new processes, generate data or correlations, or make sample quantities of products for market development. Normally, in universities we teach scale-up elements that work well on paper, but too little time and space is spent on showing students and designers why the industrial processes are failing on their first attempt. Teaching the unsuccessful cases alongside novel scale-up techniques would allow students to study the present issues facing scale-up processes.

The conventional scale-up method has been the core of certain industrial successes when the size of the pilot plant units was large enough to allow for appropriate extrapolation using "a rule of three" with some failures. However, the limitation of monetary resources forced engineers to substantially decrease the size of the units, thus creating a knowledge gap between the pilot plant units and the industrial scale one leading to process failure. Given the current digitalization outlook and abundant access to computational resources, a fresh look at scale-up is merited. For successful extrapolation to the industrial scale, this gap must be filled by knowledge gained from the existing simulation and digitalization tools. This is consistent with the ongoing transformation of traditional practices with the latest smart technologies in the fourth industrial revolution (Industry 4.0).

Reporting unsuccessful cases with the conventional scale-up method would improve documentation in chemical engineering and it would help significantly to share knowledge with educators and expand process scale-up standards. Furthermore, this information could enhance the performance of industrial process simulators. In general, students learn more lessons from unsuccessful cases. It is also very beneficial for industry and academia since it could lead to training highly qualified personnel for the industry and minimize the costs related to scale-up errors. Integrating the unsuccessful experiences into the classroom would help students better use existing knowledge and benefit from experiences gained in the industry.

The process design concepts and guidelines provided in the conventional scale-up method are clearly important, and several excellent books and references are available to address the issue and will be cited in the book where applicable. However, the primary concern of this book is to provide a more comprehensive treatment of failures through an iterative scale-up method. In the iterative scale-up, one directly moves from a laboratory scale to an industrial scale. Depending on the conditions, coming back to either the laboratory scale, the *pilot scale*, *cold scale*, or *process*

https://doi.org/10.1515/9783110713985-202

modeling scale to represent the digital version of the plant is essential for the technical and economical risk identification and reduction, which improve the key characteristics of the industrial-scale design. This happens through successive back-and-forth iterations until satisfactory conditions are met in terms of uncertainties clearance for the whole process. Furthermore, the piloting costs can be largely reduced by possibly outsourcing the pilot experimentation or, if needed, a proper down-scaled pilot can be designed and built. The new pilot, which will focus on the main areas of concern, would be less costly and more suitable for answering the questions and concerns raised during the design.

The new scale-up method requires to a large extent better use of laboratory experiments, a down-scaled pilot, process modeling and simulation (e-pilot), and process feasibility tools. This idea forms the core of the book. The contents of this book can easily be adapted to design a course, and additional workshops can be developed for students to show why chemical processes fail to a large extent. Furthermore, the book demonstrates how the iterative scale-up can improve process extrapolation and reduce costs and timing in the chain of process design and patenting. The iterative method would create a paradigm shift in process scale-up since companies may be more selective in the future about choosing processes to pilot as resources become scarcer. This book will not cover all materials in scale-up; however, its main aim is to provide more details on iterative scale-up, which replaces or complements the conventional scale-up method, provide examples where design errors were made, and present a clear methodology and solution to minimize these errors. The book is divided into three integrated parts. Part 1 covers the basics and concepts of iterative scale-up. In Part 2, enabling tools and technique are presented, which could be helpful in developing case studies. In Part 3, six complete case studies and three open cases on chemical and biochemical industries are introduced to some extent to show the applicability of iterative scale-up in process exploitation.

The book is primarily intended to serve students, those starting their career, university professors, and practitioners, especially in the process, chemical, petrochemical, biochemical, mechanical, mining, and metallurgical industries. However, other engineers, consultants, technicians, and scientists concerned with various aspects of industrial design and scale-up may also find it useful. It can be considered as a complementary book to process design for senior and graduate students as well as a hands-on document for engineers at the entry level and practitioners. The content of this book can also be taught in intensive workshops to environmental process industries. The book is intended to:

1. review the literature on scale-up failures by providing industrial examples;
2. define uncertainties and constraints facing the conventional scale-up method;
3. introduce the iterative scale-up method;
4. show the latest concepts, tools, and methodology used for process scale-up;
5. highlight the necessity of down-scaled piloting;

6. explore the importance of process modeling and simulation in process extrapolation; and
7. provide examples with the iterative scale-up method.

The content of this book has formed over many years through the concentrated research and industrial efforts of the editors and authors, primarily the first editor. We gratefully acknowledge the authors of various chapters of the book for their contribution in preparing the examples and case studies. Some of our graduate students and industrial colleagues have greatly contributed to some important results presented in this book and we express our gratitude to all. Finally, we should emphasize that much remains to be done in this area, and the application of the iterative scale-up method is expected to increase in the future.

Continuation of the work

As mentioned throughout the book, the iterative scale-up method is an innovative, dynamic, and rather challenging technology development activity. We will continue our nonstop efforts through workshops and collective work with the help of engineers, practitioners, and scientists to bring the concept into full practice and produce more case studies and results in the coming years. This will greatly expedite the process design transition to the digital age and enrich digital twins that make process scale-up attractive and less expensive, shortening the commercialization process at the same time. You can always reach us using the following forums to share your thoughts, suggestions, or feedback or obtain updates on the continuous progress.

LinkedIn: https://www.linkedin.com/company/iterative-scale-up
https://www.linkedin.com/in/sotudeh
Website: https://pearl.polymtl.ca/iterative-scale-up

April 2021

Jamal Chaouki (jamal.chaouki@polymtl.ca)
Polytechnique de Montréal, Canada
Rahmat Sotudeh-Gharebagh (sotudeh@ut.ac.ir)
University of Tehran, Iran

About the Editors

Dr. Jamal Chaouki is currently a full professor of chemical engineering at Polytechnique (P.O. Box 6079, Station Centre-Ville, Montréal, Canada, email: jamal.chaouki@polymtl.ca). He has been teaching process design and reaction engineering courses for over 30 years. His positions include: director of the Bio-refinery Center and technical program director of the 8th World Congress of Chemical Engineering, process and technology consultant to several large companies in Canada, Morocco, Europe, and the United States. His major research interests include the application of fundamental approaches to problems of industrial relevance in process design and fluidization. He holds a B.Eng. degree in chemical engineering from France's ENSIC de Nancy, plus an M.S. and a Ph.D. in reaction engineering from Canada's Polytechnique de Montréal. He has also been a visiting scientist at TOTAL. Professor Chaouki has 40 patented technologies generating significant royalties, has authored 20 books and book chapters, and more than 850 publications in major international journals, conferences, and technical reports. He has also mentored 60 postdoctoral fellows, and 100 doctorate and master students. He is the founder and Editor-In-Chief of Chemical Product and Process Modeling published by Walter de Gruyter GmbH, Germany, and a member of the Canadian Academy of Engineering.

Dr. Rahmat Sotudeh-Gharebagh is currently a full professor of chemical engineering at the University of Tehran (P.O. Box 11155–4563, Iran; email: sotudeh@ut.ac.ir). He teaches process modeling and simulation, transport phenomena, plant design, and economics and soft skills. His research interests include computer-aided process design and simulation, fluidization, and engineering education. He holds a B.Eng. degree in chemical engineering from Iran's Sharif University of Technology, plus a M.Sc. and a Ph.D. in fluidization engineering from Canada's Polytechnique. He has been an invited professor at Qatar University and Polytechnique de Montréal. Professor Sotudeh has more than 300 publications in major international journals and conferences, plus 4 books and 47 book chapters. He is the cofounder and Editor-In-Chief of *Chemical Product and Process Modeling* published by Walter de Gruyter GmbH, Germany, a member of the Iranian Elite Foundation, and an expert vitness on the oil industry with the Iranian Expert Vitness Organization.

https://doi.org/10.1515/9783110713985-203

Contents

Rahmat Sotudeh-Gharebagh, Jamal Chaouki

1 Conventional scale-up method: challenges and opportunities

Abstract: This chapter covers the application of scale-up in chemical process industries, which has been the core of certain industrial successes where the size of the pilot plants was large enough to allow for an appropriate extrapolation of the processes. The design and operation of the pilot plant are therefore central to its success. Nonetheless, the limitation of monetary resources has forced companies to substantially decrease the size of pilots and that has created a knowledge gap between these units and the industrial-scale units leading to process failure. There are, however, some challenges and opportunities surrounding the conventional scale-up. Statistical data reported in this chapter show the importance of taking a fresh look at pilot plants and ways to replace them for future scale-ups. The proper design and use of scaled-down pilot plants, process simulation, and a return to lab experiments would be considered alternative solutions to pilot plants in the path of conventional scale-up. With the information reported in this chapter and the power of information technology, it seems, in some cases, we can avoid piloting before design, and other tools can be employed for the benefit of scale-up to a large extent.

Keywords: scale-up, pilot plant, start-up index, failure, extrapolation, scaled-down pilot, process simulation

1.1 Introduction

The history of scale-up goes back to canned food consumption in the British Royal Navy in 1845 [1]. Sometime later, an outbreak of food poisoning was reported due to the use of larger cans where the heating was not sufficient in killing the bacteria in the middle of the cans. Using a conventional definition, scale-up in chemical engineering means the migration of a process from the laboratory scale to the pilot or industrial scale through similarities, process design, and scaling rules. Several excellent books and documentation are available in the literature to address the chain of the scale-up process systematically, to present various techniques for obtaining scaling rules, dimensionless groups and relations on fluid flow, heat and mass transfer, and chemical

Rahmat Sotudeh-Gharebagh, School of Chemical Engineering, College of Engineering, University of Tehran, P.O. Box 11155-4563, Tehran, Iran, e-mail: sotudeh@ut.ac.ir
Jamal Chaouki, Department of Chemical Engineering, Polytechnique Montréal, 2500 Chemin de Polytechnique, Montréal, Québec, H3T 1J4, Canada, e-mail: jamal.chaouki@polymtl.ca

https://doi.org/10.1515/9783110713985-001

reactors, and to provide typical parameters for design and geometric scale-up [2–11]. In some literature, scaling rules are classified as proportional similarities, for example, geometric similarity with dimensions, mechanical similarity with material deformations, thermal similarity with temperatures, and chemical similarity with concentrations [12]. Perry's *Chemical Engineering Handbook* provides a detailed analysis of dimensionless groups, scaling law, and relationships and could be very instrumental in scale-up [13]. One must be careful about the dependent or independent parameters to be scaled, for example, kinetics can be directly extrapolated to other types of reactors because it depends only on the nature of the support and active phases of the catalyst [10], while hydrodynamics, heat transfer, fluid flow, and so on can be dependent on the scale.

The chemical process design has progressed rapidly since Ernest E. Ludwig first published his famous three-volume series on *Applied Chemical Process Design for Chemical and Petrochemical Plants* some 60 years ago followed by two further revisions some 20 years later. These provide design procedures on suitable equipment to practicing engineers for selecting an application. With Ludwig's passing, Coker has continued the work with a significantly expanded and thoroughly updated version of Ludwig's works and he developed programs and made enormous contributions [14–16]. Various equations, correlations, and scaling laws are reported in the books and are of great importance for the designer. The conventional scale-up method has therefore been the core of certain industrial successes in building large plants in the chemical process industry (CPI) and various failures during the last decade helped improve the design enormously. However, in successful scale-ups, the sizes of the pilot units were large enough to allow for an appropriate extrapolation; otherwise, they cannot serve confidently to demonstrate the scalability of a process.

The pilot plant experiments should provide more than a single data point. The experiments should also help generate more data on the critical effects of important parameters on scale-up. Today, the conventional scale-up method is facing two major dilemmas: first, the process designs are becoming very tight in order to minimize the expenditures of the industrial project; and second, business owners have been pushing the designer to substantially decrease the size of the pilots or replace it with an e-pilot due to the limitation of monetary resources. These two dilemmas have created a knowledge gap in scale-up and a new approach is needed to fill this gap. With the new scale-up, the use of existing correlations or scaling rules should be revised in order to expand their applicability to process design, and proper scaling rules should be proposed to be consistent with changes in the type of piloting or its replacement with other means. This chapter therefore provides an overview of the conventional scale-up method by providing some details on the basics, start-up issues, challenges, opportunities, and learning points.

1.2 Terminology

As in other engineering fields, specialized terminology is used in scale-up as defined below:

CAPital EXpenditure (CAPEX): Funds used by a company to acquire, maintain, and upgrade physical assets, such as technology, equipment, plant, property, and buildings. It is an expense a firm incurs to create a benefit in the future.

Chemical process industries (CPIs): Industries in which the feedstocks are chemically/physically converted into finished products. These broadly include: the traditional chemical industries, both organic and inorganic; the gas, petroleum, and petrochemical industries; fine chemical and pharmaceutical industries; and a series of allied industries in which chemical processing takes place.

Conventional scale-up method: Migration of a process from the laboratory to the pilot and/or industrial scale. In some literature, the term of "single-train" scale-up is also used and refers to the conventional scale-up method.

Independent project analysis (IPA): A benchmarking and consulting firm devoted to the empirical research of capital investment projects to maximize business value and project effectiveness. (https://www.ipaglobal.com)

Industrial disaster: A disaster rooted in the products or processes of an industry as a result of an accident, incompetence, or negligence. This could lead to significant damage, injury, or loss of life.

Iterative scale-up: Migration of a process from the laboratory to industrial scale, and coming back to either the laboratory scale, pilot scale, cold scale, or process modeling scale is essential to improve the key characteristics of the industrial-scale design.

OPerating EXpenditure (OPEX): Expenses required for the day-to-day functioning of a business.

Pilot scale (pilot plant): Small-scale system, which is operated to study the behavior of a process before using it on a large industrial scale. This is an important system and central for risk identification before the production takes place at the full industrial scale.

Process design: Designing and sequencing of unit operations for desired physical and/or chemical conversions/purifications of feedstocks into products. It is instrumental to the process scale-up and the summit of chemical engineering, bringing together all the relevant knowledge and components.

Process simulation: A model-based representation of industrial processes and unit operations in a software or simulator.

Process simulation scale: Digital scale system is introduced in this book to study the behavior of a process with multiscale modeling and simulation tools to iteratively improve a large industrial-scale design. This is an important scale and essential for risk identification to provide a sound and firm design at the industrial scale.

Start-up index: The ratio of actual start-up time to planned start-up time (time estimated for the scale-up) of a project.

Technology readiness levels: A method for estimating the maturity of a technology during the acquisition phase, which was developed at NASA during the 1970s and based on a scale from 0 to 9, with 9 being the most mature technology.

1.3 Basics

1.3.1 General consideration

The conventional scale-up method is carried out in a sequential manner: starting at the laboratory scale, followed by the pilot scale, and ending at the industrial scale as shown in Figure 1.1.

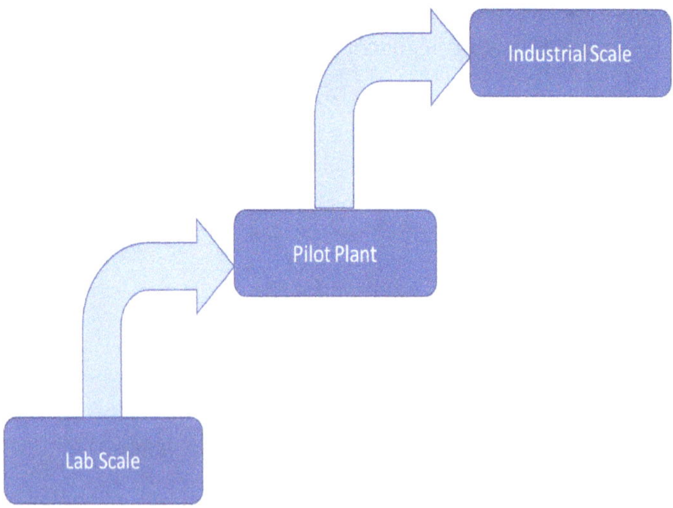

Figure 1.1: The conventional scale-up method.

Generally, during the process scale-up, design knowledge is generated to transfer the idea into a successfully commercialized scale-up. Literature reviews, consultations, laboratory experiments, piloting, modeling, and simulations can shape the repository of design knowledge and serve to identify and minimize scaling risks to acceptable levels for the successful commercial scale. Scale-up may need experimentation at two different scales, though not necessarily at the targeted scale. However, the main objectives are to identify criteria that remain constant during scale-up [17]. Building the knowledge repository would be trivial from the viewpoint of those working in the industry, but failing to do so may increase the failure of the industrial project to very high level according to the analysis of more than 1,000 industrial cases reported in the literature [18].

There are usually many risks involved when scaling up from a laboratory scale or a pilot scale to an industrial scale. Generally, one or more supplementary or intermediate experiments will be required. The problem is to identify the risks and corresponding actions needed at each stage and to acquire the required information at a minimum cost and with as little delay as possible. At least two types of intermediate experiments exist: laboratory- and pilot-scale experiments that simultaneously study the physicochemical phenomena within a process. A pilot plant can theoretically range in size from a laboratory unit to a facility approaching the size of a commercial unit [19] as shown in Figure 1.2.

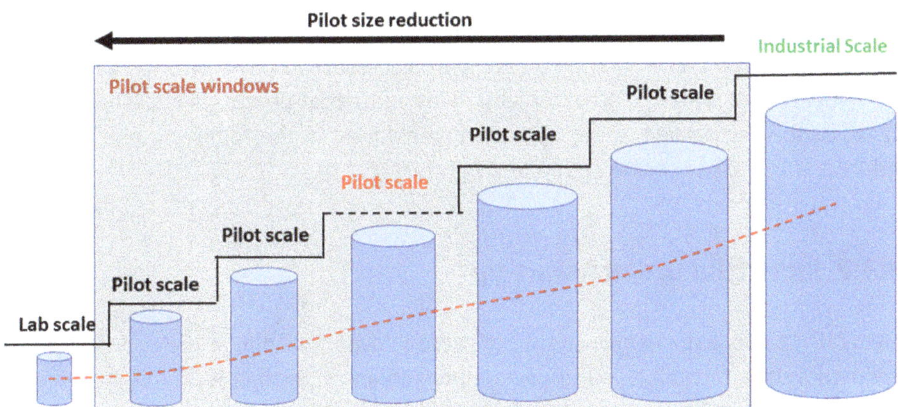

Figure 1.2: Pilot-scale windows from lab scale to industrial scale.

The proper size of a pilot plant is a trade-off between the cost and value of the obtainable data and should meet the following criteria:

- It should be as large as possible for easy extrapolation.
- It should be as small as possible to minimize the expenditures.

Pilot sizing may vary for various industries: for example, in bulk chemicals, the scale-up ratios from a pilot plant to a full capacity exceed 10,000 and the investment range is from $100 to 250 million [2]. Table 1.1 shows the typical scaling factors for the CPI for traditional gas, liquid, and biomass processes [20].

Table 1.1: Typical scaling factors for CPI.

Unit	Traditional CPIs (gas–liquids)	Biomass handing (solid)
Laboratory	0.001–0.1	0.01–0.1
Pilot plant	1	1
Industrial	10,000–30,000	1,000–5,000

Hurdles to scaling are unknown a priori and when the selected size of the pilot cannot answer all the questions, it must be completely changed, or a larger size must be built to respond to any unanswered questions. Piloting remains, however, a key point and depending on the industry, various decisions can be undertaken on the size, but the pilot plant should really be customized for the process need. If there is no firm justification to use the pilot as a design tool, this may be avoided or replaced with down-scaled pilot units or other tools.

The conventional scale-up method has many disadvantages related to the pilot step since it is the loose, uncertain, and costly element of the scale-up path. Indeed, the value of the pilot-scale test is not well established. Its size is blindly determined and one does not know what parameters are necessary to measure during the pilot-scale operation, which is possibly why many industrial processes fail [1] or are delayed at the early stage of the operation as detailed in the following sections that address start-up issues and equipment failures.

1.3.2 Industrial project start-up

Generally, there is no serious control over the relevancy of pilot-scale experiments to industrial scale for reducing the uncertainties to an acceptable level due to size and cost constraints. The majority of projects designed in this way was delayed and faced cost overruns when measured against the initial estimates by scale-up. Figure 1.3 shows the start-up index, defined as the ratio of actual to planned start-up times, for industrial-scale plants with various feedstocks. The index equal to one in this figure means that the actual start-up meets the scale-up (planned start-up time) closely. The data shown in Figure 1.3 covers 40 solid processing units owned by 35 companies in plants constructed in the United States and Canada from 1965 to 1983 [21].

In this figure, substantial deviations can be identified in the start-up performance of solid processing units compared with their liquid or gas counterparts. Solid processing units take longer times to start-up and to achieve the desired production due to the poor understanding of hydrodynamic parameters sensitive to process scale [10]. It seems that the presence of solids is always a challenge making only small scale-ups possible in contrast to fluid processes as shown in Table 1.1, where a scale-up of 1 over

10,000 is common. Scale-up is easy if there is no coupling between the chemical reaction and hydrodynamics, which might be the case in gas/liquid systems [9]. According to Bell [22], solid behavior in scale-up can be addressed through empiricism and modeling. These tools can help somehow, but insufficient to eliminate pilot plants. Bell has also reviewed some of the solid processing unit operations and highlighted relevant scale-up issues. The use of information from vendors and other third-party suppliers for scale-up was also discussed in competitive or regulated industries. However, current practice in scale-up should be improved to use the power of existing data, tools, and experimental techniques.

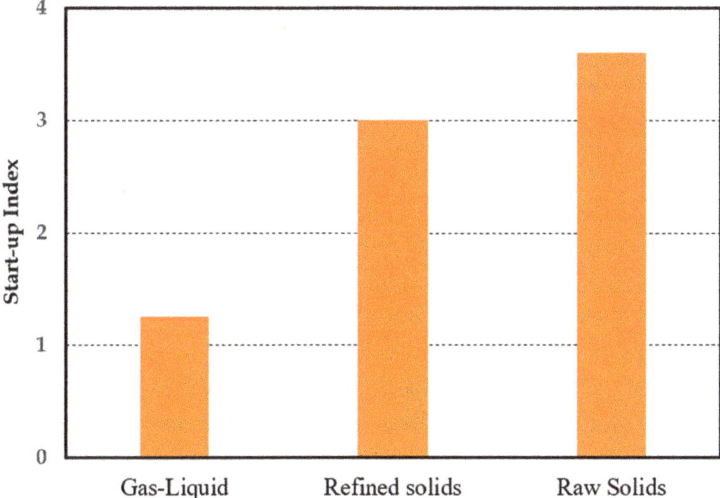

Figure 1.3: Start-up index (planned versus actual start-up times) for various feedstocks with data from [21].

Most scale-up problems found in solid processing units are related to complex hydrodynamics, and small pilot units were generally unable to address the uncertainties associated with these processes. Figure 1.4 shows the start-up index for processes with new steps ranged from simple to complex. As seen in the figure, as the number of steps increased, a big delay in the start-up time of the projects can be observed, which again points to the importance of process extrapolation. Based on the limited amount of information reported in the literature, major start-up problems are as follows: in-accurate material and energy balance, poor design of waste handling facilities, corrosion and erosion, solid processing, separation, transport, and instrumentation [21]. This would lead to under-sizing or over-sizing the main process and heat transfer equipment and delaying the start-up. It is worth mentioning that the start-up index, which is a ratio of actual to planned start-up times, is a good indication of failure if it goes far from index = 1 leading to higher costs and longer actual start-up times.

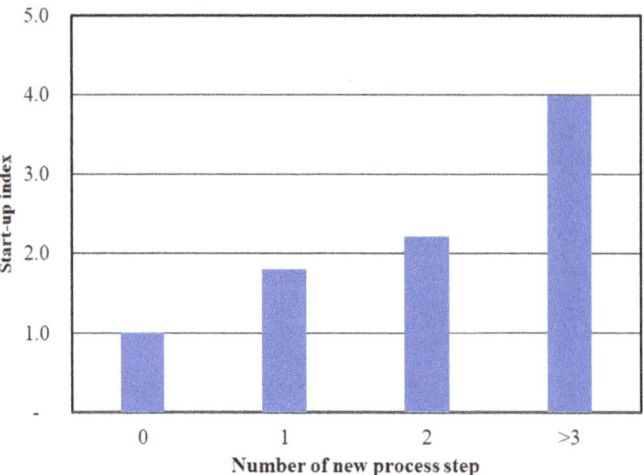

Figure 1.4: Start-up index as a function of process steps (data from [21]).

Large start-up indexes in Figures 1.3 and 1.4 show the huge start-up delay in various plants. Based on these two figures and the analysis of many industrial projects with or without a pilot plant, Merrow concluded that a successful start-up needs an integrated pilot [2] under the following conditions:

- With more than four new unit operations involved.
- With processes containing one new unit operation and a complex recycle stream.
- With processes containing a novel solid handling operation or particulate materials feedstock.

In this guideline, the new process unit operation refers the unit that has not been operated before at the industrial scale to the same process. The most common mistake is that a unit operation is considered established, due to its previous usage with other applications, but the assumption is wrong. Merrow [2] believes that if these conditions are not carefully met and a scaled-down pilot plant is not built, the project will have larger start-up time than usual. This again shows the importance of right targeting in the pilot plant construction. For other cases, however, an integrated pilot plant is not necessary, but the start-up time would be much shorter with pilot plants.

Nontechnical aspects could also play an important and critical role in start-up and scale-up. It is worth mentioning that 65% of project failures are due to softer aspects, such as people, organizations, and governance [23], and should be taken into consideration in the projects as well. Therefore, the human component is critical in the delayed start-up and not all responsibilities can be placed on piloting or not piloting. Start-up leader and personnel play a key role. According to Lager [24], a qualified start-up leader is largely responsible for 90% of the success and his/her support team should quickly analyze problems occurring during the start-up and provide the advices required.

It seems that for solving start-up issues and minimizing the risks to a large extent, the current literature is in favor of constructing a down-scaled pilot plant and using computational tools. The traditional pilot plants may be avoided with minimal risks when upgrading the process for the same product and raw materials, provided that such avoidance does not cause a financial risk to the company. Harmsen has nicely focused on industrial process scale-up and provides comprehensive details with risk identification and reduction for process equipment and integral process scale-up [2].

1.3.3 Process equipment failures and outlook

The larger start-up index shown in the previous section can be somehow related to equipment failures, which subsequently end with process failures. Statistics show that a start-up that slips behind schedule can add about US $350,000 per month to the capital costs of an average plant based on a sample of 39 solids processing plants, but the actual costs of a delayed start-up can be much higher [21]. Kleiber has also reviewed in detail the failure related to various equipment used in the process [9]. Most start-up failures originate from the failure of the individual equipment due to their unknown behaviors rather than the process design failures only. In fact, the failure of a major piece of the equipment or the widespread failures of smaller pieces of equipment, such as pumps or valves, can greatly affect the start-up in terms of time and costs. The most frequent equipment failures and their percentage contributions are shown in the following figure [21].

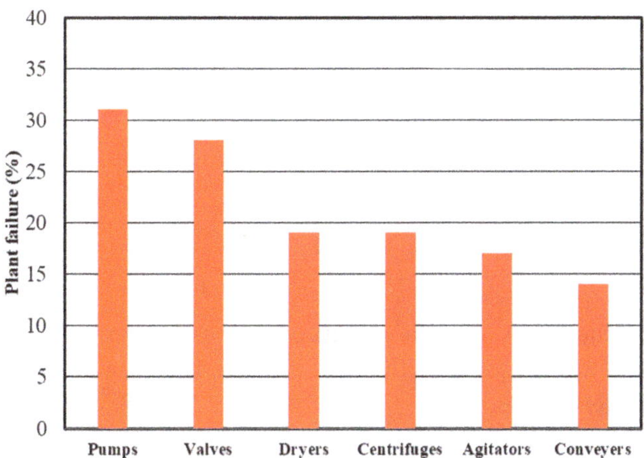

Figure 1.5: Process equipment failures with data from [21].

It is important to mention that sealing, incomplete shut off, changes in feed or material types, and handling different materials than those originally intended are the main sources of failures. Total percentage failure in the data was more than 100% as several plants had more than one type of equipment failure [2]. Other instruments, for example, flow meters, samplers for slurries, or pH meters, were also sources of failures in some cases, but these were not the major delaying factors creating fewer problems than expected [21].

It seems that by minimizing the use of equipment responsible for failure as mentioned in Figure 1.5, we may succeed in minimizing failures. Process intensification can be used as a tool for this purpose. The integration should start by identifying essential functions that should be merged into one single unit operation by changing the process conditions so that they fit into the same operating window. This would minimize the use of small equipment units. The most common functions in CPIs include mixing, transport phenomena, chemical reactions, and separation. The distillation columns, pump, and extruder are among the pieces of equipment whose functions can be combined, for example, a reactive distillation column combines reactions and separation; mixing and reactions can simultaneously occur in a pump or extruder; and dividing wall columns combines separation with heat transfer in one single unit. The process intensification can reduce the number of pieces of equipment enormously leading to lower investment, energy and maintenance costs, and a reduction of failure risks due to equipment. However, it has its own limitations since merging complex units may add to the complexity of the unit operations. It remains part of the ongoing research in academia and industry.

1.3.4 Conventional scale-up for process equipment

Major single-train conventional scale-up methods are classified into five categories as follows:
- Brute force
- Model-based
- Empirical
- Hybrid
- Dimensionless

The methods mostly follow the scheme shown in Figure 1.1. These are extensively reviewed by Harmsen [2] and we only focus on their key characteristics highlighting some of their potential and weaknesses.

1.3.4.1 Brute force

In this method, the scale-up is made through a pilot plant, where we keep all critical factors similar. In practice, this method can only be applied for a few cases due to inevitable changes in at least one or two key factors in most cases. One good example using this method is a multitube unit operation, where the number of tubes is increased while the flow rate and tube dimension are kept similar. This is a very reliable scale-up method due to minimal amount of basic design data required; however, one should evaluate the feed distribution hydrodynamics in a "cold model" or with reliable industrial data. Generally, engineers, managers, and business owners can easily understand this method.

1.3.4.2 Model based

In this method, the performance of the unit operation subjected to scale-up is predicted by a mathematical model. The model should simultaneously integrate thermophysical properties, reaction kinetics, and hydrodynamics over the unit operation. The model validation is compulsory through experimental data, originated from the pilot with multiple points, and from a "cold model," at much larger dimensions than the pilot. With the "cold model," the residence time distribution, mass and heat transfer, and mixing rate can be verified or validated with additional experiments. This scale-up method is suitable for the bulk chemical industry [2] .

1.3.4.3 Empirical

In this method, the unit operations are designed and operated with different scales, that is, bench scale, microreactor, pilot plant, and development and demonstration plant. At each scale, various performance data are measured and if needed operating parameters are adjusted for a desired performance. By plotting the experimental data versus scales, extrapolations are made to commercial scales. Similar scale-up experiences available in the company could create some degree of scale-up confidence since at various scales, a huge amount of data is generated. The statistical techniques along with the design of experiments and data analytics are also very useful in this scale-up. However, this method is quite costly, time-consuming, and not very reliable since some critical scale-up parameters may not become easily known. In general, if the brute-force and model-based approaches are not applicable to scale-up due to interplay between the hydrodynamics and reactions, this may be considered as an alternative [2].

1.3.4.4 Hybrid

This method combines the empirical and model-based scale-up, and this allows the chances of success to be highly increased. The simulation is used to predict the performance of the unit operation and the validation is made with the experimental data obtained at various scales, that is, laboratory and pilot scales. If the differences are significant, the model can be adjusted or fine-tuned with part of the data and the rest of the data can be used for validation. This method is used when hydrodynamic uncertainties surround the scale-up.

1.3.4.5 Dimensionless

In this method, dimensionless numbers are defined for all critical parameters at the test scale and their values are kept the same at the industrial scale. This method has not been used in the industry since one cannot correctly capture all critical parameters in the dimensionless numbers and, in fact, there is no validation technique to ensure that nothing is missed. Generally, this scale-up method, proposed ~60 years ago, is nearly impossible to communicate to stakeholders with no chemical engineering knowledge. Even chemical engineers and mid-managers do not endorse this method due to the lack of verification and validation. Zlokarnik provided a detailed description of the method in his book [4].

1.4 Challenges

The main challenge with the conventional scale-up method is related to its middle chain or pilot plant. As mentioned earlier, there are significant uncertainties in the size of the pilot plant and, in most cases, the objective of designing and building the pilot plant before the design of full-scale unit is not clear from the scale-up point of view. The pilot plant may serve various objectives, including marketing purposes. However, based on the details presented in this chapter, it seems that in most cases, the pilot plant can be skipped before the design and, as reported in the literature, a scaled-down pilot plant seems very adequate for the scale-up purpose, for example, in the CPI, the most common source of design errors are the result of decisions not to use a scaled-down pilot for a new process or an old process with a new feedstock [18]. These decisions however should be taken after the design. Therefore, in most cases, the pilot plant on a conventional scale does not provide satisfactory answers to the uncertainties if the sizes are too different from the industrial scale. Unfortunately, the limitation of monetary resources has forced the size of pilot plants to be substantially decreased and that has created a knowledge gap between these units and the industrial-scale units leading to process failure at ridiculous rates. The current digitalization outlook and abundant access to computational resources merit a fresh look at scale-up, and therefore, for successful extrapolation to the industrial scale, this gap must be filled by knowledge gained from the existing simulation and digitalization tools. This is consistent with the ongoing transformation of traditional practices with the latest smart technologies in the fourth industrial revolution (Industry 4.0). Analysis of a large number of projects by IPA in 1995 has showed that fully integrated pilots are essential whenever the process is complex or contains heterogeneous or particulate solids [18]. The construction of a fully integrated pilot, which normally happens after the initial design and includes all processing steps in the same fashion as the commercial design at a proper size, is time-consuming and expensive. Therefore, the right choice of the pilot plant is very important and should be targeted very well since it may account for up to 25% of the cost of the commercial-scale units [18]. Such a huge cost for the pilot plant means that the process is not economical. Some pilot facilities are still running as execution begins [18]. This is also the most serious killer of the project.

1.5 Opportunities

1.5.1 Plug-and-play look to scale-up

In an analogy to the plug-and-play approach in information technology, some people are strong believers of the modular design, which could dramatically change the working environment for the process industry, if successful. This approach may massively impact the availability, constructability, maintenance, start-up time, downtime, and Health, Safety and Environment (HSE) in the process industry. It is a dream now, but with a tremendous effort, this could be the solution in the near future through equipment and process intensification [25]. Of course, we can see signs of progress, for example, microreactors are also used in fine chemical and pharmaceutical industries. These reactors are easily adapted to laboratory scales as they require small quantities of feed and claim to reduce the time and costs required for new process development. Their scale-up is rather easy due to a lack of heat transfer limitations, and it can be performed by increasing the number of units using the same dimensions and input data. The approach is useful for the design of new processes by validating the experimental results with simulation data as it provides laboratory, pilot, and industrial scales, all in one intensified unit. Fink and Hampe [26] reported that this technology downsizes the miniplant by a factor of ~100 and reduces the feed down to 10–50 mL/h. Therefore, the CAPEX and OPEX are substantially reduced and the minimal species hold-up minimizes safety and security risks. However, although some elements of microplants, like microreactors, have been available now for decades, they are still far from ready for industrial applications. This is probably due to concerns about their operability, which is why not this technology should not be an option. More time is still needed for their acceptance in the CPI.

1.5.2 Fresh look to scale-up

The IPA[1] produces an annual benchmarking of projects, including those in CPIs, completed by their member companies. The company has a database of more than 10,000 projects. Their analyses show that since 1993 more than 50% of the projects have been disasters. A disaster is defined when more than 30% of cost growth and/or more than 38% of schedule slippage occurs. At the top of the list, it mentions that less than 39% of projects were successful during the first year of operation [25]. The failure in big or megaprojects is also significant. These are the projects with budgets exceeding US $1 billion each [27].

Ernst and Young evaluated 365 megaprojects from the oil and gas industry with CAPEX over US $1 billion each in upstream, LNG, pipelines, and refining and demonstrated that 64% of the projects were facing cost overruns while 73% were reporting a time delay [23]. These findings are in line with observations made by IPA-2011 industry study, which showed that 78% of upstream megaprojects challenged cost overruns or delays, a significant decline from 2003, where 50% of the projects were late or significantly over-budgeted [23]. They have mentioned that the failure to appropriately consider design along with other issues is considered a key challenge in the projects with a detrimental effect in the subsequent project development. Therefore, the conventional scale-up method would be partly responsible for these repeated failures due to the need to develop accurate and compact designs at minimal cost. Merrow concluded that after 2002, nonoil and gas development projects maintained a success rate of 50%, while the success of oil and gas megaprojects decreased to 22%. Of the 78% project failures, the average cost overrun was 33% and schedule slip was 30% [27]. Merrow then highlighted three major factors related to the high megaproject failure rate. They were the effectiveness of the project development and scale-up, leadership, and the tendency of businesses to adopt aggressive schedules.

Based on the *AIChE* survey from various companies [28], it is not obvious whether companies would prefer not to use pilots at all or rely instead on process simulation models developed from lab-based experimentation. Companies may, however, be more selective in the future on choosing processes to pilot as resources become scarcer. Statistics showing tremendous failure rates for industrial projects, scarcer monetary resources, and the excellent computational facilities available today all indicate that scale-up merits a fresh look for the successful extrapolation of ideas to the industrial

1 https://www.ipaglobal.com

scale and this is the reason why a new book with a promising approach is needed for industrial scale-up through the iterative approach. It is also important to mention that modular design is also the choice to combine the laboratory, pilot, and industrial scales into one integrated scale, but it is too far from industrial realization in near future.

1.6 Learning points

This section provides selected scale-up failures and learning points with examples. It is important to mention that scale-up can be complex in even simplest situations [17], and all details and simple issues should be carefully examined as detailed below:

- **Feed distribution:** In multiphase flow systems, feed distribution is of prime importance upon moving from pilot to industrial scale. Based on the experience reported by Harmsen, the industrial-scale trickle-bed reactor showed a very poor selectivity as compared with the pilot plant. The trickle-bed reactor uses the downward movement of a liquid and the co-current or counter-current movement of gas through a fixed bed of catalyst. This is a simple reactor type for catalytic reactions in CPI. The technology provider was unable to identify the cause of the poor selectivity and paid the penalty for that. Reviewing the distributor design showed that maldistribution in the liquid phase occurred upon start-up. By installing the correct distributor, the selectivity target was met [2].
- **Runaway reactions:** There is a common root cause associated with most chemical incidents due to a lack of required knowledge either at the laboratory, pilot plant, or industrial plant facilities, for example, the United Kingdom Gas Board has decided to label the gas in one of their major pipelines with sulfur hexafluoride to detect leaks [1]. Unfortunately, operators did not know that a mixture of methane and sulfur hexafluoride explodes as violently as a mixture of methane and oxygen. Fortunately, the problem was identified in time and the plan was abandoned. Such problems have been reported in the literature [1] in multiple instances. Knowing the property of materials and chemical reactions during the knowledge buildup would significantly help the scale-up. In another accident, a runaway reaction occurred on a continuous plant where the design temperature of the reactor was 200 °C but it reached 600 °C [29]. This was due to poor documentation during the design and an unreliable technical handover between managers.
- **Recycle stream:** Due to several issues, the recycle stream must be included in the industrial scale, although it is as important in the laboratory or pilot plants. A recycle stream may contain a contaminant leading to a runaway reaction at high concentrations. In one incident, the recycle stream was routinely analyzed until managerial changes occur. Unfortunately, a few months later, an explosion occurred [1]. Analysis showed that the operation of pilot plant does not provide

reliable knowledge to address the impact of impurities or inert buildup in industrial plants within the recycle stream [3].

- **Miscommunications:** Communication or soft skills may also play a key role in the scale-up process, for example, the use of the term "scale-up" may create some confusion since in various industries engineers might think differently. In two separate meetings with pharmaceutical and petrochemical industries, Harmsen reported that the petrochemical company considered the scale-up a huge task taking many years, while in the pharmaceutical companies, it is considered a small task [2]. From the petrochemical company's point of view, scale-up includes full EPC[2] of the pilot plant, but for the pharmaceutical industry, the exiting and multipurpose pilot plant, which is a mini-plant, is used with no concern for passing the tedious pilot plant design and construction. Therefore, one single term may have different meanings and before making any decisions, a common technical language should be used.
- **New equipment:** Procuring new equipment or processes from providers should be verified to ascertain whether the providers have the required knowledge on scale-up and design, for example, the Shell™ GTL[3] plant in Malaysia, with an investment of US $20–21 billion and 52,000 workers, experienced corrosion and sealing problems due to the use of many prototype pieces of noncore equipment [30]. It is important to mention that most engineering companies rely on the supplier's know-how regarding equipment, which is mostly based on in-house experiences. However, this approach may or may not be the right path for the project under consideration [3].
- **Equipment:** Most start-up failures originate from widespread failures of smaller pieces of equipment, such as pumps or valves, as shown in Table 1.1 [21]. A change in design, such as a jacketed vessel to a pump-around loop, would be an improper choice in scale-up and may lead to failure [3].
- **Procedures:** These have tremendous importance on the success of the scale-up, for example, the recipe developed by a researcher in the laboratory was successfully applied in a large-scale fermenter, but it failed in another application. By sterilizing the fermenter with a live steam injection instead of using chemicals, the ammonia was probably stripped from the medium and the problem was solved [2]. In another example, sudden mixing caused a runaway reaction due to agitator failure [29]. An aromatic hydrocarbon, 5-*tert*-butyl-*m*-xylene, was being nitrated in a 2 m^3 vessel by the slow addition of a mixture of nitric and sulfuric acids. After several hours, the operator noticed that the temperature had not risen as usual. The agitator had stopped, so he switched it on again, but wisely decided to leave the area. Five minutes later, the vessel exploded, and the contents ignited. In

2 Engineering, procurement, and construction (EPC)
3 Gas to liquids (GTL)

such conditions, it was possible to prevent the incident by installing a trip so that the agitator failure isolates the supply of incoming liquid.

- **Scale-up:** It seems that the most fundamental lessons learned from the Shell™ GTL plant in Malaysia relate to scaling from a pilot plant to industrial scale. Besides the problems with the selectivity and stability of the catalyst, the reactor scale-up from less than ten tubes in the pilot scale to a multitubular fixed-bed reactor of tens of thousands of tubes in the industrial-scale reactors was a big challenge. It seems the pilot they used was not too useful during the scale-up and this again shows that the piloting is not very well targeted in most industrial applications [31]. In another case, Stevens reported that a design flaw in the plant flare stack in Australia costs the project an additional $900 million and a 6-month delay before the plant came on stream [32].
- **Short test durations:** In laboratory-scale experiment and probably in pilot plants, the short test duration does allow the effect of chemical reactivity in corrosion, scaling, and the component buildup to be easily visible. Therefore, ongoing research and development activities are a must during the scale-up to minimize uncertainties related to invisible conditions.
- **Unrecognized changes:** A lack of knowledge among operators and personnel may create enormous problems on scale-up, for example, in a gelatin plant, a sudden fall in the quality of the product was reported [1]. Two consultants were asked to investigate the problem. They went to the foreman of the plant and asked if anything had recently changed, and apparently nothing had changed. However, one of the consultants observed a rusty vessel basket and questioned its purpose. The foreman replied that a basket full of hydrogen peroxide was added to each batch, but since the basket was rusty, he had replaced it with a new one. Although the linear dimension of new basket was only 25% big in size, the foreman did not realize the volume doubling. This finding, which seems quite simple, was the cause and upon correcting it, the problem was solved. Thermal characteristics, such as heat transfer issues due to the reduced ratio of area to volume, are also reported as the source of failure [3].
- **Operation mode:** Uncontrolled modification in scale-up may create serious problems, for example, in a process scale-up from the laboratory to commercial scale, a company has changed its operation mode from semibatch to batch. About 20% of the batches showed temperature excursions based on the given instructions, but the operators were able to manage it. The company then decided to increase the size of the reactor from 5 to 10 m^3 and reactants input by 9%. The proportion of batches with temperature excursions rose to half, and ultimately the operators were unable to manage the batch. The reactor manhole cover was blown off, and the ejected material ignited leaving nine injuries [33]. A change from a batch- to a continuous-flow reactor leading to serious failures was also reported in the literature [3].

- **Validation of simulation results:** The use of the process simulator may mislead the engineer. Any simulation model should be properly validated before it can be used for the design process due to component properties, thermophysical and transport property models, equipment hydrodynamics, and reaction complexity. After passing the validation stage, the simulation results can be used for scale-up.
- **Safety considerations:** An increase in the equipment size may lead to failure as mentioned in the GTL case study earlier. In industrial equipment, storage of a larger volume of chemicals may lead to a temperature runaway [13], a significant buildup of impurities due to unknown reactions, and other things may increase the possibility of a catastrophic explosion or fire [3], for example, in a fertilizer plant in Texas, a routine fire turned into a disaster by diffusing to ammonium nitrate storage tanks leading to explosion
- **Regulatory:** This includes raw materials, process, products, and by-product disposal [3].
- **CAPEX–OPEX analysis:** An economic analysis must be conducted to ensure that a scale-up project is really of value. The development of this analysis at an early stage will avoid the risk of going ahead with a financially nonattractive project. Normally, with a minimal amount of bench-scale data and engineering common sense and shortcut methods, values of CAPEX and OPEX can be determined with sufficient accuracy [3] as briefed in Chapter 2.

Knowledge of past accidents, failures, errors, and disasters could help the scale-up engineers and scientists to put more energy on understanding the effects of chemical properties, impurities, thermodynamics, chemical reactions, mixing, and the human aspects. It is also worth mentioning that every failure is a learning experience, adding to our knowledge base and allowing us to perform our job better. All these issues may affect the scale-up in one way or another. In the sixth edition of the book titled, *What Went Wrong? Case Histories of Process Plant Disasters and How They Could Have Been Avoided*," the authors [1] have nicely described many of the plant accidents and their root causes. These accidents happened because of various managerial changes, conflicts, and failures: failure to follow the procedures by engineers, employees, and mid- and top managers, failure to consider previous violations, and, of course, the lack of knowledge gathered during the scale-up. An analysis of 1,000 major incidents in CPI shows that only ~20% of the most frequent problems are observed in 70% of the incidents in total [1]. It is worth mentioning that most of the incidents are very simple and can be avoided with only the previous knowledge reported in the literature. Reading this book and similar books is highly recommended to engineers and scientists involved in the scale-up process and would help in the identification and minimization of risks during process development and extrapolation.

1.7 Conclusion

The work of IPA has illustrated that the current scale-up process can easily deal with gas and liquid processing units while serious challenges face in the particulate processes. The use of pilot plants with uncertain sizes may increase perceived costs and seemingly slow the scale-up process before industrial design as they increase costs and time investment. It is true that the value of pilot plant is often not appreciated until a failure has occurred, but companies may be more selective in the future on choosing processes to pilot as resources become scarcer. There are plentiful evidences that solid processing units are more difficult to scale-up since they are tedious to characterize or fully understood. On the other hand, process interactions with various internals within unit operations, process equipment, and instrumentations can be irregular. Excessive confidence has been placed on suppliers or third-party design firms without clear procedure to check whether they meet the standards and regulatory specifications. Mathematical modeling has high potential and would be applied for many problems along with other tools. The current practice in scale-up is failing to minimize the gap between scale-up and start-up and it only provides incremental progress on some fronts. However, the path forward is clear, and it seems that the scale-up efforts should continue on many fronts using mathematical simulation, down-scaled pilot plants, re-visiting laboratory data, and the continuous updating of a design knowledge repository for the identification and minimization of design risks to an acceptable level.

References

[1] T. Kletz and P. Amyotte, What went wrong? Case histories of process plant disasters and how they could have been avoided, Sixth edn, Cambridge: Elsevier, 2019.
[2] J. Harmsen, Industrial process scale-up; a practical innovation guide from idea to commercial implementation, Second edn, Cambridge: Elsevier, 2019.
[3] J. M. Bonem, Chemical projects scale up: how to go from laboratory to commercial, Cambridge: Elsevier Inc, 2018.
[4] M. Zlokarnik, Scale-up in chemical engineering, Wiley-VCH, 2006.
[5] S. M. Hall, Rules of thumb for chemical engineers, 6th edn, Elsevier Inc, 2018.
[6] J. Worstell, Scaling chemical processes: practical guides in chemical engineering, Butterworth-Heinemann, 2016.
[7] J. Shertok, The art of scale-up: the successful chemical practitioner's guide to creating a profitable process, Red Bank: Absolutely Abby Press, 2018.
[8] J. R. Backhurst and J. H. Harker, Process plant design, Heinemann Educational Books, 1973.
[9] M. Kleiber, Process engineering; addressing the gap between studies and chemical industry, Walter de Gruyter GmbH, 2016.
[10] J.-P. Dal Pont, Process engineering and industrial management, London: ISTE Ltd, 2012.
[11] G. Malhotra, Chemical process simplification: improving productivity and sustainability, John Wiley & Sons, Inc, 2011.

[12] J. M. Lorenzo and. F.. J. Barba, Eds., Scaling-up Processes: Patents and Commercial Applications, in Advances in food and nutrition research, Vol. 92, María-José Ruiza Francisco J. Martí-Quijala†, Elsevier Inc., 187–223, 2020.
[13] D. W. Green and M. Z. Southard, Perry's chemical engineers' handbook, McGraw-Hill Education, 2018.
[14] A. K. Coker, Ludwig's applied process design for chemical and petrochemical plants, 4th edn, Vol. 1, Gulf Professional Publishing, 2007.
[15] A. K. Coker, Ludwig's applied process design for chemical and petrochemical plants, 4th edn, Vol. 2, Gulf Professional Publishing, 2010.
[16] A. K. Coker, Ludwig's applied process design for chemical and petrochemical plants, 4th edn, Vol. 3, Gulf Professional Publishing, 2013.
[17] J. P. Clark, Practical design, construction and operation of food facilities, Academic Press, 2009.
[18] E. W. Merrow, Industrial megaprojects, concepts, strategies, and practices, New Jersey: John Wiley & Sons, Inc, 2011.
[19] R. P. Palluzi, Pilot Plants, in ULLMANN'S encyclopedia of industrial chemistry, Wiley-VCH Verlag GmbH, 1–10, 2014.
[20] Biofuels commercialisation, November 2012. [Online]. Available: http://ocw.ump.edu.my/pluginfile.php/6770/mod_resource/content/1/Relevant%20Paper.pdf. [Accessed July 2020].
[21] E. W. Merrow, Estimating startup times for solids-processing plants, Chemical Engineering, 24, 89–82, 1998.
[22] T. A. Bell, Challenges in the scale-up of particulate processes – an industrial perspective, Powder Technology, 150, 60–71, 2005.
[23] Ernst and Young, Spotlight on oil and gas megaprojects, 2014. [Online]. Available: https://www.ey.com/Publication/vwLUAssets/EY-spotlight-on-oil-and-gas-megaprojects/$FILE/EY-spotlight-on-oil-and-gas-megaprojects.pdf, Accessed on July 2020. [Accessed July 2020].
[24] T. Lager, Startup of new plants and process technology in the process industries; organizing for an extreme event, Journal of Business Chemistry, 9(1), 3–18, 2012.
[25] H. Bakker, Management of projects: a people process, May 2008. [Online]. Available: https://www.napnetwerk.nl/download/?id=2781. [Accessed July 2020].
[26] H. Fink and M. J. Hampe, Designing and Constructing Microplants, in Microreaction technology: industrial prospects, Heidelberg: Berlin, 2000.
[27] E. W. Merrow, Oil and gas industry megaprojects: our recent track record, Oil and Gas Facilities, 38–42, 2012.
[28] D. Jack, Pilot plants, 2009. [Online]. Available: https://www.chemicalprocessing.com/assets/wp_downloads/pdf/pilot-plants-special-report-ContTech.pdf. [Accessed July 2020].
[29] T. Kletz, Lessons from disaster, how organizations have no memory and accidents recur, Houston: GULF PUBLISHING COMPANY, 1993.
[30] T. Van Helvoort, R. Van Veen and M. Senden, Gas to liquids – historical development of GTL technology in Shell, Amsterdam: Shell Global Solutions, 2014.
[31] L. Carlsson and N. Fabricius, From Bintulu shell MDS to pearl GTL in Qatar, 2005. [Online]. Available: http://www.ivt.ntnu.no/ept/fag/tep4215/innhold/LNG%20Conferences/2005/SDS_TIF/050139.pdf. [Accessed July 2020].
[32] M. Richardson, D. Borg and A. Wood, The Impact of Modular Construction on Liquefied Natural Gas (LNG) Megaprojects, in Perspectives in project management: a selection of masters degree research projects, A. Wood and R. Rameezdeen, Eds., Newcastle upon Tyne, Cambridge Scholars Publishing, 291, 2018.
[33] Morton International, Inc, Chemical manufacturing incident, 1998. [Online]. Available: https://www.csb.gov/assets/1/20/morton_report.pdf?13798. [Accessed July 2020].

Jamal Chaouki, Rahmat Sotudeh-Gharebagh

2 Iterative scale-up method: concept and basics

Abstract: This chapter covers the basics, tools, and applications of the iterative scale-up method in chemical process industries. In this approach, the piloting, which is central to the conventional scale-up method, will be removed from the scale-up path as the chemical and biological process industries avoid operating uncertain pilot plants due to their minimal benefits, high operational anxieties, and rather significant costs. This leads to direct pre-design of an industrial plant based on the existing data generated from the literature, initial laboratory, bench, and small pilot plants. At this stage, uncertainties will be judiciously identified and wisely reduced to an acceptable level in various iterations through process modeling and simulation, small-scale runs and/or down-scaled piloting within the frame of the feasibility analysis. This approach to scale-up, however, remedies the shortcomings of the conventional scale-up method and fills the knowledge gap created in the wake of the pilot plant removal. The chapter includes the basics of the iterative scale-up method, tools and techniques to minimize the uncertainties and introduce the description of the case studies to be included in the book. With the information reported in chapters one and two, and the power of information technology, it seems we can easily avoid the costly, tedious, and time-consuming piloting before design with the close collaboration of the chemical process industry. The iterative scale-up method would create a paradigm shift in process design and more cases would be available in the future to support its wide applicability. We hope this scale-up can be integrated into design courses to train a new generation for its industrial implementations.

Keywords: scale-up, piloting, scaled-down pilot, process simulation, iterative, conventional

2.1 Introduction

As mentioned in Chapter 1, the chemical, mining, and biological (CMB) process industry is being forced to reckon with a rapid change in which scale-up may look dramatically different than it has in past decades. Nowadays, the conventional scale-up

Jamal Chaouki, Department of Chemical Engineering, Polytechnique Montréal, 2500 Chemin de Polytechnique, Montréal, Québec, H3T 1J4, Canada, e-mail: jamal.chaouki@polymtl.ca
Rahmat Sotudeh-Gharebagh, School of Chemical Engineering, College of Engineering, University of Tehran, P.O. Box 11155-4563, Tehran, Iran, e-mail: sotudeh@ut.ac.ir

https://doi.org/10.1515/9783110713985-002

method faces significant challenges due to ambiguity in determining the proper size of the pilot plant, its high capital expenditure (CAPEX)[1] and operating expenditure (OPEX)[2], and of course an increasing doubt in its suitability to generate confident scaling laws for minimizing the scale-up risks. Based on the editors and authors' expertise, the pilot plant should be either removed or moved to a proper place after the design and upon evaluation of other enabling tools explained in this book; if there is a need to integrate a scaled-down pilot plant or any experimental subsets, we can then consider building a proper pilot plant. This could consequently serve for various applications and if any future changes must be envisaged in the plant, we can test the idea in the integrated pilot plant and then apply the results with minimal risk to the entire plant. This is consistent with current practices in the CMB process industry where the operation of large-scale pilot plants is largely suspended, and other means are under consideration to replace them.

Very limited information is reported in the literature for industrial design with no piloting. Lonza, a Swiss chemical company, used process simulation for the entire industrial design of a dividing wall column to be employed in multipurpose production [1]. A new approach was introduced to overcome the frequently occurring convergence problems, which enhances the robustness of the simulation. Since the launch of their column, several different processes were successfully using the equipment. It is worth mentioning that the process was designed and erected without piloting, solely on basis of modeling and simulation. On the other hand, process concept is directly transferred from simulation scale to the plant scale. Lonza tried to avoid piloting in many cases, even for rather complex unit operations, like extractive distillation [1–5]. The entire unit was designed by simulation and only a few key laboratory data were used, which cannot be simulated, to eliminate uncertainties. They were solid precipitation, foaming, and product decomposition. Tanthapanichakoon [6] introduced simple strategies based on his experience and validated technical references which can be used to expedite process and product development. This can increase the success rates of research and development projects through non-piloting. In the algorithm presented in this paper, it was mentioned that process simulators, process and equipment scale-down studies, design software, CFD simulation, and pinch analysis can ease making such decision. A guideline was also given for a decision to skip piloting on common unit operations and processes.

Successful simulation requires a reliable fluid package to accurately describe the physical and transport properties of the components in the mixture. This can sometimes be a bottleneck as very often the necessary data is not readily available due to a lack of molecule inclusion in the simulator databases. There has not been

1 **Ca**pital **ex**penditures (CAPEX)
2 **Op**erating **ex**penses (OPEX)

any piloting approach reported in multiple studies that has largely reduced costs and time to market [2]. Instead, the necessary validation of thermodynamic model was conducted with a single lab-batch distillation. In another successful study, Kockmann et al. [4] reported the scale-up concept from a single-channel microreactor extended from process development to industrial production. The team at Lonza AG designed and tested these microreactors and performed lab studies of pharmaceutical reactions with a successful transfer to commercial production using a simple correlation based on the hydraulic diameter over typical range of flow rate. This approach led to a consistent scale-up for single-channel microreactors avoiding laboratory to pilot plant studies. Heckmann et al. [5] reported successful process transfer from laboratory to manufacturing scale by omitting intermediate piloting stages in the organic fine chemicals field. They have performed the process fine-tuning directly in the plant. A prior piloting option was not chosen since the amount of information obtained from piloting campaigns rarely justified the time and resources allocated. Based on his extended experience, Berg [7] mentioned that one must think about scale-down rather than scale-up for process development. He proposes to first start with the end in mind by defining business success, including process economics and a high-level design of the commercial process, and then iterate the design to meet success targets and critical-to-quality (CTQ) processes and products. CTQ is an attribute of a product or process that is critical to quality with a direct and significant impact on its actual quality. However, they hint at some views found in the literature, which support the need for a new approach to scale-up.

In the novel scale-up technique, the small-size test units at a laboratory, bench, or mini plant will be built as necessary and/or kept as in-house small-scale units in operation before design. Removing or displacing the pilot plant from the scale-up path could create some concerns due to the knowledge gap in the wake of pilot plant removal. However, in the concept detailed in this chapter, we can easily remedy these shortcomings in order to minimize costs and improve the quality of scale-up in the sense that most technical and economical questions have been answered before the design. However, our interest was to acquire statistical information from benchmarking firms to show the facts and data on current practice in scale-up. A few figures and data reported in Chapter 1 confirmed the novel scale-up idea proposed in this book, but it seems a huge amount of unpublished information is available and we hope companies will decide to release this information for the purpose of using in design books to train the next generation of scale-up engineers and designers. Specifically, information on the following aspects would be of interest:
- Share of process design in failure
- Actual versus planned CAPEX and OPEX
- Contribution of soft skills to industrial failure
- Typical timeline from idea to commercialization
- Start-up time of the projects with piloting or non-piloting
- Percentage of projects passing through pilots in chemical process industry

- Start-up time of the projects (planned versus actual) for various feedstocks
- Industrial failure in terms of costing and possible damages due to non-piloting
- Start-up time of the projects based on the number of new process steps/unit operations
- Typical expenditures of the projects at different stages from development to pilot, design, and start-up

Apparently, due to confidentiality and IP[3] issues, international companies and benchmarking firms have no intention of publicly reporting the failure data of industrial projects and the extent of pilot plant usage in commercialization sequences. However, the general understanding from the data reported by independent project analysis (IPA) [8–11], personal communications with industrial experts, and our own experience show that the industrial projects are failing at incredible rates. On the other hand, our investigation in the Web of Science (WOS)[4] database showed a decline in the number of papers published in last ten years where pilot studies have been reported. Figure 2.1 shows the trend of papers indexed in WOS with the three keywords, "scale-up," "pilot plant," and "process simulation." These keywords are the main characteristics of the novel scale-up. The figure shows few publications on pilot plant studies due to the fact that extremely high costs are associated with generating pilot plant data. E-pilot, to be introduced in the next chapter, would provide an alternative solution for this shortcoming.

As shown in Figure 2.1, the results are promising where a significant increase can be seen in the scale-up activities in the wake of the decrease in "pilot plant" papers. The gradual changes in the trend of "process simulation" papers can also be observed. We conclude from the three curves in this figure that the pilot plant activities are about to be replaced with process simulation and we will see an increased and promising application of multi-scale process simulation in design. This along with some statistical data reported in the literature [8–11] highly support the iterative scale-up method. However, we still hope that the companies decide to release their experiences in using process simulation and data analytics tools instead of large-scale pilots for the sake of education. This would convince the new generation of designers and scientists to move to the more accurate and less costly iterative scale-up method in the coming years. With these attempts, more promising results can be foreseen soon. However, close collaboration is needed during the transit period to move to next stage. In a nutshell, the following transition route is proposed from the conventional scale-up method to the iterative scale-up method:

3 Intellectual property
4 https://www.webofknowledge.com, accessed on August 2020.

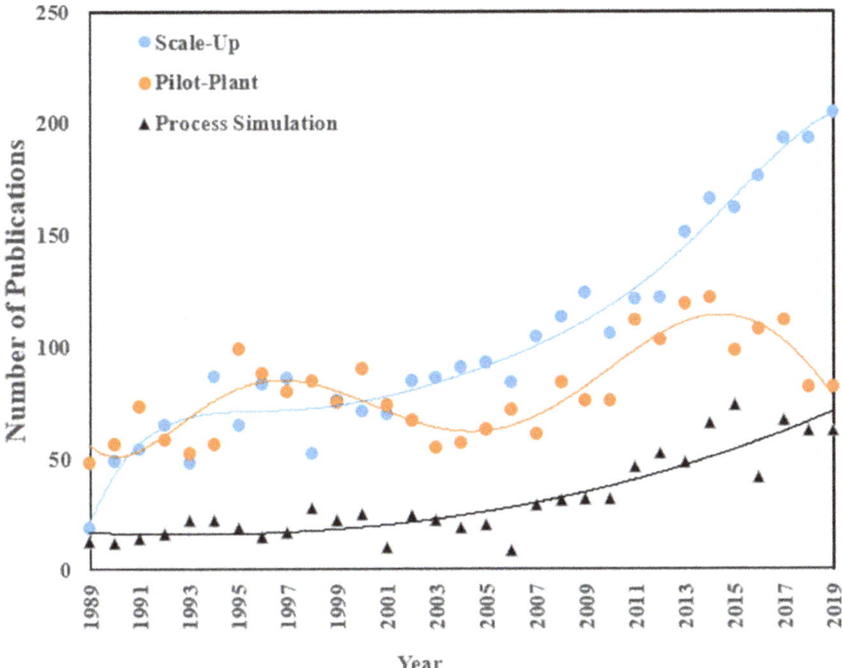

Figure 2.1: Web of Science publication last 30 years on main keywords of the iterative scale-up method.

a) Large-scale pilot plants can be excluded from the scale-up path. However, the small-scale in-house pilot plants, bench-scale, laboratory apparatus, and third-party pilot plants, which we call an all-in-one category here as "small-scale units" can always be used for generating basic data during the iterative scale-up method since they may be less costly and time-consuming, easily operable and flexible as well.
b) It is important to investigate all means effectively to build the complete knowledge repository for design. Therefore, after providing enough basic design data from literature and small-scale units, the preliminary design will be completed leading to the identification and listing of all important and critical uncertainties.
c) These uncertainties will be iteratively minimized to move to the end of the uncertainty zone as detailed in the following section. During the iteration, various powerful and enabling techniques are used to evaluate the economic and environmental benefits of minimizing the risks to acceptable levels.

Since large-scale piloting is usually excluded in the novel scale-up process, its cost, which is normally huge, can be saved as less expensive, and more suitable techniques are applied for answering the questions and concerns raised during the process design stage. Consequently, the iterative scale-up method can improve process extrapolation, drastically reduce costs and timing in the chain of process design, and decrease the uncertainties related to scale-up as shown in a few studies [1, 2]. In addition, as the new scale-up is a highly multidisciplinary, this would enhance collaboration and knowledge exchange among researchers involved in different steps of the project as highlighted in the book.

In this method, design, construction, and start-up of a novel industrial-scale process are conducted without prior research and development at the pilot plant scale. The method is sometimes used by start-up companies for processes containing one major piece of equipment for unit operation [12]. Technology providers of special unit operations use this method for a new process application, but it is not common. There is, however, a strong belief in the industry that this method always fails. A good example of this failure reported in the literature [12] is *Shell's* gas-to-liquids (GTL) technology in Bintulu, Malaysia, where the company used the design of certain unit operations – not core to GTL – from prototypes rather than proven technologies. Considering all these facts and the information provided in Chapter 1 and this chapter, we strongly believe that the chance of success for the iterative scale-up method is too high to pursue. The iterative scale-up would be easy to:

- propose new solutions for anomalies
- find the bottlenecks through early PFD[5]
- provide the opportunity for early patent filing
- decrease the uncertainties related to scale-up
- develop a fast and more economical process design
- eliminate many mistakes and more likely to work efficiently
- detect anomalies, that is, HSE, feedstock, impurities, and so on as early as possible
- build the right pilot plant in terms of size, cost, and measurements to be done.

The above features of the iterative scale are organized in this book into three integrated sections as follows:

Part I: Basics and concepts. This section includes Chapter 1 and this chapter. In Chapter 1, major issues of the conventional scale-up method were explained with statistical data and the importance of soft skills in industrial projects is highlighted. Some failures were also reported, which could help designers, engineers, and students in their training. In this chapter, we mostly focus on the basics and structure of the iterative scale-up method technique and explain the key tools and techniques. A guideline is also proposed for developing case studies with these enabling techniques.

5 **P**rocess **f**low **d**iagram (PFD)

Part II: Enabling tools and technologies. This part includes Chapters 3–5 where we explain the enabling tools and techniques to be used during the scale-up process to fill the knowledge gap created in the face of scale-up removal. These techniques, which are instrumental to scale-up, are briefly explained in this chapter and will be discussed in detail in the subsequent three chapters.

Part III: Case studies. In this part which covers Chapters 6–11, six case studies are reported showing the challenges of the conventional scale-up method and the progress and applicability of the iterative scale-up method to process exploitation. The summary of key learning points of these cases will be presented in this chapter.

2.2 Terminology

As in other fields, specialized terminology is used by the scale-up community. Definitions of the following terms may be helpful for those new to the field:

Capital expenditure (CAPEX): Funds used by a company to acquire, maintain, and upgrade physical assets, such as technology, equipment, plant, property, and buildings. It is an expense a firm incurs to create a benefit in the future.

Chemical process industries (CPIs): Industries in which the feedstocks are chemically/physically converted into finished products, they broadly include the traditional chemical industries, both organic and inorganic, the gas, petroleum and petrochemical industries, the fine chemical and pharmaceutical industries, and a series of allied industries in which chemical processing takes place.

The conventional scale-up method: Migration of a process from the laboratory to the pilot and/or industrial scale. In some literature, the term of "single-train" scale-up is also used, which refers to the conventional scale-up method.

Critical to quality: It is the quality of a product or service in the eyes of the customer.

Independent project analysis[6]: A benchmarking and consulting firm devoted to the empirical research of capital investment projects to maximize business value and project effectiveness.

The iterative scale-up method: Migration of a process from the laboratory to industrial scale. Coming back to either the laboratory, pilot scale, cold scale, or process modeling scale is essential to improve the key characteristics of the industrial-scale design.

6 https://www.ipaglobal.com

Key performance indicator (KPI): It is a measurable value that demonstrates how effectively a business is achieving key objectives. Organizations use KPIs to evaluate their success at reaching targets.

Life cycle assessment (LCA): It is a method used to evaluate the environmental impact of a product through its life cycle from the processing of raw materials, manufacturing, distribution, use, and recycling, to final disposal.

Manufacturing efficiency index (MEI): A new approach in measuring manufacturing performance.

New product development (NPD): In business and engineering, NPD covers the complete process of bringing a new product to market. Its central aspect is product design, along with various business considerations.

Operating expenditure: Expense required for a day-to-day functioning of a business.

Pilot scale (pilot plant): Small-scale system which is operated to study the behavior of a process before using it on a large industrial scale. This is an important system and central for risk identification before production takes place at the full industrial scale.

Process design: Designing and sequencing of unit operations for desired physical and/or chemical conversions/purifications of feedstocks into products. It is instrumental to process scale-up and the summit of chemical engineering, bringing together all the relevant knowledge and components.

Process simulation: A model-based representation of chemical, physical, biological, and other industrial processes and unit operations in a software.

Process simulation scale (e-pilot): A digital-scale system, which is introduced in this book, to study the behavior of a process with multiscale modeling and simulation tools to iteratively improve a large industrial-scale design. This is an important scale and central for risk identification to provide a sound and robust design at the industrial scale.

Technology readiness levels (TRLs): A method for estimating the maturity of a technology during the acquisition phase, which was developed at NASA during the 1970s and based on a scale from 0 to 9 with 9 being the most mature technology.

Other terms are introduced and defined as needed in the text and individual chapters.

2.3 Basics

As explained in Chapter 1, the conventional scale-up method starts from the laboratory unit to the pilot plant and then to industrial design. In the iterative scale-up method, we disregard this path and directly move from a laboratory-scale to an industrial-scale design. At this stage, with a preliminary design, all potential technical problems are uncovered, and the aim is to solve these problems with an acceptable accuracy by various tools and techniques through iteration. Therefore, depending on the conditions, the utilization of new data generated from either small-scale tests, modeling, or data analytics may be essential to improve the key characteristics of the design. It is only at this stage that the pilots could be considered and the decision should be made on its size, subset, operation, and way to generate useful data as the objective of this step is to reveal any ambiguities. With each iteration, more knowledge is generated, and more risks can be identified, assessed, and mitigated to an acceptable level so that the industrial design improves. However, risk identification for a new process is very tedious, because not all details are always readily available. If a certain piece of information is not available, then it may be identified as an unknown. For certain risks though, even that information may be lacking. Therefore, risk identification is carried out several times during the iterative scale-up for project extrapolation. The final solution can be obtained through the successive back and forth progressive iterations until satisfactory conditions are met. This would reorient research with a more realistic focus in terms of CAPEX, OPEX, and LCA. The idea as shown in Figure 2.2 was developed by Jamal Chaouki and well received by the industry and scientific communities [13].

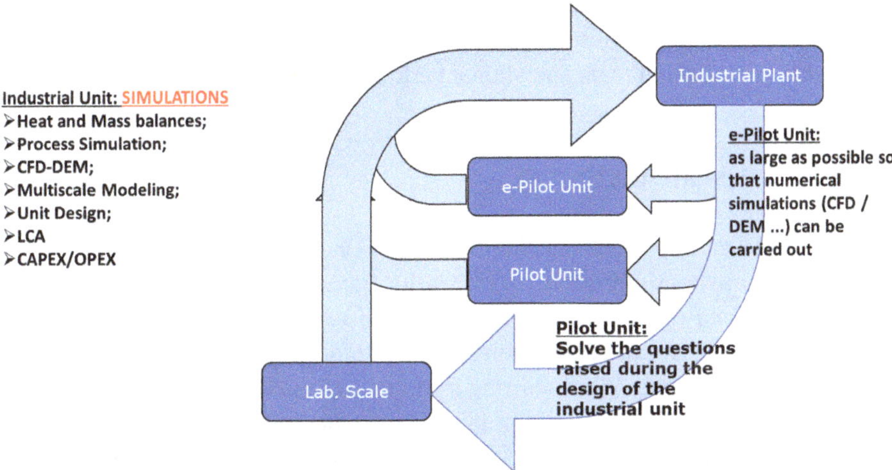

Figure 2.2: Schematic of the iterative scale-up method.

In the following subsections, the main elements of the iterative scale-up method will be briefly explained.

2.3.1 Literature review

Before starting any design tasks, preliminary data on process and products can be gathered from various handbooks, encyclopedias, data bases, the latest literature, consultation activities and brainstorming, industry and technical reports, patent analysis, standards, and market analysis to verify the initial idea. As shown in the following Table 2.1, the success rate for the initial idea is about 0.03%, which means that of every 3,000 ideas generated, only one project is actually commercialized successfully [7].

Table 2.1: Success rate in idea to commercialization chain (data from Stevens and Burley [7]).

Idea	Submitted	Projects	Large chance	Major chance	Launch	Commercial
3,000	300	125	9	4	1.8	1
0.03%	0.3%	0.8%	11%	25%	60%	100

Therefore, this is a critical stage in preparing the foundation for transferring ideas into successful commercialization. At this stage, gross profit analysis [14] can also be performed to determine which pathway one should follow in the design tree. This stage is very important in building the design knowledge and achieving the selected path among potential solutions. A complete basic technical data repository of a project is fundamental to its operational performance and overall excellence [15].

2.3.2 Small-scale tests

Small-scale tests may include laboratory tests, small-size in-house pilot plants, bench-scale units, and third-party pilot plants. Berg believes that one must scale-down the planned commercial process to design laboratory- and pilot-scale processes [7]. These tests can be the main source of information at the beginning or in the iterative design loop. When starting from scratch and after preparing the available literature data, the single-point laboratory experiments are used as proof of principle to verify if the new idea of the technology works in first instance. However, after the preliminary design and during the scale-up when uncertainties become known, one may iterate/circulate several times from industrial design to small-scale tests to address certain risks and improve the design. Normally, obtaining the data from these sources is less expensive and could largely help to minimize design uncertainties. This will be achieved by completing CAPEX and OPEX for individual unit operations, which helps us to be selective in directing experimental efforts on the laboratory scale to remove uncertainties of more expensive equipment.

2.3.3 Industrial design

At the first iteration, the whole process is predesigned based on the information gathered during the literature review and small-scale tests on leading unit operations and the uncertainties surrounding the scale-up will be identified and listed. The uncertainties may originate from the individual unit operation where many dependent and mostly non-linear process variables interact and play a role. The following figure shows the interaction of various leading phenomena. The key point is to decide what parameters of a given phenomenon have the greatest impact in a process step [7] through CAPEX and OPEX analysis.

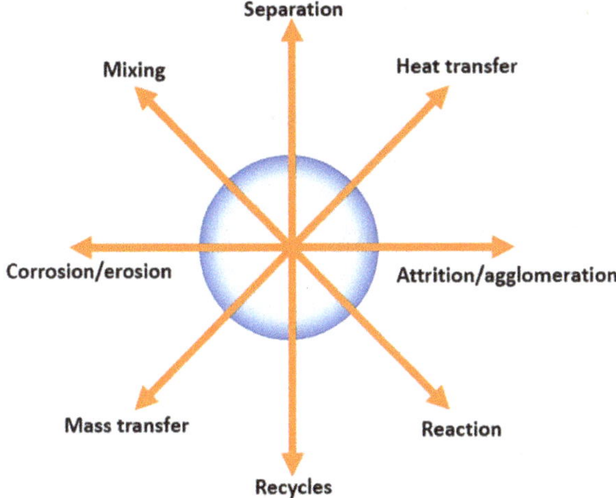

Figure 2.3: Interaction of various leading phenomena in the iterative scale-up method.

Apparently, a preliminary flowsheet can be developed with simple kinetics and hydrodynamics and key elements for the process can be identified for mixing, reactions, separations, and possible operating conditions. Herein, the initial draft of technological, environmental, and economics design risks are identified. Available process design and simulation tools can be used for this purpose. As the risks are listed, the scale-up goes into iteration through small-scale tests, and various enabling tools and techniques in order to remove the risks depending on economic feasibility and environmental impacts. Upon successful iteration, the design knowledge repository can be shaped and updated, which could serve to identify and minimize scaling risks to acceptable levels to be successful at the commercial scale. Therefore, the iterations will gradually refine the design, minimize the risks, and continue until a feasible solution is achieved. In a nutshell, at this stage, generally available technology risk assessment tools are chosen to evaluate risks associated with the scale-up. This is normally conducted by multidisciplinary experts and senior management where we identify risks, access the probability of occurrence and impact, and develop a mitigation plan. During the iteration, we must ensure whether the plan is implemented properly. This is common practice proposed in the literature [7]. However, international companies may use various tools to assess the technology. Total™ uses PROQUAL[7] as a standardized methodology for incorporating advances into projects in five phases. Phase D of this tool is used for uncertainties analysis aiming to access the remaining uncertainties and define requirements for additional data, tests and pilots, (i.e., research and development plan) [16, 17]. It has the following subsets in order to:

- assess major uncertainties by reviewing main risks, low maturity items, and gaps
- resolve or reduce uncertainties through further studies, data acquisition, tests, and pilots
- reduce potential risk and uncertainty to an acceptable level
- evaluate research and development plan probability of success
- define project risk mitigation guideline, and plan.

These steps can be easily applied to process iteration to improve the iterative scale-up method.

2.3.4 Modeling and data analytics

Multiscale modeling and data analytics are two invaluable tools in the iteration scale-up. Modeling and simulation allow the use of condensed knowledge of chemical engineering in an automated way to easily complete the design with existing documentation, equations, and correlations. However, historical tracking and event data and case-based reasoning from the exiting large-scale units could also be used to improve the design. The low cost of data storage and powerful algorithms on big data analytics would make this happen in scale-up. Coupling these two leading tools would expedite the rapid development of the iterative process.

2.3.5 Pilot plants

A pilot plant should be a scaled-down version of the commercial plant and not a scaled-up version of the laboratory apparatus [18]. It is worth mentioning that in certain cases, conventional pilot plants may have configurations, which can highly differ from the industrial units. This may add uncertainties to large-scale units when

7 **Pro**cess for technology **qual**ification (PROQUAL)

the information from these pilot plants are used for design. If needed, well-targeted scaled-down piloting can be employed as a more suitable source for obtaining design data and it can consistently help to improve the process conditions during the iterative design, start-up, operation, and NPD. The data reported by IPA shows that the start-up index for the processes where a pilot is not built is ~2.2 higher than where pilot is built [15]. The operability of the process with piloting is 33% better than non-piloting cases [15]. This shows the importance of right scaled piloting when it is appropriate.

2.3.6 Technology readiness level

A method for estimating the maturity of a technology developed at NASA during the 1970s is used to scale various steps of the iterative scale-up method as shown in Table 2.2. Harmsen [12] also used the TRL classification as an efficient way of assessing an idea or process concept. However, for the purpose of the iterative scale-up method, our classification slightly differed from his classification.

Table 2.2: Technology readiness level (TRL) definition for the iterative scale-up method.

Level	Description	Remark
0	Process idea statement from literature and market analysis	Idea
1	Experimental proof at laboratory	Start of iteration
2*	First commercial-scale design	Risk identification and listing
3*	Prefeasibility analysis	Risk evaluation
4*	Risk identification and reduction (lab experiments, process modeling, and data analytics or piloting)	Risk reduction
5*	Feasibility analysis	
6*	Revised commercial design	
7	First commercial plant in operation	Precommissioning
8	Preparing for full operation	Commissioning
9	Full operation with all specifications	Deployment

*Levels 2–6 are completed simultaneously through multiple progressive iterations to make sure the uncertainties decrease to acceptable levels.

2.4 Enabling tools and techniques

Proper application of enabling tools and techniques are central in the iterative scale-up for design improvement. As shown in Figure 2.4, by integrating these tools, the design is progressively improved during the iteration in a cone-shape fashion to reduce the uncertainties. This is the method that must be followed for the projects.

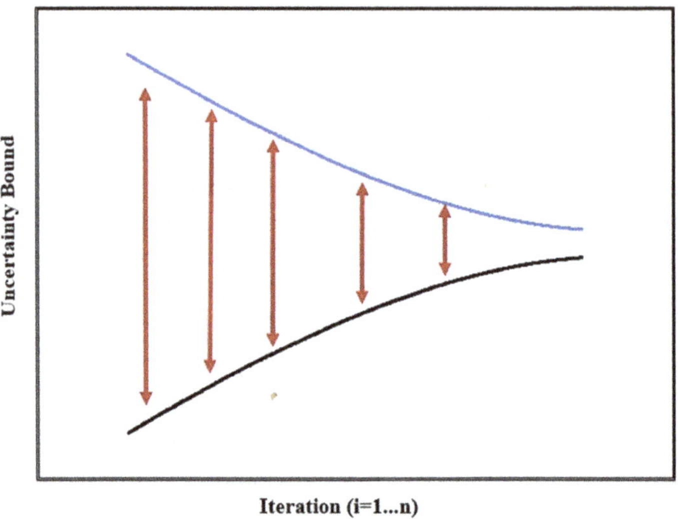

Figure 2.4: Uncertainty cone in the iterative scale-up method.

2.4.1 Modeling and data analytics

2.4.1.1 Process modeling and simulation

Rapid growth in process simulation is attributed to incredible advances in hardware and software. As shown in Figure 2.5, the high computation speeds and low storage costs that are available today show how with affordable costs, we can use computation tools, equipment/process modeling and simulation tools, mathematical software, and spreadsheets to ease the design process.

This is mostly important in the iterative scale-up method as it uses these tools in detecting, reducing, and minimizing the design uncertainties. However, the key point here is to find the proper process for the verification and validation of simulation results. A chapter is devoted to providing more details on enabling modeling and simulation tools for process exploitation with industrial process simulators, compartmental modeling, CFD-DEM techniques, online software, or any other means to predict the performance of the unit operations along with verification and validation techniques.

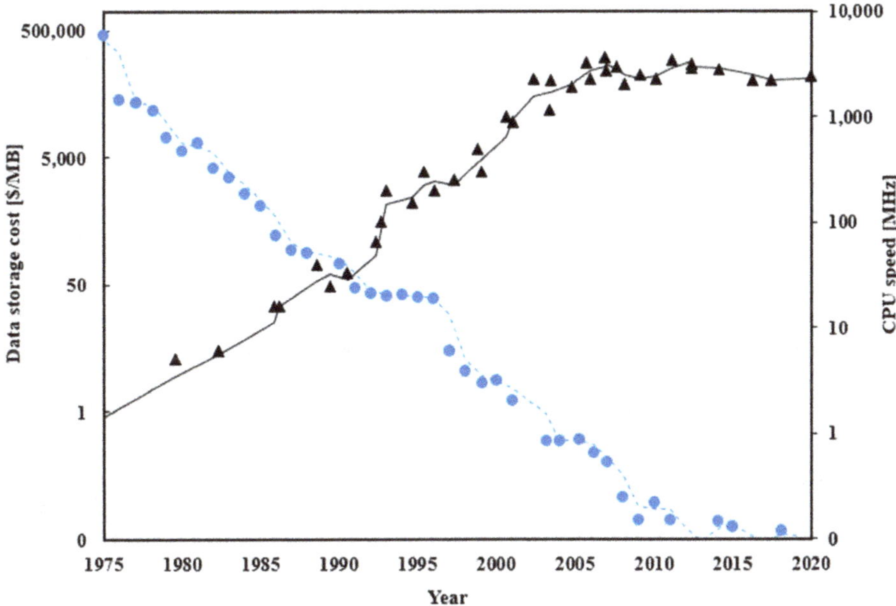

Figure 2.5: Storage costs and CPU speed versus time using the data from literature [19, 20].

A survey of literature shows the wide application of simulation tools in the industry to meet the technical, economic and environmental needs. Selected examples are shown in Table 2.3.

Table 2.3: Successful industrial applications of process simulators.

Challenge	Solution	Company	Main benefits
Debottlenecking gas processes [21].	HYSYS® and Integrated Aspen™ Exchanger Design & Rating (EDR)	**Petrofac**	20% capacity increase with payback in less than a month. Improved understanding of process constraints and scale-up.
Optimize HDPE[8] plant. Reduce side reactions to prevent off-specs [22].	Used Aspen Polymers™ and process data from plant historian and final product MWD[9] to accurately predict outputs	**Qenos**	Completed plant trials for a new product 6 months ahead of schedule Saved $135,000 per year.

8 **H**igh-**d**ensity **p**oly**e**thylene (HDPE)

9 **M**olecular **w**eight **d**istribution (MWD)

Table 2.3 (continued)

Challenge	Solution	Company	Main benefits
Troubleshoot and revamp the naphtha splitter column unit [23].	Using the column analysis capability in Aspen HYSYS® and its integration with Aspen EDR	**Tupras**	40% increase in column capacity 15% reduction in utility consumption (~$500,000 saving per year) Payback in less than one year
Convert low-value fuel oil into high-value products. [24]	Used Aspen HYSYS® to reconfigure the refinery by developing plant digital twins of hydrocracker, CDU[10]/VDU,[11] etc.	**Saudi Aramco**	100 MBD[12] capacity increase Evaluate refinery reconfigurations Debottleneck existing units
Fouling of the heat exchange system [25].	Simulation of fouling behavior using Aspen HYSYS® and Aspen EDR	**INEOS**	Save $4 million per annum Minimize maintenance costs & downtime and ensures reliable operations
Troubleshooting petrochemical plant [26].	To effectively troubleshoot the benzene/toluene units, using Aspen Plus	**Reliance**	Increased production Savings of $2.4 million per annum
Increase plant capacity and reduce energy consumption site wide [27].	Using Aspen Plus for debottlenecking Using APEA[13] and pinch analysis to reduce excessive energy usage.	**LG Chem**	15% increase in plant capacity Successfully integrated and sequenced streams to save energy
Develop a zeolite membrane to traditional ethanol dehydration [28].	Use Aspen Custom Modeler to model an emerging technology.	**Hitachi Zosen**	Considered CAPEX and OPEX while simultaneously exploring designs 18% saving in the conceptual design
Optimization of natural gas facilities [29]	Debottlenecking processing facilities to identify operating conditions using an integrated HYSYS® digital twin model	**YPFB Andina**	$280 million in one year 18% total capacity increase

10 Crude **d**istillation **u**nit (CDU)
11 Vacuum **d**istillation **u**nit (VDU)
12 Thousand **b**arrels per **d**ay (MBD) (Latin “M” meaning thousand)
13 Aspen **p**rocess **e**nergy **a**nalyzer (APEA)

Table 2.3 (continued)

Challenge	Solution	Company	Main benefits
Integration of streams and equipment blocks using multiple property packages in a single simulation [30].	Create a digital twin of the entire SRU,[14] including recycle streams using HYSYS with Sulsim	**Fluor**	Multiple O_2 models for furnace Prediction of sulfur degassing system Rapid output of information into Excel
Reduce cost overruns in estimating methods and application [31]	Use of Aspen Capital Cost Estimator with a deployment and training plan.	**Linde Group**	Improve accuracy of actual costs Support modular design strategy Reduce capital cost project overruns
Increasing capacity in sulfur production [32]	Use ASPEN Sulsim™ to simulate case study, conduct sensitivity analysis and define a window of operating conditions	**Siirtec Nigi**	Minimal equipment changes Increase capacity from 50 to 90 tons per day

As shown in Table 2.3, advanced calculation tools and sophisticated methodologies can greatly assist the designer at various stages.

2.4.1.2 Cold flow modeling

Thanks to the low cost of data storage and high computation speeds to which we have access today, it is possible to simulate large industrial units with some simplifications. This modeming approach is called cold flow modeling in general and the CFD is one of its subsets. If these models can be properly verified and validated with experimental data from the cold flow units, we can use them for design purposes with confidence. The cold flow units, normally operated at low operating conditions, are used to study the hydrodynamics of complex unit operations or equipment in the commercial design. This is extremely useful for obtaining large-scale data at a relatively moderate cost, which otherwise cannot be obtained easily. This is particularly of interest and may be the single choice for particulate or complex processes due to current limitations in their scale-up knowledge. The data from the cold flow unit is extremely useful in validating large-scale CFD[15] models of the commercial unit and the validated model, as a base case, can be then used to study various parameters and changes in geometry and flow rates, and to perform case studies and sensitivity analysis. A validated CFD model can

14 Sulfur recovery unit (SRU)
15 Computational fluid dynamics (CFD)

be used to optimize the geometry of a cold unit for a desired performance. Furthermore, critical issues can also be evaluated with the validated model, such as the residence time distribution, premixing and non-premixing effect of chemicals, mass transfer, heat transfer, gas and liquid holdups, and pressure drop [12]. These are valuable enabling tools, which make it possible to significantly reduce the scale-up risk and uncertainties during the iterative scale-up method.

There are, however, some limitations on cold flow units, such as the ones for CFD models related to bubble and droplet break-up and coalescence, agglomeration, and thermal or catalytic attrition. For example, if a gas–liquid mixture is present in the real industrial unit, since we cannot use the real components in the cold unit, water is used to represent the liquid phase and air/nitrogen is used to represent the gas phase. Other simplifications would be adding chemicals to create breakage, coalescence, and agglomerates in the system and study the corresponding effect. This is, of course, unrealistic. For liquid–liquid systems, one may use water/alkane mixtures to create and study the droplet formation, breakage, and coalescence. The above limitations are acknowledged, but since the higher size for cold flow units can be used, this is an excellent tool to study the hydrodynamics of a complex system, which otherwise cannot be evaluated, even in pilot plants. In Chapter 3, the potential and limitations of CFD and CFD-DEM[16] models will be discussed in detail for the simulation of complex processes.

2.4.1.3 Data analytics

Various technologies are available today for the collection, storage, processing, and visualization of data sets, such as the massive, structured, and unstructured data sets covering experimental, historical, and operational data. In the chemical process industry, with the development of data collection tools, tremendous process datasets are being generated from the laboratory, equipment, unit operations and processes through sensors and actuators. To generate insights from these collected datasets, different machine learning, deep learning, data mining, and time series algorithms can be employed on top of distributed computing frameworks for batch and stream processing purposes.

Most of these datasets are not frequently examined and 88% of chemical industry managers consider data analytics necessary to maintain a competitive advantage within 5 years [33] based on the 2016 Global Industry 4.0 Survey. Therefore, data analytics is emerging across all branches of process engineering affecting manufacturing and business, and process design, control and monitoring. In the near future, data-driven scale-up will enrich design practices and teaching in chemical engineering.

16 **C**omputational **f**luid **d**ynamics–**d**iscrete **e**lement **m**ethod (CFD-DEM)

A survey of literature shows the wide applications of data analytics in the chemical process industry to meet the current demands. Selected examples of the successful application of data analytics in chemical engineering are shown in Table 2.4.

Table 2.4: Successful industrial applications of data analytics.

Challenge	Solution	Company	Main benefits
Optimize HDPE plant and reduce side reactions to prevent off-specs [22]	Used Aspen Polymers™ and process data from plant historian and final product MWD to accurately predict outputs	**Qenos**	Completed plant trials for a new product 6 months ahead of schedule Saved $135,000 per year.
Production increase [34]	Data analytics and generation of KPIs[17] using Aspen InfoPlus.21	**Infineum**	10% global increase in MEI[18] 2,000 ton production increase per annum
Early failure predictions [35]	Compressor failure 35 days in advance using Aspen Mtell to avoid an emergency shutdown	**Diversified Energy Company**	$30 M potential savings Reduce maintenance costs Improve HSE performance

2.4.2 Pilot plants

As mentioned earlier, the normal piloting is the main element of the conventional scale-up method, but its design, construction, and operation take more time and huge costs. Nowadays, project owners must swiftly bring a new product or process to the market and they end up either shortening or even omitting pilot plant runs. If the pilot plants are run for a limited time only, however, the build-up of trace elements, catalyst decay, fouling, corrosion, and ageing effects cannot be easily observed, which leaves scale-up uncertainties unanswered and renders the pilot useless. Therefore, in the iterative scale-up method, the normal pilot plants are skipped from the scale-up path, and instead scaled-down pilot plants and cold flow units are used in the iteration loop as necessary to improve design as explained below:

The analysis of a large number of industrial projects by benchmarking firms and the opinion of experts showed that scaled-down piloting is essential whenever the process is complex consisting of more than four new steps and/or recycle streams, or it contains particulate/crude solids [7, 9]. However, in less complex projects, only the relevant subsets are piloted and not the whole process. For the process containing two to four new processing steps, the piloting is a legitimate decision to reduce potential risks in view of the total time schedule and it leads to a shorter start-up time of

17 **K**ey **p**erformance **i**ndicators (KPIs)
18 **M**anufacturing **e**fficiency **i**ndex (MEI)

the industrial unit [8, 36]. It is important to mention that the projects not passing to the piloting stage according to the Merrow criteria, can also be successful. In this section, we provide more details on scaled-down piloting, which is an integrated pilot plant and the scaled-down version of the industrial design. It normally runs continuously and long enough to indicate reliability as mentioned by Marton [36]. This type of piloting, which includes all processing steps and recycle streams, is expensive and its construction is timely. Therefore, it is only applied when the other enabling technologies introduced in this chapter and detailed in the book are unable to provide the solution to ambiguities. This point is also mentioned in the literature if one skips normal pilot plants in the process scale-up [12].

Harmsen [12] provided the following list of piloting purposes in scale-up:

- To discover the unknown parameters, for example, fouling, corrosion, erosion, catalyst decay, azeotrope formation, foaming, and particulates formation
- To validate the process flowsheet, to find optimal operating windows, and to check the build-up of trace components through various data taken at multiple points in the pilot plant
- To test construction materials on corrosion at similar conditions to those of the commercial-scale design
- To validate process control and operating procedures
- To provide new product samples to customers to test the performance
- To train operators of the commercial plant.

From his view, for all the above purposes, the pilot plant, which could be relatively large, should contain the same process steps, unit operations, recycle stream, and conditions of the industrial scale unit and should be contracted if it is really needed either fully integrated or not. It is worth mentioning that the size of a pilot must also consider the quantity of products to be tested either by the company itself or third parties. The following table shows the size of these types of piloting based on benchmarking data [15].

Table 2.5 shows that to obtain the proper solution from the pilot plant, the size should be rather large if the resources allow. It is worth mentioning an important quotation from Kate Threefoot, DuPont Research Fellow [7]: "You always build a pilot plant, it's just that sometimes it's at the commercial scale." This shows the importance of employing large-size pilots.

Table 2.5: Typical scaling factors for scaled-down piloting (source of data [15]).

Unit	Gases and liquids	Homogeneous solids	Run-of-mine
Pilot	1	1	1
Industrial	500–1,000	100–200	10–25

2.5 Guideline to develop the iterative scale-up method

Based on the authors' expertise and the case studies reported in this book, the following 12-step guideline as shown in Table 2.6 is proposed to develop the iterative scale-up method case studies. We hope that with the use of this guideline, more successful and emerging designs can be carried out saving hundreds of lives due to incidents, failures, and disasters in the chemical process industry.

Table 2.6: Guideline to develop the iterative scale-up method case studies.

Step	Description
1.	Identify and prepare preliminary data for multiple solutions from literature, consultations, industry, technical reports, patent analysis, and standards (product and process).
2.	Develop a preliminary PFD for these solutions, and list all unit operations and operating conditions (conduct lab or small-scale experiments, if necessary).
3.	Identify potential HSSE[19] issues.
4.	Survey local/international markets and decide on industrial-scale capacity
5.	a) Perform unit operations design with simple or revised (i = 2, . . . , n[20]) hypothesis. b) Perform a life cycle assessment of the process.
6.	Prepare and simulate SFD[21] using industrial process simulators, for example, Aspen Plus, HYSYS™, and Unisim a) Use simple or revised (i = 2, . . . , n) kinetics and hydrodynamics. b) Prepare detailed material and energy balance, sizing and costing.
7.	Perform CAPEX-OPEX analysis using, that is, APEA,[22] spreadsheets, ... as explained in Section 2.6.
8.	Refine solutions, minimize risks to acceptable levels through: a) Revision of literature and lab data or generation of additional lab data b) Perform multi-scale modeling, that is, TPM,[23] TFM,[24] CFD, DNS,[25] CFD-DEM, and so on. c) Prepare the e-pilot. d) Conduct compartmental modeling. e) Run cold flow test. f) Use open pilot plant data.

19 **H**ealth, **s**afety, **s**ecurity and **en**vironmental (HSSE)
20 Scale-**up** **i**teration **n**umber (SUIN)
21 **S**imulation **f**low **d**iagram (SFD)
22 **A**spen **p**rocess **e**conomic **a**nalyzer (APEA)
23 Two-**ph**ase **m**odel (TPM)
24 Two **f**luid **m**odel (TFM)
25 **D**irect **n**umerical **s**imulation (DNS)

Table 2.6 (continued)

Step	Description
9.	Repeat step 5 and propose scenarios to determine remaining risks.
10.	Consider building in-house down-scaled pilot plant (proper size, subsets) if necessary, to obtain additional and enlightened data for removing remaining ambiguities and risks.
11.	Finalize the design if risks are clarified and minimized, otherwise, go to step 12. a) Conduct the profitability analysis with known methods to determine feasible solutions. b) Optimize the key economic objectives of the process. c) Prepare full feasibility and go to detailed design.
12.	Iterate (i = 2, . . . , n) and return to step 3 with revised hypothesis. Continue all steps until conditions of step 11 are fully satisfied.

2.6 Cost estimation

Cost estimation is a critical element of any the iterative scale-up method project as in Table 2.6. To evaluate a scale-up project, it is compulsory to estimate both CAPEX and OPEX. This section does not cover a complete review of either CAPEX or OPEX estimating procedures, but it provides some concepts that can be used in scale-up evaluation. This is because managerial decisions in companies are normally based on counterbalancing CAPEX and OPEX. Many types of estimates are available for these issues. Table 2.7 shows five methods based on the classification made by the American Association of Cost Engineers [37] with their range of accuracy [38].

As shown in Table 2.7, each estimate requires progressively more detailed technical and economic data and preparation time. However, for the iterative scale-up method, level one to three would be sufficient and with the existing information in the literature, we can estimate the CAPEX and OPEX to an acceptable accuracy since more precise data are not readily available during the scale-up. However, as we reach the final point in the iterative scale-up method, we can then move to levels 4 and 5 with definitive and detailed estimates that require detailed and accurate data, calculations, and quotations.

Table 2.7: Type of estimates with their range of accuracy.

Level	Estimate	Accuracy (%)
1.	Order of magnitude	±30 to 50
2.	Study estimate	±20 to 30
3.	Preliminary estimate	±10 to 25
4.	Definitive estimate	±5 to 15
5.	Detailed estimate	±2 to 5

Table 2.8: Commonly used computer-aided tools for cost estimation.

Tool	Developer	Application	Source
Aspen Process Economic Analyzer (APEA)	ASPEN TECHNOLOGY	CAPEX, OPEX, and Profitability Analysis	https://www.aspentech.com/en/products/pages/aspen-process-economic-analyzer
CHEMCAD	CHEMSTATIONS	Economics Module	https://www.chemstations.com/
SuperPro Designer	INTELLIGEN INC	Cost Analysis and Economic Evaluation	https://www.intelligen.com/competencies/cost-analysis/
CostOS Estimating	AVEVA	Cost Analysis and Economic Evaluation	https://www.aveva.com/en/products/costos-estimating/
Cost estimator	MCGRAW-HILL	Equipment Capital Costs	http://www.mhhe.com/engcs/chemical/peters/data/
Capital Cost Estimator	DWSIM	Capital and Operating Costs	https://dwsim.inforside.com.br/new/
Costing Spreadsheet	VANDERBILT UNIVERSITY	Cost estimation for purchased equipment	https://www.vanderbilt-chil.com/equipment-costing-spreadsheet
Matches	MATCHES	Process Equipment Cost Estimates	https://www.matche.com/equipcost/Default.html

The software can help significantly to perform the cost estimation when the historical data or vendor's databases are not available. A variety of commercial simulators has their own embedded tools and databases to estimate major process equipment costs. Moreover, some freeware tools are also available to perform CAPEX and OPEX analysis. Table 2.8 summarizes the list of selected tools and simulators for cost estimation.

All the software, simulators, and spreadsheets can be easily used for CAPEX and OPEX estimation. Among them, McGraw Hill provided an online and hands-on cost estimator tool to estimate the purchased cost of equipment with minimal design data based on information provided by Peters et al. [39]. The costs are for January 2002 and should be updated based on the CE Index shown in this chapter. The purchased costs of the following equipment can be estimated from this online tool:

1. Agitators, autoclaves, bayonet heaters, blenders, blowers
2. Centrifuges, chutes and gates, columns and connections
3. Compressors, condensers, solid conveyors, crushers, cutters, disintegrators
4. Drivers, dryers, dust collectors, ejectors, electric motors, evaporators
5. Expanders, extruders, fans, filters, furnaces, grinders, heat exchangers
6. Heaters, hoists, immersion heaters, insulation, kettles, kneaders, mixers

7. Mixing tanks, packed columns, packing, piping, pulverizers, pumps
8. Reactors, separators, storage tanks, trays(distillation), turbines, valves
9. Vaporizers and vibrating screens

CAPEX and OPEX estimates of a proposed plant are continuously carried out during the development of the iterative scale-up method based on available technical and cost data.

2.6.1 CAPEX analysis

CAPEX or the fixed capital investment of a plant [40] is the fund required for engineering, procurement, and construction of a plant. It is the capital required to provide all the depreciable facilities. The facilities may be divided into the battery limits and auxiliary facilities. The battery limits include all manufacturing and processing equipment. The auxiliary facilities are the storage areas, administration offices, utilities, and supporting facilities. The CAPEX estimation of facilities is a critical part of an iterative scale-up method. Peters et al. [39] detailed seven methods for CAPEX estimations as follows: detailed item estimate, unit cost estimate, percentage of delivered-equipment estimate, approximation by Lang factor, power factor or capacity ratio, investment per unit of capacity, and turnover ratio. Coker [41, 42] introduced the capacity ratio, factored cost estimates, and detailed factorial cost estimates to estimate the cost of the plant and equipment. In this section, for the iterative scale-up method, we present three selected methods to calculate the CAPEX for equipment and processes.

2.6.1.1 Capacity exponent method

It is often necessary to calculate the cost of a piece of equipment when there is no available cost data for a given size or capacity. If we know the cost of a piece of equipment or plant with a given capacity, the cost of the same piece of equipment or plant with a new capacity can be calculated from

$$C_2 = C_1 \left(\frac{Q_2}{Q_1}\right)^m \tag{2.1}$$

where C_1 is the cost of a piece of equipment or plant at original capacity “1”; C_2 is the cost of a piece of equipment or plant with new capacity “2”; Q_1 is original capacity; Q_2 is new capacity; and m is cost exponent (or capacity factor).

The value of m depends on the type of equipment or plant. It is generally considered to be 0.6, the well-known six-tenths rule [39]. This value can be used to obtain a rough estimate of the expenditures, if there is insufficient data to calculate the index (m) for the size of the equipment or plant required. It is worth mentioning that in some literature, m is referred to be as the power sizing component. Its value is usually less than 1. Coker has also reported the capacity exponent for process units ranging from 0.6 to 0.8 [41]. The value of m typically lies between 0.5 and 0.85 depending on the type of plant; however, a care is required for each time it is used to estimate a situation. Table 2.9 lists values of m for various types of equipment.

Cost indices should be used to bring the historical cost data to a desired year. If the cost (C_1) and index (I_1) at time "1" in the past are known, the cost at time "2" (C_2) can be determined from the following equation with the known index at time "2":

$$C_2 = C_1\left(\frac{I_2}{I_1}\right) \tag{2.2}$$

The indices are used to give a general estimate, but no index can account for all the factors. Many different types of cost indices are published in the literature, such as the Chemical Engineering Plant Cost Index which is often referred to as the CE index, Marshall and Swift Cost Index (M&S), Intratec Chemical Plant Construction Index (IC), and Nelson–Farrar Indexes (NF) [41]. CE is a composite index for US CPI published monthly in the *Journal of Chemical Engineering*. The journal also publishes the Marshall and Swift Index (M&S Equipment Cost Index) [44]. Table 2.10 shows the Overall CE index from 2001 to 2019. In this table, indices from 2000 to 2016 were taken from Turton et al. [45] and indices from 2017 to 2019 were accessed through the *Chemical Engineering* magazine [46].

Table 2.9: Typical capacity exponents for equipment [43].

Equipment	Exponent (m)
Turbo blowers and compressor	0.5
Evaporators/rectangular tanks	0.5
Heat exchangers	0.65–0.95
Spherical tanks/towers, constant diameter	0.7
Pumps/piping	0.7–0.9
Reciprocating compressor	0.75
Electric motors	0.8
Towers, constant height	1.0

Table 2.10: Overall CE index from 2001 to 2019 [45, 46].

Year	Index	Year	Index
2000	394	2010	551
2001	394	2011	586
2002	396	2012	585
2003	402	2013	567
2004	444	2014	576
2005	468	2015	557
2006	500	2016	542
2007	525	2017	567.5
2008	575	2018	603.1
2009	521	2019	607.5

2.6.1.2 Lang factor method

The purchased cost of a piece of equipment, free on board is quoted by a supplier and may be multiplied by a factor of 1.1 to give the approximate delivered cost. In this factorial method for estimating the total installed cost of a process plant, a combination of materials, labor, and overheads is considered. The CAPEX of a plant facility can be estimated using the Lang factor (F_L) method [47] as follows:

$$CAPEX = F_L \sum_{1}^{n} C_i \tag{2.3}$$

where n is the number of major pieces of equipment and $\sum_1^n C_i$ is the sum of the purchased costs of all the major pieces of equipment.

Lang factors may vary widely from about 2 to 10 depending on the process scale and type, materials of construction, location, and degree of innovation and technology [48]. Table 2.11 shows the recommended Lang Factors which vary from 3 to 6 for different industrial processes.

As seen in this table, a rising trend in Lang factors is visibleshowingrelatively less expenses for major pieces of equipment due to more focus on safety, energy savings and instrumentation [49]. A significant difference can be also seen between solids and fluid processes, which was attributed to the amount of piping associated with the main equipment where more fluids are involved [47]. However, these values should be used as a general guide; the proper factor is best calculated from a firm's own cost data. In an industry where the cost of major pieces of equipment is dominant, a lower Lang factor can be expected. Table 2.12 shows the Lang factors used

Table 2.11: Lang factors published by Lang in 1948, Peters et al. in 2003 and Towler and Sinnott in 2012 [49].

Types of the plants	Lang factor (CAPEX)			Working capital investment [39]
	1948 [47]	2003 [39]	2012 [44]	
Solid processing	3.1	4	4.55	0.7
Mixed solids–fluid processing	3.63	4.3	6.05	0.7
Fluid processing	4.74	5	6.0	1

in selected plants, which somehow differ from those reported in Table 2.10 due to cost of major pieces of equipment.

Table 2.12 shows that the Lang factor for the food industry is much lower than those reported in the chemical industry due to high equipment costs and the use of stainless steel materials [49].

W.E. Hand proposed different factors by extending the Lang method for each type of equipment (columns, vessels, heat exchangers, and other units) as shown in

Table 2.12: Lang factors used in selected plants for CAPEX.

Plant type	Lang factor	Remark
Food industry [49]	1.8	Without contingencies
Yeast production [49]	1.41–2.72 (average = 1.82)	Including contingencies
Biogas plants [48, 50]	1.79 1.81	Large scales, sub-Saharan African countries 20 medium-large scales, South Africa
On-farm oil production [51]	1.46–3.02	Low process complexity
Mechanical pressing [51]	1.65–2.26	Low-to-medium process complexity
Solvent extraction [51]	4.–4.21	High process complexity
Biomass to hydrogen [52]	3–5.4	Base equal to 4
Biomass to hydrocarbons [53]	4.31	
Processing of biobased Resources [54]	2.7–6	Typical value equal to 4.57
Chemical, refinery, and petrochemical [55]	3.38–6.86 (average = 5.12)	data base of over >250 projects ranging from 10 to 250 million
Chemical process industry [56]	3..4–5.2	
Chemical process industry [57]	2.5–9.53	

Table 2.13: Hand factors developed by W. E. Hand [58].

Equipment type	Factor
Fractionating columns	4.0
Pressure vessels	4.0
Instruments	4.0
Pumps	4.0
Heat exchangers	3.5
Compressors	2.5
Miscellaneous	2.5
Fire heaters	2.0

Table 2.13. In this extension, equipment is grouped in various categories [39] and summed to give the battery limits installed cost [58].

A more detailed list was compiled by W.F. Wroth, which consists of installation factors as shown in Table 2.14 for a selected group of equipment. It is worth mentioning that Lang and Hand use purchased equipment costs while Wroth uses delivered equipment costs [58].

Table 2.14: Categorized Lang factors for selected equipment [58].

Equipment	Factor
Blender	2.0
Blowers and fans	2.5
Compressors	2.0–2.3
Ejector	2.5
Electrical motors	3.5
Furnace	2.0
Heat exchanger	4.8
Instrument	4.1
Pumps	5–7.0
Refrigeration	2.5
Tanks	2–4.1
Towers	4.0

2.6.1.2.1 The Lang factor for pilot plants

It seems that a limited value for the Lang factor is reported for pilot and demonstration plants. Table 2.15 shows the Lang factors for pilot to medium plants.

The value of the Lang factor shown in this table should be used as a general guide; the proper factor is best calculated from a firm's own cost data during scale-up. However, in view of its large impact on CAPEX and OPEX, it is important that the Lang factor be as accurate as possible.

Table 2.15: Lang factors for pilot, small, and medium plants.

Plant type	Lang factor	Remark
Small-/medium-scale biogas plants [48]	2.63–3.04	2.63 gives more accurate estimates
Leaching and solvent extraction [59]	4.8	
Fish processing plants [60]	1.59–3.3	For various countries
Biopharmaceutical facilities [61]	3.3–8.1	
Pharma facilities [62]	3.1–4.74	
Biotech facilities [62]	4–8	
Pilot plants [57]	7.53–9.53	Value of individual equipment item <10,600

2.6.1.3 Multiplying factors method

In this method, with known delivered equipment costs (E), the CAPEX of facilities can be estimated by the following equation [39]:

$$\text{CAPEX or OPEX} = E\left(1 + \sum_{1}^{n} f_i\right) = E \times f_{\text{CAPEX or OPEX}} \tag{2.4}$$

where f_1, f_2, f_3, . . . , f_n are multiplying factors for piping, electrical, and indirect costs and are determined based on the process used. This method is commonly used for preliminary and study estimates [39] with the expected accuracy in the range of ±20–30. Table 2.16 shows the multiplying factors.

For methods described in this section, the estimation of individual equipment costs is the critical part. Besides the data available in the repository of the companies, third-party data providers, vendors, or suppliers, and various softwares are also available to estimate the purchased cost of the individual equipment.

Table 2.16: Multiplying factors for estimating CAPEX and OPEX based on delivered equipment cost [39].

Equipment	Multiplying factors (fraction of delivered equipment)		
	Solid processing	Solid–fluid processing	Fluid processing
Purchased equipment delivered	1	1	1
Equipment installation	0.45	0.39	0.47
Instrumentation	0.18	0.26	0.36
Piping	0.16	0.31	0.68
Electrical	0.10	0.1	0.11
Buildings	0.25	0.29	0.18
Yard improvement	0.15	0.12	0.1
Service facilities	0.40	0.55	0.7
Engineering and supervision	0.33	0.32	0.33
Construction expenses	0.39	0.34	0.41
Legal expenses	0.04	0.04	0.04
Contractor's fee	0.17	0.19	0.22
Contingency	0.35	0.37	0.44
f_{CAPEX}	**3.97**	**4.28**	**5.04**

2.6.2 OPEX analysis

The components of OPEX or total production cost are the fixed and variable costs, plus general expenses as shown in Table 2.17 on an annual basis. In academia, OPEX are usually estimated as a percentage of capital costs, often a nominal 10%. However, a professional practice might be different. Various components of the routine OPEX are summarized in the following table. We recommend using the typical values given in this table as a guideline only.

The following points are important to mention about the table:

1. **Operating costs include fixed and variable operating costs**: The fixed operating costs are independent, while the variable operating costs are dependent on the quantity produced. The division between these costs is somewhat arbitrary.
2. **Nonroutine OPEX:** This includes shutdowns, turnarounds, and other nonrecurring costs which are normally excluded from the routine OPEX [65].
3. **Raw materials**: The raw materials are the essential materials required to manufacture the product. Their quantities can be obtained from the material and energy balance calculation made by process simulators (see Chapter 3 for more

Table 2.17: Summary of the estimation basis for routine OPEX on annual basis.

Cost	Typical value [63]	Typical value [39]
Variable costs (A) ($/y)		
1. Raw materials (note 1)	Material and energy balance calculations	10–80% of OPEX
2. Utilities	Material and energy balance calculations	10–20% of OPEX
3. Miscellaneous operating materials	10% of item 5	
4. Shipping and packaging	Normally negligible	
Fixed costs (B) ($/year)		
5. Maintenance (labor and materials)	5–10% CAPEX	2–10% of CAPEX
6. Operating labor	From manning estimates <15% OPEX	10–20% of OPEX
7. Laboratory costs	20–23% of item 6	10–20% of item 6
8. Supervision	20% of item 6	10–20% of item 6
9. Plant overhead	50% of item 6	
10. Capital charges	10% CAPEX	Depends on method of calculation
11. Insurance	1% CAPEX	0.4–1% of CAPEX
12. Local taxes	1–2% CAPEX	1–4% of CAPEX
13. License fees and royalties	1% CAPEX	0–6% of OPEX
Direct production costs (A ± B) ($/year)	Sum of items 1–13	66% of OPEX
General expenses (C)		
14. Sale expense	20–30% of direct production cost	15–25% of OPEX
15. General overhead		9.5–11.5% of sales [64]
16. Research and development		
Annual OPEX (A + B + C) ($/year)	Sum of items 1–16	
Annual production rate (kg/year) (P)	From process flow diagram	
Annual OPEX/unit of product ($/kg)	(A + B + C)/P	

details) based on a given process flow diagram. These are multiplied by the operating hours per year to obtain the annual requirements. Their prices are best acquired through quotations from potential suppliers, but in the iterative scale-up method, the market price of various commodities in different regions can be obtained from the literature or market price providers, for example, Independent Commodity Intelligence Services [66], formerly Chemical Market Reporter, European Chemical News, and Industry Updates [67]. However, the cost of the raw material is normally the largest part of the OPEX.

4. **Utilities:** The utilities cover power, gas, steam, water, compressed air, cooling and process water, and effluent treatment if not calculated separately. Their

quantities can be obtained from the material and energy balance calculation performed by process simulators (see Chapter 3 for more details) based on a given process flow diagram. The prices can be acquired from Company records, if available, or from local authorities. They will depend on the primary energy sources and the plant location. The range given in Table 2.17 can be used to make preliminary estimates.

5. **Operating labor:** The manpower costs may be calculated from an estimate of the number of shift (usually five working shifts) and annual working days based on experience with similar processes and local expenses. Various methods were proposed in the literature to calculate the manpower. The manning chart represented by the Wessel equation proposed in 1952 [58] gives a method of estimating the number of man-hours required for a production rate of 2–2,000 tons/day:

$$\mathrm{Log}(\mathrm{Y}) = -0.783\,\mathrm{Log}(\mathrm{X}) + 1.252 + \mathrm{B} \tag{2.5}$$

where Y is the operating hour/ton per processing unit; X is the plant capacity, tons/day; and B is the process constant (0.132 for a batch with minimum labor requirement and 0 for operations with average labor requirements, and −0.167 for a well-instrumented continuous process).

2.7 Challenges

The main challenge with the iterative scale-up method is related to confidentiality regarding the release of data from the exiting practices. Due to the lack of reliable information on the conventional scale-up method and their link with industrial failures, the designer largely believes that the pre-design pilot plants should be in place even though they know obtaining satisfactory answers to the uncertainties is not possible to the full depth and dimension if the sizes are too far from the industrial scale. It will take a while to overcome these barriers since the full acceptability of the iterative scale-up method and collaborative efforts are needed for this purpose. In fact, companies may be more selective in the future on choosing processes to pilot as resources become scarcer [7], but if alternative and less challenging solutions are proposed, this could attract their attention. At the same time, the enabling technologies also have some limitations in dealing with micro-meso and microscale problems. These problems may slow the development of the iterative process, but since there is a strong desire to use the potential of digitization in the process industry, the future will be much more promising.

2.8 Conclusion

Companies will be more selective in the future about choosing the conventional scale-up method for their designs as resources become scarcer. This opens new avenues in scale-up, called the iterative scale-up method. In this chapter, the basics and concepts behind this scale-up along with enabling tools and technologies are introduced. Some case studies on chemical and biochemical industries as detailed in subsequent chapters in this book are also shown to highlight the applicability of the iterative scale-up method in process exploitation. This practice and paradigm shift would ease filling the gap between the conventional scale-up method time and start-up time of projects while reducing the design and implementation and operational costs in the process deployment. We believe that with this clear concept, the scale-up efforts will be pushed on many fronts to create value in the chemical process industry and promote robust designs in the future leading to improved safety in industrial operations.

References

[1] D. Staak, T. Grützner, B. Schwegl and D. Roederer, Dividing wall column for industrial multi purpose use, Chemical Engineering and Processing: Process Intensification, 75, 48–57, 2014.

[2] T. Grützner, C. Schnider, D. Zollinge, C. S. Bernhard and N. Künzle, Reducing time to market by innovative development and production strategies, Chemical Engineering and Technology, 39(10), 1835–1844, 2016.

[3] D. Staak and T. Grützner, Process integration by application of an extractive dividing-wall column: an industrial case study, Chemical Engineering Research and Design, 123, 120–129, 2017.

[4] N. Kockmann, M. Gottsponer and D. M. Roberge, Scale-up concept of single-channel microreactors from process development to industrial production, Chemical Engineering Journal, 167(2–3), 718–726, 2011.

[5] G. Heckmann, F. Previdoli, T. Riedel, D. Ruppen, D. Veghini and U. Zacher, Process development and production concepts for the manufacturing of organic fine chemicals at LONZA, CHIMIA, 60(9), vol. 60, no. 9, 530–533, 2006, 2006.

[6] W. Tanthapanichakoon, Accelerating process and product development, Chemical engineering, pp. 48–54, February 2013.

[7] D. A. Berg, Technology Evaluation, 2018. [Online]. Available: https://chemeng.queensu.ca/courses/CHEE400/TechEval.pdf. [Accessed August 2020].

[8] E. W. Merrow, Estimating startup times for solids-processing plants, Chemical Engineering, 24, 89–82, 1998.

[9] E. W. Merrow, Industrial megaprojects, concepts, strategies, and practices, New Jersey: John Wiley & Sons, Inc, 2011.

[10] E. W. Merrow, Oil and gas industry megaprojects: our recent track record, Oil and Gas facilities, pp. 38–42, 2012.

[11] Ernst and Young, Spotlight on oil and gas megaprojects, 2014. [Online]. Available: https://www.ey.com/Publication/vwLUAssets/EY-spotlight-on-oil-and-gas-megaprojects/$FILE/EY-spotlight-on-oil-and-gas-megaprojects.pdf, Accessed on July 2020. [Accessed july 2020].

[12] J. Harmsen, Industrial process scale-up; a practical innovation guide from idea to commercial implementation, Second edn, Cambridge: Elsevier, 2019.
[13] J. Chaouki, Scaling up step by step, in The University of Toronto Connaught Industry Alliance Symposium on Carbon Capture and Conversion, Toronto, 2016.
[14] University of Colorado Boulder, Gross Economic Profit Analysis, 2012. [Online]. Available: https://www.youtube.com/watch?v=0GfES6dgg_8. [Accessed August 2020].
[15] A. Marton, Commercialization of New Technology, 2010. [Online]. Available: https://www1.eere.energy.gov/bioenergy/biomass2010/pdfs/biomass2010_track3_s1_marton.pdf. [Accessed August 2020].
[16] TOTAL, FSU single train-ProQual kick off meeting, Total, 2013.
[17] Total's E&P division, Creating energy, 2012. [Online]. Available: https://www.bomobile.total.com/uploads/gestion_media/2012-03-Createurs_EN.pdf. [Accessed August 2020].
[18] D. Gertenbach and B. L. Cooper, Scaleup issues from bench to pilot, in AIChE National Meeting, Nashville, 2009.
[19] K. Rupp, 42 Years of Microprocessor Trend Data, Feburary 2018. [Online]. [Accessed August 2020].
[20] J. C. McCallum, Disk Drive Prices 1955+, 2020. [Online]. Available: https://jcmit.net/diskprice.htm. [Accessed August 2020].
[21] Aspen Technology, Petrofac Improves Process Design Accuracy by Debottlenecking Gas Processes Increasing Capacity by 20%, 2020. [Online]. Available: https://www.aspentech.com/en/-/media/aspentech/home/resources/case-study/pdfs/at-05802-petrofac-cs-2020-0302.pdf. [Accessed August 2020].
[22] Aspen Technology, Inc., The Innovation Fast Track: Saving Time and Money in New Product Development, 2017. [Online]. Available: https://www.aspentech.com/en/-/media/aspentech/home/resources/case-study/pdfs/at-03648-cs-qenos-the-innovation-fast-track.pdf. [Accessed August 2020].
[23] Aspen Technology, Inc, Control Column Performance with a Plant Digital Twin Using Aspen HYSYS®, 2020. [Online]. Available: https://www.aspentech.com/en/-/media/aspentech/home/resources/case-study/pdfs/at-05656-tupras-case-study.pdf. [Accessed August 2020].
[24] Aspen Technology, Inc., Saudi Aramco Increases Refinery Capacity by 100,000 Barrels/Day Using Plant Digital Twin, 2020. [Online]. Available: https://www.aspentech.com/en/-/media/aspentech/home/resources/case-study/pdfs/at-05658-cs-saudi-aramco_final.pdf. [Accessed August 2020].
[25] Aspen Technology, Inc, European Refiner Tackles Heat Exchange Issues and Saves Millions in the Process, 2018. [Online]. Available: https://www.aspentech.com/en/-/media/aspentech/home/resources/case-study/pdfs/at-03864_ineos.pdf. [Accessed August 2020].
[26] Aspen Technology, Inc, Petrochemical Plant Troubleshoots Using Plant Digital Twin Built with Aspen Plus® and Saves $2.4M USD Annually, 2020. [Online]. Available: https://www.aspentech.com/en/-/media/aspentech/home/resources/case-study/pdfs/at-5657-reliance-cs.pdf. [Accessed August 2020].
[27] Aspen Technology, Inc, Major Korean Chemical Producer Significantly Increases Plant Capacity and Reduces Energy Usage, 2016. [Online]. Available: https://www.aspentech.com/en/-/media/aspentech/home/resources/case-study/pdfs/11-8248-cs-lg-chem-final-(1).pdf. [Accessed August 2020].
[28] Aspen Technology, Inc., Mid-Size EPC Reduces Cost of Innovative Ethanol Dehydration System By 18% with Aspen Custom Modeler, 2015. [Online]. Available: https://www.aspentech.com/en/-/media/aspentech/home/resources/case-study/pdfs/11-7753-cs-hitachi-zosen.pdf. [Accessed August 2020].

[29] Aspen Technology, Inc., Production Optimization of Natural Gas Pipelines & Production Facilities Using Performance Engineering, 2017. [Online]. Available: https://www.aspentech.com/en/-/media/aspentech/home/resources/case-study/pdfs/at-05713-cs-ypfb.pdf. [Accessed August 2020].
[30] Aspen Technology, Inc, Fluor Achieves Significant Time Savings in SRU Simulation, 2020. [Online]. Available: https://www.aspentech.com/en/-/media/aspentech/home/resources/case-study/pdfs/at-05712-fluor-case-study.pdf. [Accessed August 2020].
[31] Aspen Technology, Inc, Global Engineering Organization Improves Bids and Estimates with Aspen Capital Cost Estimator, 2015. [Online]. Available: https://www.aspentech.com/en/-/media/aspentech/home/resources/case-study/pdfs/11-7040-cs_linde_f.pdf. [Accessed August 2020].
[32] Aspen Technology, Inc., Increasing Capacity in Sulfur. Production Using Sulsim™ Modeling, 2018. [Online]. Available: https://www.aspentech.com/en/-/media/aspentech/home/resources/case-study/pdfs/at-04709-cs_siirtec.pdf. [Accessed August 2020].
[33] R. Geissbauer, J. Vedso and S. Schrauf, Industry 4.0: Building the digital enterprise, 2016. [Online]. Available: https://www.pwc.com/gx/en/industries/industries-4.0/landing-page/industry-4.0-building-your-digital-enterprise-april-2016.pdf. [Accessed August 2020].
[34] Aspen Technology, Inc., Infineum Increases Production by 2,000 Tons per Year With Global Information Management System, 2018. [Online]. Available: https://www.aspentech.com/en/-/media/aspentech/home/resources/case-study/pdfs/at-04085-cs-infineum.pdf. [Accessed August 2020].
[35] Aspen Technology, Inc, Digital Transformation with Predictive Maintenance Drives Cost Savings, 2019. [Online]. Available: https://www.aspentech.com/en/-/media/aspentech/home/resources/case-study/pdfs/at-05064-ss-hydrogen-compressor.pdf. [Accessed August 2020].
[36] A. Marton, Getting off on the right foot: innovative projects, IPA Newsletter, 3(1), 1–4, 2011.
[37] AACE, Inc., COST ESTIMATE CLASSIFICATION SYSTEM – AS APPLIED IN, 2005. [Online]. Available: https://www.costengineering.eu/Downloads/articles/AACE_CLASSIFICATION_SYSTEM.pdf. [Accessed August 2020].
[38] M. Gerrard, Guide to capital cost estimating, 4th edn, IChemE, 2000.
[39] M. S. Peters, K. D. Timmerhaus and R. E. West, Plant design and economics for chemical engineers, McGraw-Hill Education, 2003.
[40] C. Zhang and M. M. El-halwagi, Estimate the Capital Cost of Shale-Gas Monetization Projects, DECEMBER 2017. [Online]. Available: https://www.aiche.org/resources/publications/cep/2017/december/estimate-capital-cost-shale-gas-monetization-projects. [Accessed March 2021].
[41] A. K. Coker, Ludwig's applied process design for chemical and petrochemical plants, 4th edn, Vol. 1, Gulf Professional Publishing, 2007.
[42] A. K. Coker, Fortran programs for chemical process design, analysis, and simulation, Gulf Professional Publishing, 1995.
[43] Institution of Chemical Engineers, A New Guide to Capital Cost Estimating, London: Institution of Chemical Engineers, 1977.
[44] G. Towler and R. Sinnott, Chemical engineering design: principles, practice and economics of plant and process design, 2nd edn, Butterworth-Heinemann, 2012.
[45] R. Turton, J. Shaeiwitz and D. Bhattacharyya, Analysis, synthesis, and design of chemical processes, Fifth edn, Pearson Education, Inc, 2018.
[46] Chemical Engineering magazine, 2019-chemical-engineering-plant-cost-index-annual-average, [Online]. Available: https://www.chemengonline.com/2019-chemical-engineering-plant-cost-index-annual-average/. [Accessed March 2021].

[47] H. J. Lang, Simplified approach to preliminary cost estimates, Chemical Engineering, 55, 112, 1948.
[48] B. Amigun and H. Von Blottnitza, Capital cost prediction for biogas installations in Africa: lang factor approach, Environmental Progress & Sustainable Energy, 28(1), 134–142, 134 April 2009.
[49] M. Van Amsterdam, Factorial Techniques applied in Chemical Plant Cost Estimation: A Comparative Study based on Literature and Cases, MSc Thesis Work, TU Delft, Nethelands, 2018.
[50] B. M. Nagel, An Update on the Process Economics of Biogas in South Africa based on Observations from Recent Installations, University of Cape Town, 9 February 2019.
[51] M.-H. Cheng, B. S. Dien and V. Singh, Economics of plant oil recovery: a review, Biocatalysis and Agricultural Biotechnology, 18(101056), 1–10, 2019.
[52] U.S. Department of Energy Hydrogen and Fuel Cells Program, Hydrogen Production Cost Estimate Using Biomass Gasification: Independent Review, September 2011. [Online]. Available: https://www.hydrogen.energy.gov/pdfs/51726.pdf. [Accessed March 2021].
[53] National Renewable Energy Laboratory and Pacific Northwest National Laboratory, Process Design and Economics for the Conversion of Lignocellulosic Biomass to Hydrocarbons via Indirect Liquefaction: Thermochemical Research Pathway to High-Octane Gasoline Blendstock Through Methanol/Dimethyl Ether Intermediates, March 2015. [Online]. Available: https://www.nrel.gov/docs/fy15osti/62402.pdf. [Accessed March 2021].
[54] I. Lewandowski, Bioeconomy: shaping the transition to a sustainable, biobased economy, Switzerland: Springer International Publishing, 2018.
[55] T. E. Wolf and P. E. Sugar Land, Lang Factor Cost Estimates, 2013. [Online]. Available: http://prjmgrcap.com/langfactorestimating.html. [Accessed March 2021].
[56] D. E. Garrett, Chemical engineering economics, New York: Van Nostrand Reinhold, 1989.
[57] D. Brennan and K. Golonka, New factors for capital cost estimation in evolving process designs, Chemical Engineering Research & Design, 80(A6), 579–586, September 2002.
[58] D. Green and M. Z. Southard, Perry's chemical engineers' handbook, 9th edn, New York: McGraw-Hill Education, 2018.
[59] F. Arroyo, C. Fernández-Pereira and P. Bermejo, demonstration plant equipment design and scale-up from pilot plant of a leaching and solvent extraction process, Minerals, 5, 298–313, 2015.
[60] FAO, Economic engineering applied to the fishery industry, [Online]. Available: http://www.fao.org/3/v8490e/v8490e05.htm. [Accessed March 2021].
[61] S. Farid, Process economics of industrial monoclonal antibody manufacture, Journal of Chromatography B, 848, 8–18, 2007.
[62] T. Pereira Chilima, F. Moncaubeig and S. S. Farid, Estimating capital investment and facility foot print in cell therapy facilities, Biochemical Engineering Journal, 155(107439), 1–14, 2020.
[63] R. K. Sinnott, Coulson & Richardson's chemical engineering: chemical engineering design, Fourth edn, Vol. 6, Elsevier Butterworth-Heinemann, 2005.
[64] W. D. Seider, D. R. Lewin, J. D. Seader, S. Widagdo, R. Gani and K. Ng, Product and process design principles: synthesis, analysis and evaluation, John Wiley & Sons, May 2016.
[65] R. B. Hey, Performance management for the oil, gas, and process industries: a systems approach, Cambridge: Gulf Professional Publishing, 2017.
[66] Independent Commodity Intelligence Services, ICIS: Petrochemicals, Energy, and Fertilizers Market Information, [Online]. Available: https://www.icis.com/explore/. [Accessed March 2021].
[67] Chemical Week, European Chemical News and Industry Updates, [Online]. Available: https://chemweek.com/Europe. [Accessed March 2021].

Reza Zarghami, Rahmat Sotudeh-Gharebagh, Bruno Blais,
Navid Mostoufi, Jamal Chaouki

3 Process extrapolation by simulation

Abstract: In this chapter, the importance of simulation in the design and operation of chemical processes, as well as its role in scaling studies, is briefly discussed. The chapter includes the role of simulation in iterative scale-up and the basics of the multiscale simulation approach to improve process extrapolation. Conventional process simulators mostly use ideal models to simulate the whole process. More detailed unit simulators with higher resolution abilities, like computational fluid dynamics (CFD) codes, have also been developed to capture more complex process phenomena. Compartment model (CM) is then introduced to approximate complex phenomena for which CFD requires a higher computational cost. The ability of CM, or even the hybrid CFD–CM, to simulate large-scale phenomena allows them to be used as e-pilots, instead of real pilots, in the scale-up studies. Both CFD and its alternative, the simplified CM, must be verified and validated for being properly used as an e-pilot. Verification and validation methods are explained, followed by the introduction of uncertainty quantification methods for quantifying the various sources of uncertainty within the model and related simulation.

Keywords: scale-up, e-pilot, CM, CFD, modeling, simulation, verification, validation, uncertainty

3.1 Introduction

The need to change the scale of a developed chemical process in laboratory size is a general challenge in chemical process industries. The conventional approach to scale-up is based on the principle of similarity from laboratory to pilot, demonstration, and finally full industrial scale. The main objective of scale-up in process design is to

Reza Zarghami, School of Chemical Engineering, College of Engineering, University of Tehran, P.O. Box 11155-4563, Tehran, Iran, e-mail: rzarghami@ut.ac.ir
Rahmat Sotudeh-Gharebagh, School of Chemical Engineering, College of Engineering, University of Tehran, P.O. Box 11155-4563, Tehran, Iran, e-mail: sotudeh@ut.ac.ir
Bruno Blais, Research Unit for Industrial Flows Processes (URPEI), Department of Chemical Engineering, Polytechnique Montréal, C.P. 6079, succ. Centre-ville, Montréal, Québec, H3C 3A7, Canada, e-mail: bruno.blais@polymtl.ca
Navid Mostoufi, School of Chemical Engineering, College of Engineering, University of Tehran, P.O. Box 11155-4563, Tehran, Iran, e-mail: mostoufi@ut.ac.ir
Jamal Chaouki, Department of Chemical Engineering, Polytechnique Montréal, 2500 Chemin de Polytechnique, Montréal, Québec, H3T 1J4, Canada, e-mail: jamal.chaouki@polymtl.ca

https://doi.org/10.1515/9783110713985-003

identify the related criteria (similarity rules) for transferring information between different scales, from the laboratory scale to the full commercial scale. Such information generally includes scale-dependent phenomena that behave differently at different scales. For example, mass and heat transfer phenomena are scale-dependent, while chemical reaction and thermodynamics are scale-independent and do not follow the rules of similarity. However, most industrially important reactions occur in heterogeneous systems in which mass and heat transfer between phases often take place in addition to the reaction. Conversions in different reactors with differing scales are identical only if both the transfer processes and chemistry were similar. Therefore, a reaction may have various conversions at different reactor scales. Scaling is typically achieved through conducting experiments in three main scales: laboratory, pilot, and demonstration. In the laboratory experiments, mostly the scale-independent phenomena are evaluated and some physical properties (e.g., densities, viscosities, and specific heat) are determined. The pilot experiments are designed for a particular purpose related to an industrial plant. Pilot-scale experiments represent an essential step in the investigation of a process toward a commercial scale. The pilot experiments take into account both scale-dependent and scale-independent phenomena within the postulated industrial and operating constraints. A pilot scale uses the laboratory-scale information and provides necessary information about the process with a lower resolution at a higher scale. Scale-up problems are investigated through pilot-scale experiments. The pilot scales should be designed with reasonable costs in mind. The demonstration scale is based on a modest scale (e.g., one-tenth of the full scale). Experiments at the demonstration level are very costly, but in most cases inevitable.

Today, the conventional scale-up method faces two major problems: the limitations of the dimensional similarity to provide direct and detailed quantitative information and the limited ability to minimize project costs by decreasing the size of the pilot or changing its type. On the other hand, there are examples of process design experiences that have failed to achieve their goals, despite running experimental pilots (see Chapter 2 for details). A new scaling approach is therefore needed to address these problems. Using basic conservation laws, along with the laws of thermodynamics and rate of chemical reactions to mathematically describe the process, can mitigate these shortcomings. With this new scale-up method, the use of existing correlations or scaling rules is combined with the proper mathematical models to predict the changes in the size and type of the scaled-down pilot or even consider the model as an e-pilot. Modifying the scale is accompanied by changes in the spatial sizes and may involve changing the model resolution. The method, therefore, should be adopted at the scales at which data are available, and results are needed. In this context, scaling associated with a simulation tool is the process of showing how and to what extent the simulation validated in one or several reduced scale experiments can be applied with sufficient confidence to the real process at higher scales.

3.2 Terminology

Like other fields, specialized terminology is used by the simulators community. The definitions of the following terms may be helpful for those new to the field:

Computational fluid dynamics: Computational fluid dynamics (CFD) is a computational technique used in the simulation of complex fluid flows when a high resolution is required.

Compartment model: The compartment model (CM) is a low computational cost model used to capture typical hydrodynamic features of the system.

Iterative scale-up: Migration of a process from the laboratory to industrial scale and getting back to either the laboratory, pilot, cold model, or process simulation stage for improving key characteristics necessary for the industrial-scale design.

e-Pilot: A numeric model that represents a real pilot plant and can be used instead of a pilot in scaling studies.

Pilot: A small-scale unit that is employed to study the behavior of a process before using it on a large industrial scale.

Model: A model is a mathematical description of a system.

Modeling: Modeling is the process of building a model.

Simulation: Simulation is the act of using a model to study the behavior and performance of a system.

Simulator: A simulator is software used for the simulation of a chemical process.

Verification and validation: Verification and validation (V&V) are the initial steps to assess the accuracy and reliability of modeling, programming, and simulation phases.

Uncertainty quantification: Uncertainty quantification (UQ) is the quantitative characterization of uncertainties in modeling, programming, and simulation steps.

3.3 Simulation in iterative scale-up

Simulation is the enabling technique in iterative scale-up, as mentioned in Chapter 2. In some cases in the industry, the design, construction, and start-up of industrial-scale processes are performed without prior research and development at the pilot plant scale. This approach is used by start-up companies for processes containing one major piece of unit operation [1] with further trial and error during the operation to minimize risks. However, simulation can be used in process or equipment design to reduce costs and time to market. Simulation can help to establish the process concept and generate the basic data on the streams, their compositions, sizing, and rating of the main equipment. Limited information is reported in the literature on industrial design with process simulation tools due to confidentiality issues. Nevertheless, the growth of information being released about this technique is evident. For example, the Lonza Group AG tried to use simulation tools in many cases, even for rather complex processes and unit operations, such as extractive distillation [2–6]. They designed an entire unit by simulation using very little key laboratory data to eliminate uncertainties. These uncertainties, which cannot be simulated, were on solid precipitation, foaming, and product decomposition.

Based on multiple studies, there has been no piloting approach reported that has largely reduced the cost and time to market [3]. If needed, validation of the thermodynamic model (invariant parameters) can be conducted with a single lab-batch distillation. Staak et al. [2] used process simulators for the entire industrial design of a dividing wall column to be employed in multipurpose production. A new approach was introduced to overcome the frequently occurring convergence problems, which enhanced the robustness of the simulation. Since the launch of their column, several different processes were completed in the equipment. It is worth mentioning that the process was designed and built solely based on simulation. On the other hand, the scale-up was directly applied from the simulation to the plant scale. A survey of the literature reveals the wide application of simulation tools in the industry to meet technical, economic, and environmental needs [7–11]. Selected examples are shown in Chapter 2.

With the help of simulation, the iterative scale-up can improve process extrapolation, drastically reduce costs and time in the chain of process design, and decrease the uncertainties related to scale-up with the details provided in the following sections.

3.4 Multiscale simulation approaches

While the terms *simulation* and *modeling* are often used synonymously, modeling describes the process by mathematical equations whereas simulation solves the equations and studies the performance of the process. A complete model consists of mass, momentum, and energy conservation laws coupled with phase equilibrium and chemical kinetic equations. A model is similar to, but simpler than, the real process due to imposed simplifications and approximations in its formulation [12]. With the chemical process simulation it is possible to evaluate behavior and stability to optimize and predict the performance of a chemical process. The model employed for a chemical process takes into account the process of operating conditions and physicochemical properties. A chemical process generally includes various connected units, related equipment, and their connections. Notable computational tools used for solving process models are general engineering process simulators and more detailed engineering software.

Figure 3.1 illustrates a typical chemical process and its building blocks. A chemical process can be categorized into three different levels:

- **Process, plant, or system**: the whole process or plant, for example, gas-phase polyethylene production process, fluid catalytic cracking process;
- **Component, equipment, or unit:** for example, mixer, separator, reactor, bubble column, spray dryer;
- **Local zone or region:** for example, the impeller region in a stirred tank reactor, or inlets, outlets, distributor region, or dead zone in a chemical reactor.

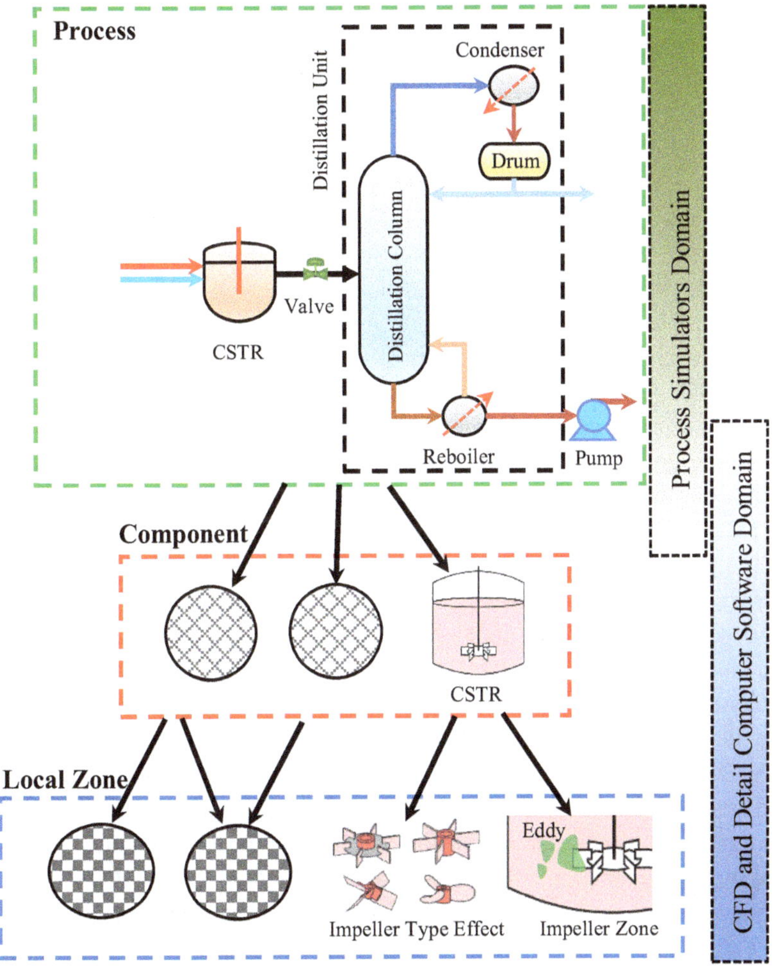

Figure 3.1: Multiscale simulation approach.

Process simulators (e.g., HYSYS® and Aspen®) are widely used for the simulation of chemical processes in steady-state and dynamic modes. These process simulators are used from the design stage to predict the whole process behavior and to optimize and control the process operation. Nowadays, specialized toolboxes are employed in process simulators to produce a preliminary design and relative cost estimation. On the other hand, more elaborate software, like unit simulators, CFD codes (commercial software, such as ANSYS® Fluent, or open-source codes, like OpenFOAM®), or computer codes for designing specific components mostly developed by designers, is developed in the detailed engineering phase. These unit simulators are used by process engineers and equipment designers to study detailed behavior (e.g., final equipment design or thorough cost estimates) within a lower scale of interest and higher resolution of the individual components or local zones. CFD has been used extensively in the simulation of complex flows when a high degree of precision is needed. The information estimated by CFD codes can also be integrated into the simulator [13, 14]. For example, Fluent can compute the flow pattern and species distribution in a stirred tank reactor and the weighted average of the field variables at the reactor outlet can be transferred to Aspen Plus®. On the other hand, stream information, physical properties, thermodynamic models, and reaction kinetic data can be sent back from Aspen Plus to Fluent.

Figure 3.2 illustrates the idea of the multiscale approach. Smaller timescale phenomena occur at corresponding lower length-scale systems. The information at the smaller (fine) length scale and timescales is passed into the larger (coarse) scales. The most common goal in using a multiscale approach is to estimate the macroscopic behavior of a process based on the information in the smaller scales (which is necessary for scale-up). The second goal of the multiscale approach is to estimate and control the finer scale phenomena while manipulating macroscopic variables, such as pressure and temperature (which is used for scale-down).

In a multiscale approach, the smaller scale models consider various more complex phenomena (e.g., fluid eddies and reaction) and their interactions in detail. These interaction details can be used with some assumptions and averaging to develop closure laws for calculating the effective interactions (e.g., conversion in the reactor) in the larger-scale models, making it possible to capture essential information required on a larger scale. Alternatively, the calculation of effective interactions can be performed using local experimental data, if available. Transferring the information from one (e.g., temporal or spatial) scale to another, however, is not an easy task and one of the greatest scientific challenges. This issue is primarily found in a chemical process where differences in scale exist in nearly all modeling and simulation stages. Models are typically developed for a specific temporal and spatial scale, yet they are often calibrated, verified, validated, and applied at larger or even smaller scales.

Large-scale components can be simulated based on the CMs or even a hybrid approach of CFD codes and CMs. The CMs are broadly applied practices in chemical engineering [15] and are an appropriate way to approximate complex phenomena with

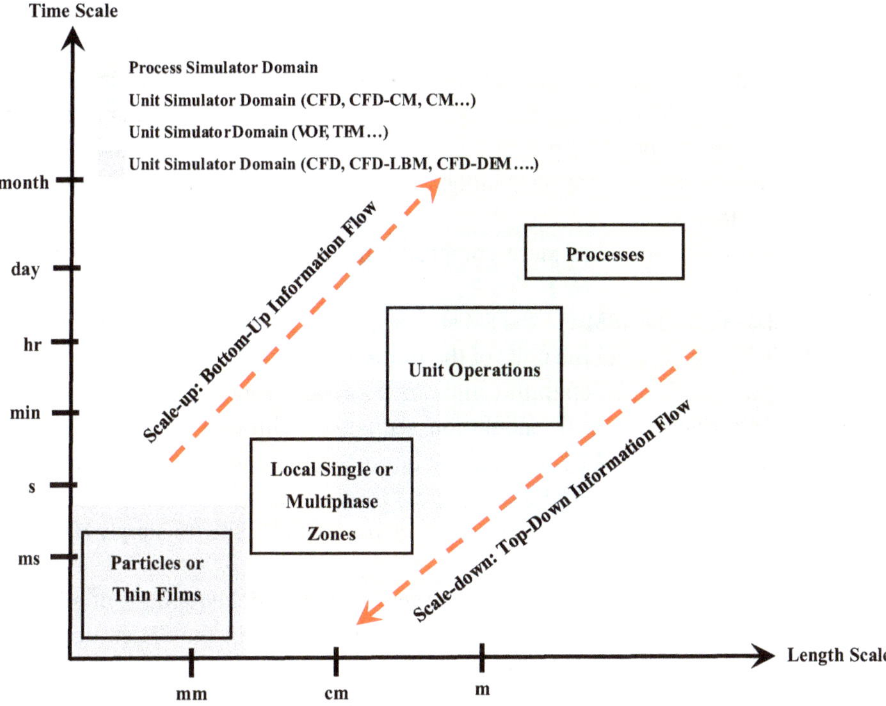

Figure 3.2: Schematic of the multiscale approach.

significantly less computational demand than CFD. The CM utilizes simple models as building blocks or nodes with low resolutions and are, in most cases, ideal. The complexity of the pure CFD models is reduced with the proper integration of CMs into CFD computer codes. Model simplification and/or reduction, therefore, allow components to be simulated on larger scales with relatively good precision and acceptable computation times, which is important for model-based scaling studies as well as building an e-pilot.

3.5 Process simulation

Process simulation is a major activity in process engineering and used in all stages of process design, operation performance, and research and development. The purpose of process simulation is to illustrate a chemical process through a mathematical model that includes the calculation of mass and energy balance equations along with phase equilibria and transport and chemical kinetic equations. Process simulation does not encompass the whole reality of a process but is an acceptable approximation of a real system depending on the level of assumptions made. Due to the difference between a model and reality, simulator users must interpret the results of the process simulation with sufficient confidence. The approximate model used in the process simulation includes some linear, nonlinear, and differential algebraic equations and their solution methods. A chemical process simulator is software used for simulating the chemical process behavior in steady-state or dynamic modes. Control of a chemical process, properties estimation and analysis, process optimization, equipment sizing, and cost estimation are other capabilities of current advanced process simulators.

Currently, there are three main computational strategies in the solvers of the simulators [16]:

- **Sequential modular (SM)**: In the SM strategy, the sequential order of calculations is established to solve all the units of the process one by one. The model equations are solved in each unit separately and move forward. A process with recycles is broken into one or several calculation sequences (through tear streams) and the solution is obtained by an iterative approach in which both units and tear streams should satisfy the convergence criteria. Examples of software using this strategy are AspenTech™ software, such as Aspen HYSYS (or simply HYSYS®) and Aspen Plus®.
- **Equation oriented (EO)**: In the EO strategy, all the model equations of all units are solved simultaneously, instead of solving each unit in the SM strategy, the EO assembles all the model equations and solves them simultaneously. Examples of software using this strategy are gPROMS, Aspen Dynamics®, ASCEND, and SPEEDUP.
- **Hybrid strategy**: In a hybrid strategy, a combination of SM and EO strategies is used. Two levels of units and flowsheet are defined in the hybrid strategy. The simulation begins with the solution of tense models at the unit level in a sequential step and then a simultaneous step starts for the solution of the linear models at the flowsheet level [17]. An example of software using this strategy is the new release of Aspen Plus.

While both SM and EO strategies have their strengths and weaknesses, the SM is normally used in steady-state simulations and the EO is usually used in dynamic simulations and optimization.

Commonly used process simulators with a significant share in the chemical process market are listed in Table 3.1.

3.6 Unit simulation

Unit simulation refers to the simulation of the internal behavior of a unit operation through numerical modeling. It allows engineers and researchers to predict the spatial and possibly temporal variation of temperature, fluid velocity, reactant

Table 3.1: Commonly used process simulators in chemical process industries.

Process simulators	Developer	Application	Operative system	License
Aspen HYSYS	Aspen Technology	Process simulation and optimization	Windows	Commercial
Aspen Plus	Aspen Technology	Process simulation and optimization	Windows	Commercial
Aspen Dynamic	Aspen Technology	Dynamic process simulation	Windows	Commercial
CHEMCAD	Chemstations	Process simulation	Windows	Commercial
DYNSIM	AVEVA	Dynamic process simulation	Windows	Commercial
gPROMS	PSE Ltd.	Advanced process simulation		
Petro-SIM	KBC	Dynamic process simulation	Windows	Commercial
PIPESIM	Schlumberger	Steady-state flow line simulation	Windows	Commercial
ProMax	Bryan	Process simulation	Windows	
PRO/II	AVEVA	Steady-state process simulation	Windows	Commercial
SPEEDUP	Aspen Technology	Dynamic process simulation	Windows	Commercial
UniSim	Honeywell	Process simulation and optimization	Windows	Commercial

concentration, and other physical variables of interest. Although the increased capacities of computer architecture have made the unit simulation more accessible than ever, it still requires a significant investment of time, money, and human resources to achieve state-of-the-art results. This is especially true when the unit operation includes fluid flows, granular flows, or other complex nonlinear physics for which the state of the art is still constantly evolving. When the physics are simple and straightforward (e.g., conduction heat transfer), simulation of the unit is relatively simple and can be used to accurately obtain information about temperature profile, heat fluxes, and other variables in a relatively small turnover time, depending on the complexity of the geometry. In practice, however, this is rarely the case. Models for industrial unit operations are generally significantly more complex than the simple heat transfer problem.

Unit simulation takes many names and forms depending on the physics, which is solved. For example, CFD refers to the simulation of the dynamics of fluid flow and is leveraged by numerous industries, ranging from aeronautics to pharmaceuticals.

The same can be said about computational electromagnetics, granular flow modeling, computational heat transfer, or computational structural mechanics. The combination of one or more of the aforementioned fields and their coupling is referred to as multiphysics modeling. Chemical processes have multiscale, multiphysics, and multiphase features in nature. With CFD, the flow of phases can be modeled by the Eulerian approach in continuum domains. This hypothesis may be true for fluid phases, but it may bring less accurate results when considering noncontinuum discrete phases (e.g., solid particles) as a continuum. To properly model the noncontinuum discrete phases, the Lagrangian approaches, such as the discrete element method (DEM), have been developed, in which the motion of individual discrete elements is tracked in space and time. The Lagrangian approach is complementary to the Eulerian approach for modeling multiphase flows and the combination of both is referred to as the Eulerian–Lagrangian approach.

Figure 3.3 illustrates the type of results that can be obtained with single-phase modeling, in this case for mixing through CFD. Figure 3.4, on the other hand, demonstrates a result for granular flow obtained with the DEM, and Figure 3.5 shows multiphase flow obtained with coupled CFD–DEM. Some popular CFD/DEM-based simulation tools are listed in Table 3.2.

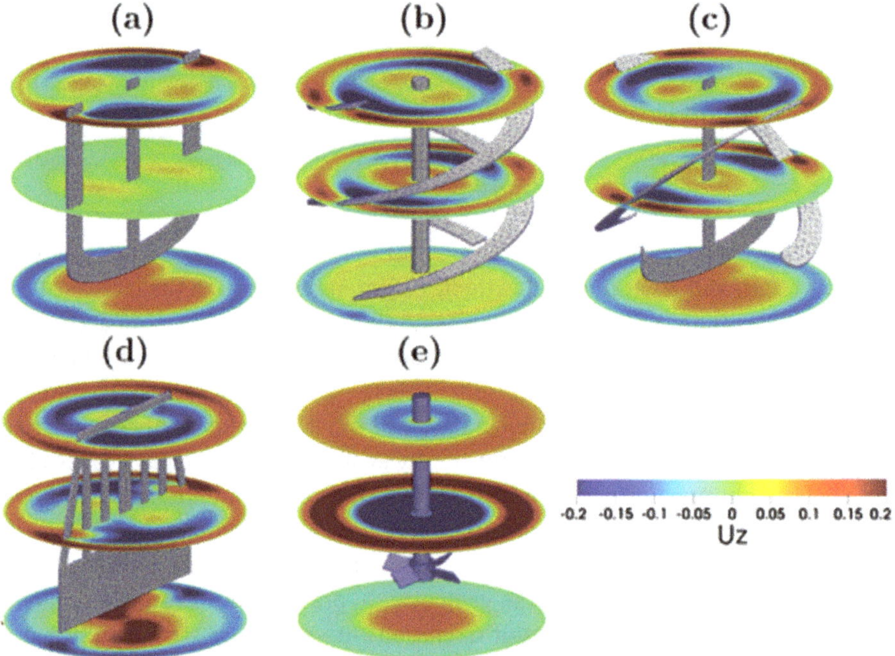

Figure 3.3: Axial flow patterns obtained for a series of industrial mixers using single-phase CFD: (a) anchor, (b) helical ribbon, (c) Paravisc, (d) Maxblend, and (e) pitched-blade turbine (these mixers can be fully simulated in the laminar regime at the industrial scale [18]).

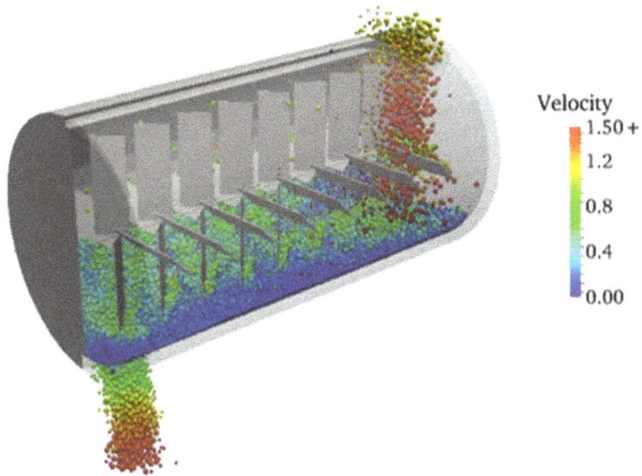

Figure 3.4: Granular flow in a continuous mixer obtained through the discrete element method (DEM) [19].

With the advent of computing power and new commercial or open-source simulation software, unit simulation is now more popular than ever. It can produce visually appealing results and present the intuitive intrinsic behavior of a unit operation. However, it is often used in situations where it is not needed. The following two scenarios may occur and are pitfalls that should be avoided:

- An existing solution already provides a more than satisfactory answer to the problem being investigated. In this scenario, an equally good answer can be obtained using unit simulation, although at a cost of additional resources and time.
- The problem is too complex or computationally intensive to be modeled accurately using unit simulation. Consequently, no satisfactory results will be obtained if the model is too inaccurate and/or the simulations take too much time (weeks or months), and/or the model requires ad-hoc patches or a sub-model to ensure that it adequately represents the reality. This is especially true for multiphase CFD, which can generate visually appealing results that sometimes may have little to no physical grounding. Thus, learning when to use numerical modeling is the key to a successful scale-up of unit operations.

There exists a large number of numerical techniques to solve the motion of the fluid, the transfer of energy, the dynamics of granular materials, and the electric and/or magnetic fields. Each numerical method, whether more general like the finite difference, finite volume, or finite element approaches, or more specialized, like the lattice Boltzmann method and kinetic Monte Carlo, can be the topic of several specialized books. Consequently, we are content to present a brief overview of

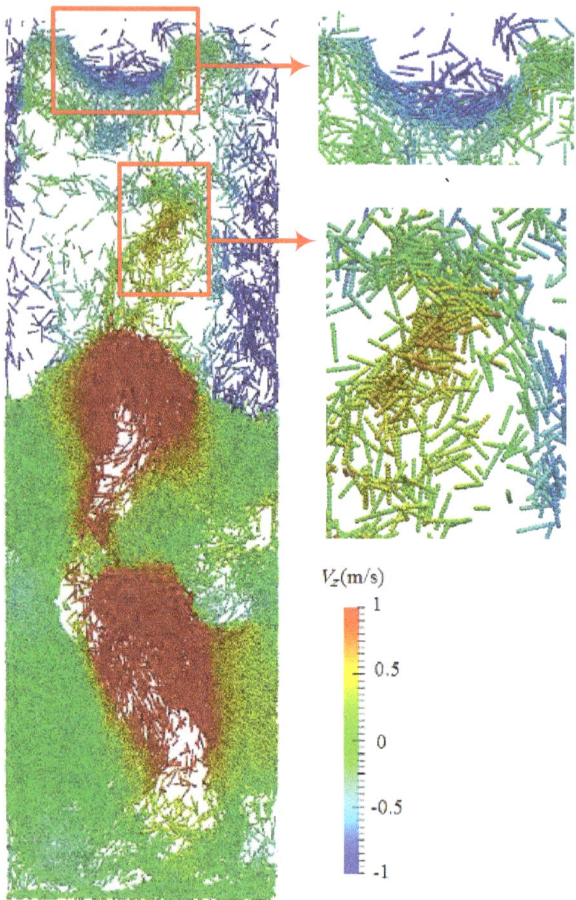

Figure 3.5: Elongated spherocylindircal particles distribution obtained through CFD–DEM simulation of a fluidized bed.

Table 3.2: Commonly used CFD/DEM simulators in chemical process industries.

Process simulators	Developer	Application	Operative system	License
ANSYS CFX	Ansys, Inc.	CFD	Windows	Commercial
ANSYS Fluent	Ansys, Inc.	CFD	Windows	Commercial
ANSYS Multiphysics	Ansys, Inc.	CFD Multiphysics simulation	Windows	Commercial
COMSOL Multiphysics	Comsol, Inc.	CFD Multiphysics simulation	Windows	Commercial

Table 3.2 (continued)

Process simulators	Developer	Application	Operative system	License
Autodesk CFD	Autodesk, Inc.	CFD	Windows	Commercial
SimScale	SimScale GmbH	CFD	Windows/Linux	Open source
FLOW-3D	Flow Science	CFD	Windows/Linux	Partially open source
PIPESIM	Schlumberger	CFD Multiphase flow simulation	Windows	Commercial
LIGGGHTS	DCS Computing	DEM	Linux	Open-source
CFDEM®coupling	DCS Computing	CFD-DEM	Linux	Open-source
Aspherix	DCS Computing	DEM	Windows/Linux	Commercial
MercuryDPM	MercuryDPM	DPM	Windows/Linux	Open-source
EDEM	EDEM	DEM	Windows/Linux	Commercial

the strategies used to solve single-phase and multiphase flow problems and highlight the key challenges associated with simulating the scale-up of each unit operation.

3.6.1 Single-phase flow

Single-phase unit operation refers to an operation in which there is a single liquid or gas as the continuous phase in motion and that can be modeled by the Eulerian approaches (e.g., volume of fluid). A continuous fluid can contain multiple species as long as they form a single-phase and, as such, they must be miscible. Turbulence is the biggest challenge encountered in single-phase unit operations. The core issue with turbulence is its multiscale character. Large turbulent structures generate smaller structures up to a certain scale at which viscosity can play a dominating dissipating effect. A typical spectrum of energy as a function of wavelength for turbulent flow is shown in Figure 3.6. This figure illustrates three general models used to simulate turbulent flows: direct numerical simulation (DNS), Reynolds-averaged Navier–Stokes (RANS) simulation, and large-eddy simulation (LES).

The direct numerical simulation aims to simulate the spectrum of the turbulent scale. Although it is generally the most accurate approach, it becomes prohibitively impossible to use as the Reynolds number increases, since the finest mesh resolution must be proportional to $L/Re^{\frac{3}{4}}$. For example, simulating the entire turbulent cascade in a vessel agitated by a stirrer the size of $L = 1$ m at a Reynolds number of 10^6 would require a mesh resolution of 30 µm and, consequently, over 10^{15} cells. This is beyond the reach of current computing capacities. Consequently, as the Reynolds number increases, DNS becomes less and less usable. RANS solves a statistically averaged form of the Navier–Stokes equation alongside a turbulence model, which aims at modeling the entire cascade. RANS solutions can be obtained relatively fast on workstation-type machines and are readily available in the majority of commercial and open-source software. However, their accuracy can sometimes be lacking, especially if the transient characteristic of the process greatly affects its outcome. LES appears as an intermediate strategy in which the largest scale of the flow is simulated, but the smaller scales are modeled either through an explicit or implicit filter.

In the chemical process industry, RANS is by far the most commonly used approach. It provides reliable solutions when fluctuations within the flow do not greatly affect its characteristics. The use of LES is recommended when transient flow structures influence the process and its use has been growing considerably in the last few decades.

Not all unit operation scale-up is carried out at a constant Reynolds number. When this is not the case, the Reynolds number generally increases as the size of the unit operation is increased. Consequently, more complex turbulent structures may be encountered. This change may not be beneficial to the outcome of the process, as these structures may lead to the creation of additional dead zones, recirculation regions, or bypasses. CFDs is highly useful in this context to predict how the change in

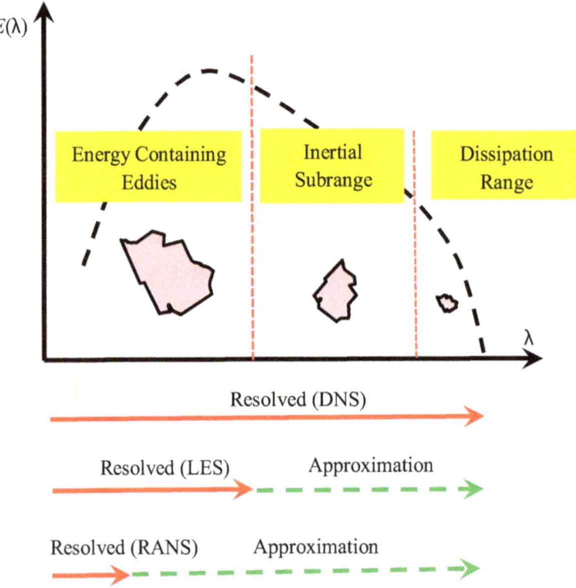

Figure 3.6: General turbulent flow models (RANS, LES, and DNS) and related approximation ranges of energy cascade.

Reynolds number can affect the characteristics of the flow. In this case, choosing the right turbulence modeling strategy is critical.

3.6.2 Multiphase flow

Multiphase flows, whether gas–solid, liquid–solid, liquid–liquid, or gas–liquid–solid, are notoriously difficult to simulate due to their multiscale nature, especially when one phase is dispersed into another. In this case, the interaction between the two phases at the scale of the dispersed phase greatly alters the macroscopic behavior of the flow. This is true for solid–fluid flows (e.g., fluidized and spouted beds), gas–liquid (bubble columns), or liquid–liquid (emulsions). Models for multiphase flows are generally characterized by the scale (macro, meso, and micro) at which the phases are modeled. In general, three families of models exist [20]:

- **Multifluid approaches**: In these strategies, all phases are considered as an interpenetrating continuum. These models are generally computationally affordable because the interfaces between the phases are not modeled and discrete entities (such as particles) are lumped into a continuous field.
- **Unresolved or hybrid approaches**: In these strategies, some phases are modeled with a higher degree of accuracy than the other. For example, within a concurrent coupling framework, the motion of each particle may be simulated by a Lagrangian method, but the flow is modeled at a larger scale by an Eulerian method. These approaches are an interesting compromise because they can simulate significantly larger systems with less loss in accuracy than the multifluid approaches. In general, they cannot simulate flow fields larger than the lab or pilot scale.
- **Resolved approaches**: In these strategies, the flow around each particle, bubble, or droplet is fully resolved. Although these strategies are by far the most accurate, they cannot simulate systems in which the ratio between the characteristic size of the interphase is significantly smaller (e.g., 100×) than the process.

Scaling-up multiphase unit operations through simulation is very difficult since the size of the dispersed phase is generally not affected by the scale. The particles used in a fluidized bed do not become larger if the size of the bed is increased and the bubbles generated in a bubble column do not necessarily become bigger if a column with a larger diameter is used. Consequently, as the size of the system increases, the ratio between the size of the system and the size of the dispersed phase increases. This leads to a prohibitively large number of cells, and generally requires the introduction of alternative modeling strategies with more modeling hypotheses. The engineer is then faced with the challenge of the model becoming so computationally demanding that it can no longer be solved or its accuracy being severely reduced due to the change in modeling strategy. Multiscale simulation is a potent solution to this problem. In this context, more accurate models that simulate the process at the interphase level are used with coarser hybrid or multifluid approaches that can be used at a larger scale. Although this requires a significant investment of time and resources, it can be a very powerful tool to scale-up unit operations.

3.7 Compartment model

Simple reactor models used in chemical process simulators are ideal. These reactors are the continuously stirred tank reactor (CSTR) and plug flow reactor (PFR). Compartmental models are an intermediate approach between ideal systemic models (e.g., CSTR and PFR) of process simulators and more detailed computer programs, like CFDs. CMs are used to approximate the mixing behavior in real reactors without using the governing 3D convection-diffusion partial differential equations. The CMs are developed based on a combination of ideal reactors or even simple, nonideal cases (i.e., dispersion model) to model the real reactor [15]. Tanks-in-series (TIS) is an example of a simple CM that uses a series of CSTRs to approximately model a CSTR with Rushton turbines. The dispersion model is used most often for nonideal tubular reactors. The measured residence time distribution (RTD) is used to adjust the number of tanks in the TIS model and the dispersion coefficient in the dispersion model. The concept of RTD was first developed by Danckwerts [21], who noted its importance in flow processes. The RTD simply shows the age, $E(t)$, of the individual molecules staying in the vessel, or more precisely, the distribution of residence times of the fluid elements. Generally, the RTD is experimentally obtained by injecting a tracer at the inlet and measuring the tracer concentration at the outlet. RTD is associated with bulk flow patterns or macromixing. However, it can also be determined by a reliable model.

Figure 3.7 shows the side capacity model, which is another simple CM for characterizing bypassing or stagnancy phenomena in a real mixed reactor [15, 22]. The real mixed reactor in the side capacity model is divided into two ideal CSTRs with different volumes. To model the bypass, for example, the small CSTR would have a residence time less than that of the large tank. The volume of the compartments in the side capacity model is adjusted in such a way that the RTD of the model is qualitatively comparable with the RTD determined from the tracer measurement.

The volume of a process component (reactor, mixing tank, etc.) in the CM approach is divided into a reasonable number of fully mixed volumes called compartments or nodes. The main challenge in the CM is to characterize the compartments' network with the appropriate number of compartments, the respective volumes, and the relative mass and energy fluxes between the adjacent compartments. The number of compartments strongly depends on the geometrical complexity and the scale of the

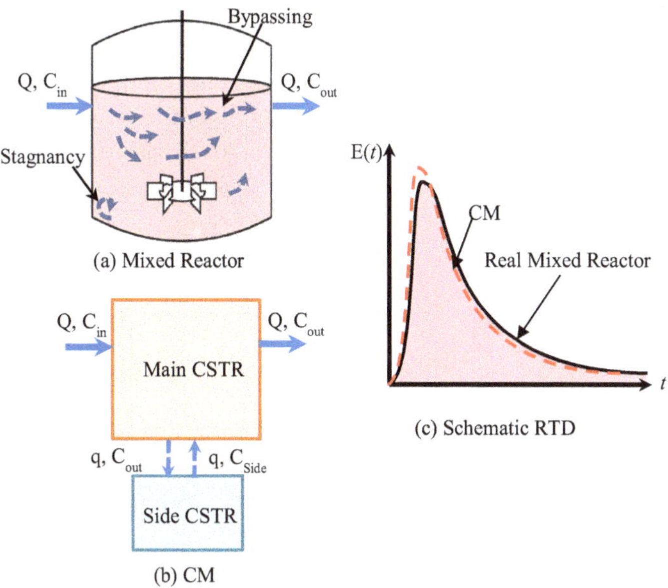

Figure 3.7: Simple compartment model, (a) schematic of a real mixed reactor, (b) side capacity as a simple CM, and (c) schematic RTD curves.

component [23]. Compartment volumes are chosen based on a macro flow behavior (e.g., RTD) in simple CMs or determined based on more details and local criteria, such as flow pattern, temperature or concentration gradients, or even local energy dissipation [24, 25] in advanced CMs.

Figure 3.8 shows two well-defined nodalization approaches to capture the flow behavior within a mixing tank equipped with standard baffles and a flat blade turbine (FBT) radial impeller. A radial flow impeller produces two circulating loops, one below and one above the impeller (Figure 3.8(a)), and mixing occurs between the two loops, but less intensely within each loop. This is an example of a real compartmentalization behavior within a stirred tank reactor with a radial flow impeller.

In advanced CMs, the proper compartments are identified through a manual or automatic nodalization approach [25–29]. Nodalization is mostly based on visual analysis of the underlying hydrodynamics. Different criteria (e.g., flow pattern) are used to detect and define compartments. For example, the overall flow pattern in a mixed reactor is used as a base criterion to detect the compartments. Figure 3.8(b) shows that a total of six compartments are identified by the manual nodalization approach based on the overall flow pattern near the FBT. On the other hand, the results of CFD simulations are processed to identify proper compartments in the automatic nodalization approach [30, 31]. In the automatic nodalization, certain important model parameters (e.g., energy dissipation and shear distribution) are selected as the base criteria for building the compartments. In this approach, the compartments are defined based on the uniformity of these model parameters (called averaging or clustering). As shown in Figure 8(c), the compartments are created by the averaging of the CFD cells, in which the value of the specified property is uniform relating to the given tolerance [26, 32, 33].

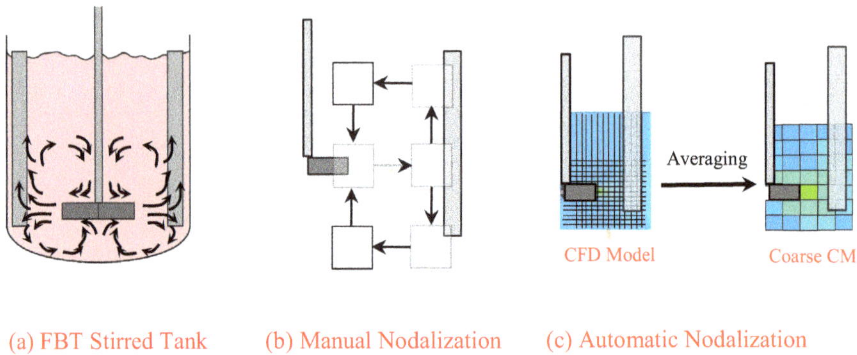

Figure 3.8: Nodalization of (a) a stirred tank with FBT via (b) manual and (c) automatic approaches.

The interactions (e.g., flow fields) among compartments can be easily computed from the results of the CFD model [34]. Since interactions among compartments are not steady and may change with time, some researchers proposed dynamic CM instead of steady state to take into account the changes in the compartmental volumes during the simulation [25]. A coarser CFD mesh is used through the CMs that can capture the hydrodynamics with proper resolution and be coupled with the reaction kinetics and/or chemical thermodynamics with a reasonable computational cost [35, 36]. On the other hand, many hybrid models based on hydrodynamics combining CFD and CMs have been developed [37]. Multiple compartments are defined in these hybrid models in which the hydrodynamic properties of each phase are similar. Recently, some newly developed hybrid CFD–CMs based on the turbulent kinetic energy dissipation rate have been used to estimate the mass transfer operation, for example, in relatively large bubble columns [38].

Both the CFD and its alternative simplified/reduced CMs should be verified and validated (see validation section). More complete and precise experimental data leads to a more accurate validation process, which increases the cost of experiments. Some researchers use heuristic hybrid CFD–CM approaches in which there is limited experimental information [39].

3.8 e-Pilot

Scaling is typically achieved through conducting experiments in a real pilot, as shown with the green dashed line in Figure 3.9. However, within a proper iterative scale-up strategy, the macroscopic behaviors can be estimated from the information in the smaller scales (direct gray dashed route in Figure 3.9). On the other hand, the smaller scale phenomena can be adjusted and controlled through manipulating macroscopic variables (e.g., pressure and temperature) for larger scales (indirect grey dashed route in Figure 3.9). Depending on the computational cost of simulating the real pilot, a valid CM, a valid CFD model, or even a hybrid CFD–CM can be used as the e-pilot of different sizes instead of the real pilot. For example, the experimental kinetic data of a lab scale fixed bed can be formulated through a coupled hydrodynamics–reaction kinetics CFD e-pilot to simulate a two-phase gas–solid reactor. The CFD is used to formulate the prediction of the flow field, then the reaction kinetic model characterizes the reaction conversion.

Unfortunately, most of the studies involving CFD/CM simulations do not report the time required for the calculations. Of course, the computational time depends to a large extent on the simulation method, the multiscale phenomena to be studied, the needed resolution for capturing those phenomena, the available computational resources (hardware configuration), and the software (serial or parallel processing). A large number of discretization methods result in a system of linear algebraic equations that need to be solved. The size of the set of equations depends on the scale of the problem, the number of cells/elements, and the discretization practice. The number of operations for the solution of a system of N equations with N unknowns employing a direct method (e.g., Gaussian elimination) is of the order of N^3. The simultaneous storage of all N^2 coefficients of the set of equations in the core memory is required. The total number of operations in iterative methods (e.g., Krylov methods) is typical of the order of $N^{1.5-2}$ per iteration cycle. However, for realistic CFD problems (3D with 1 million unknowns), iterative methods are more efficient and require significantly less memory than a direct method.

Figure 3.10 shows a very rough estimation of the computing time, in similar circumstances, at different length scales. Macro models (e.g., a pure CM e-pilot) need a very short execution time and, therefore, can be easily used in a different unit and process scales. The hybrid CFD–CM e-pilot requires a longer computing time (from hours to days) and can therefore still be used in different scales of the unit and even process. The CFD e-pilot, however, depending on the circumstances, can take from hours and days to simulate a relatively small unit scale to even months or years for the simulation of larger industrial unit scales. As an example, a rough estimation of the required number of cells to simulate a typical two-phase flow pilot reactor (with the scale in the order of magnitude of 1 m) using a mesomodel (e.g., TFM) is about 10^6. This number becomes approximately 10^9 cells for a ten times larger reactor. As the number of cells increases, the computing cost of the

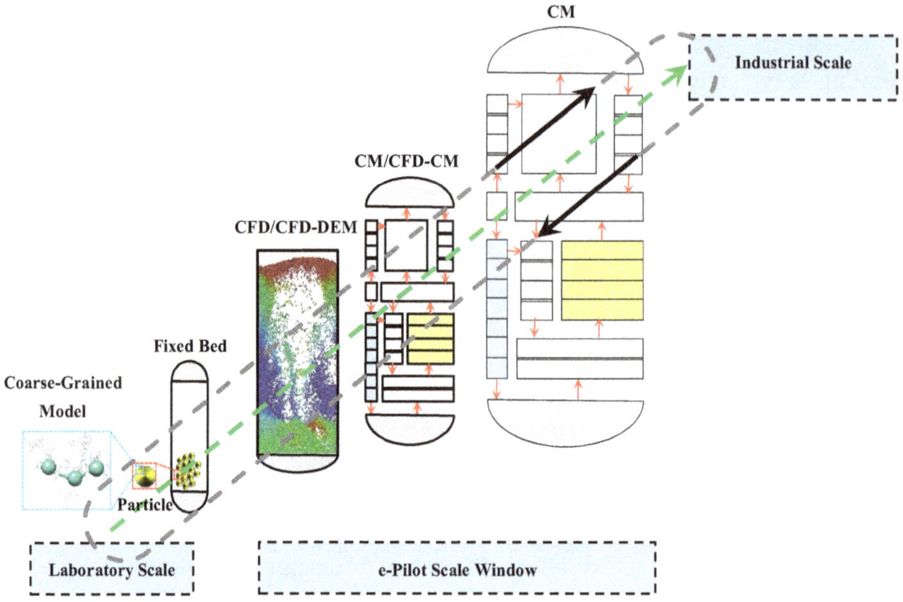

Figure 3.9: Role of e-pilot in an iterative scale-up strategy.

traditional solvers, such as the semi-implicit method for pressure-linked equations, grows significantly (e.g., [43, 44]). It is clear that different CFD models require different mesh sizes and eventually a different number of cells, even for the same length scales.

The parallel algorithms significantly decrease the computational time (depending on the speedup efficiency) by increasing the number of processors (see for example [40–42]). As an example, if a CFD simulation with commercial codes takes 16 days on a single core CPU, adding 32 parallel cores might reduce the computing time to 12 h. Increasing the number of cores from 32 to 64 might decrease the computing time by half again (from 16 h to 8 h). In other words, the speedup (executing time in serial with one processor divided by the executing time in parallel with many processors) of the parallel computing system increases by increasing the number of processors to a certain level. Considering the numerical solution of the 3D Navier–Stokes equations, an increase in the computer power by a factor of 100 should allow for an increase in the number of cells by a factor of about 10 to 100 and a decrease in the cell size by a factor of about 2 to 5. Although Figure 3.10 exhibits merely a rough approximation, it provides a good estimate of the pilot dimensions with available computing resources and needed resolution for capturing the related phenomena.

When the process is in single phase and the laminar regime, the CFD e-pilot performs the same as the full industrial scale. In the case of the turbulent flow, the

CFD e-pilot would perform like the full industrial unit if a RANS model is used. When LES is used, the full unit can be simulated, but the computational cost generally will be the limiting factor. LES simulations require significantly finer meshes which, consequently, greatly increase the simulation time (a bare minimum of 10 times the RANS simulation). Subsequently, for the single phase flow, the limitations of the CFD e-pilot depend on the desired accuracy but generally can be handled through the use of the RANS model or significantly increased computational resources.

- Macromodel: For example, CM, CFD-single phase-laminar, CFD-single phase-turbulent (RANS), and CFD-multiphase (mixture)
- Mesomodel: For example, CFD-single phase-turbulent (LES), and CFD-multiphase (VOF/TFM)
- Micromodel: For example, CFD–DEM, and CFD-LB

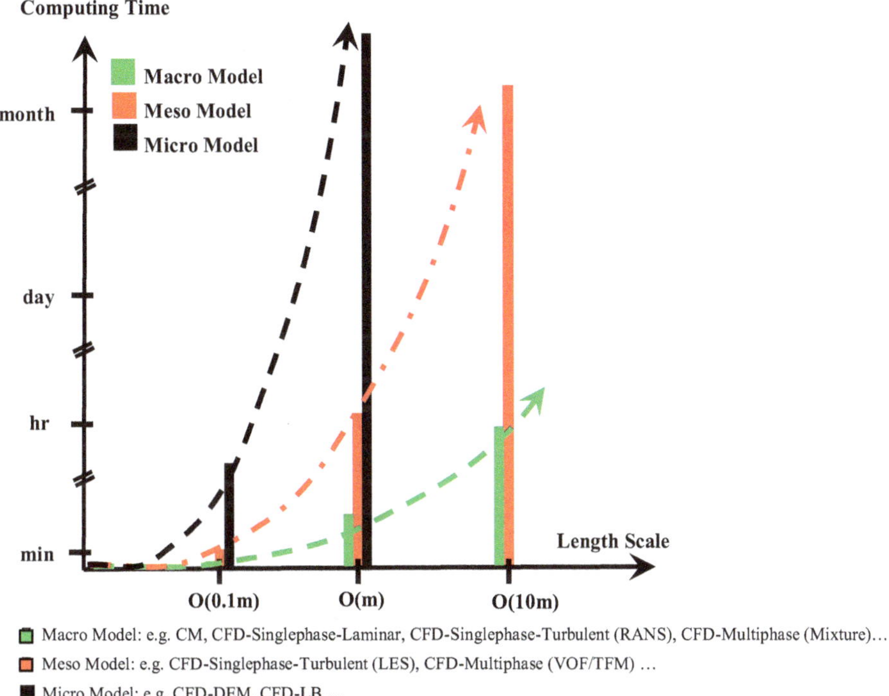

Figure 3.10: A rough estimation of computing time for different simulation methods.

Limitations for the CFD e-pilot become more pronounced when we encounter multiphase flows. For multiphase flows, the characteristic length of the multiphase interaction is at a much smaller scale than the unit operation. For example, the drag force acting on particles in a fluidized bed occurs at the particle scale. Consequently, the scale separation between the multiphase interaction and the process increases as the process is scaled-up. As a result, the computational cost quickly surpasses what is feasible to simulate. This limitation depends on the available computational capacity and the type of multiphase flow being solved for. The mixture model is a simplified multiphase model for industrial scales, and it may be a better option for them. Volume-of-fluid (VOF) and multifluid (e.g., TFM) usually provide more accurate results than the mixture model. The TFM requires a longer simulation time than mixture models, using the same grid [46]. However, TFM can usually provide more accurate results. VOF models generally require much finer grids. The VOF or level set approaches for gas–liquid or liquid–liquid flows require a sufficient number of cells (at least 10) per dispersed entity and are thus quickly limited to small scale geometries. Finally, the TFM approach does not possess any limitation, since it considers the phases as interpenetrating continua. Consequently, notwithstanding turbulence, they can be used to simulate industrial scale units. However, their accuracies are severely limited. Methods based on the DEM can rarely simulate more than 10^8 particles [45]. Approaches based on unresolved CFD–DEM share that limitation, whereas resolved CFD–DEM is limited to less than 10^6 particles.

The continuous development of computer power is boosting the CFD market in chemical engineering. For the next few decades, macroscopic simulation by process simulators and micro- or mesoscopic simulations with unit simulators will continue to play a dominant role in solving the majority of scaling problems in chemical industries. Table 3.3 shows an order-of-magnitude estimate of computing time for typical process units with currently available conventional computing resources.

Table 3.3: Rough estimation of computing time (CT) for typical process units.

Typical process units	**Dimension***	**Macromodel**		**Mesomodel**		**Micromodel**		
		Cells	**CT (h)**	**Cells**	**CT (day)**	**Cells/ particles**		**CT (month)**
Catalytic fluidized beds (FBs)	5–20	3,000	1–1.5	10^6–10^7	1–7	25×10^7	10^{16}	UF$^×$
FB incinerators/CFB combustors	20–20–50	20,000	24	10^7	7	10^9	10^{20}	UF
Spouted beds (SBs)	4–20	2,000	1	10^6–10^7	1–7	10^8	10^{10}	UF
Agitated vessel in gold industries	10–10	6,000	3	10^7	7	10^9		12
Stirred tank reactors	5–8	1,000	0.5	10^7	7	5×10^8		4
Bubble columns (BCs)	10–60	5,000	2–2.5	5×10^7	30		10^{18}	UF
Slurry reactors	10–60	5,000	2–2.5	5×10^7	30	10^9	10^{20}	UF

Table 3.3 (continued)

Typical process units	Dimension*	Macromodel		Mesomodel		Micromodel	
		Cells	CT (h)	Cells	CT (day)	Cells/ particles	CT (month)
Trickle bed reactors	5–20	3,000	1.5				
Multitubular fixed bed (1,000 tubes)	0.02–4[+]	3,000	1.5	10^{12}	360	10^{20}	UF
Thin-film reactors and distillation	4–20	2,000	0.5–1				
Pharmaceutical V-blenders	3–10	500	0.25	10^7	7	10^{15}	UF

*D(m)-H(m) or L(m)-W(m)-H(m).
+Tube size.
×Unfeasible.

3.9 Verification, validation, and uncertainty analysis

The development of any model or simulation of a process or unit operation should comprise a V&V step to assess its accuracy and performance. V&V were defined by the American Institute of Aeronautics and Astronautics (AIAA) [46] and the American Society of Mechanical Engineers (ASME) [47] as:

- Verification: Determining how a computerized implemented model accurately represents the conceptual model.
- Validation: Determining how a computerized model is an accurate representation of the real world.

V&V are two core concepts that are essential for any simulation activity. Using simulation to extrapolate or scale-up operations without a strong understanding of these concepts is dangerous since results can be obtained (and they will), but they will have an unknown accuracy at best or, at worst, will be completely erroneous. V&V has been the topic of reference books, such as Oberkampf and Roy [48] or Roache [49]. Consequently, we focus on giving brief presentations of the elements necessary for a careful V&V of a numerical model when using it to design or scale-up unit operations.

Figure 3.11 shows the role of V&V in modeling and simulation, adopted by the Society for Computer Simulation [50]. Two types of models are identified in this figure, a conceptual model and a computerized model. The conceptual model is composed of all mathematical modeling and equations that describe the reality of interest. The computerized model or code is a computer program that implements a conceptual model for the simulation purpose. As shown in Figure 3.11, the verification deals with the relationship between the conceptual model and the computerized model. Validation deals with the relationship between the computerized model and reality. Qualification deals with the relationship between the conceptual and reality and determines the modeling adequacy and shows how accurately a mathematical model can describe the reality. Qualification is related to the lack of knowledge of the underlying physics in the modeling phase [51, 52].

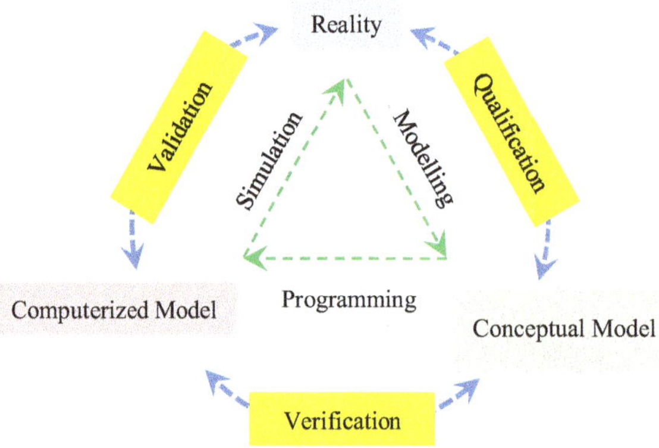

Figure 3.11: Modeling and simulation and the role of verification and validation.

3.9.1 Verification

Verification is generally carried out intuitively either when programming a new software from scratch or adding functionality to an existing model (e.g., adding a user-defined function to the software). In this case, the developer/user generally verifies that the output of the model corresponds to the expected results. For example, one may wish to implement a unique correlation to calculate the specific heat from the temperature. In this case, verifying that the model gives the correct value can be done straightforwardly by evaluating the specific heat obtained at a few given temperatures and comparing it with values calculated outside of the software in which the model is implemented. This trivial example is straightforward but should be used extensively when simple modifications are added to a model. Regretfully, such an approach cannot be used for more complex models that are based on partial differential equations (e.g., fluid mechanics, heat transfer, and advection–diffusion of chemical species). In this case, a more elaborate approach should be used.

The verification is divided into code verification and solution (calculation) verification. Code verification can be further decomposed into numerical algorithm verification and software quality assurance (SQA). Numerical algorithm verification addresses the implementation of the numerical algorithms in the computerized model and that they worked correctly. For example, the numerical algorithm verification would demonstrate that a computerized model can produce the expected convergence rate as the discretization is refined. Numerical algorithm verification is mostly conducted by comparing computational solutions with highly accurate solutions (e.g., analytical solutions). However, it is almost impossible to find analytical solutions to complex problems. Some practical approaches have been developed to generate highly accurate solutions required for code verification (e.g., the method of manufactured solutions). The emphasis in SQA, however, is on computerized model reliability and producing repeatable results on specified computer hardware, operating systems, and compilers. On the other hand, the main focus of SQR is to identify and eliminate programming and implementation errors within the software. SQA procedures are needed during the computerized model development process. The purpose of solution verification is to identify, quantify, and remove the numerical error (e.g., insufficient spatial or temporal discretization, insufficient convergence tolerance, round-off error, incorrect inputs). Three different sources of errors, blunders or mistakes are addressed by solution verification in the programming stage: The first is errors in the preparation of the input. The second is numerical errors resulting from the discretization of the conceptual model on a computer and third is errors, blunders, or mistakes in any processing of the output. The first and third sources refer to human errors exclusive of any other sources. It should be noted that the solution verification does not account for the relationship between the simulation results and the reality, which is the main subject of validation. However, the solution and its error estimation from the solution verification are employed in the validation.

We limit ourselves to the presentation of a single universal approach that can be used for ordinary or partial differential equation models: the method of manufactured solutions. All numerical models, whether they aim at solving an ODE or a PDE, have an order of convergence. For ODE, such as those that are solved in the system simulation, this order of convergence describes the error that is committed in the time discretization. For PDE, this is slightly more complex as there is an error on the spatial discretization and, possibly, on the time discretization. The order of convergence for both these errors does not have to be the same (e.g., first-order time integration scheme is generally combined with second-order accuracy in space). This order of convergence can always be demonstrated mathematically for problems that are sufficiently regular (i.e., without strong discontinuity, such as shocks) and should always be recovered. Order of convergence analysis uses the fact that the order of convergence is known and should be recovered asymptotically to verify that the numerical implementation of a discretization scheme is without error.

Without loss of generality, to evaluate the order of convergence of a simple PDE ($\nabla^2 T = S$, the heat equation for temperature T with a source term S), one needs to calculate a measure of the error of the simulation. The easiest way to measure the error is using the analytical solution and an appropriate norm, such as the L^2 norm of the error. By solving the equation on successively refined meshes, the evolution of the error as a function of the discretization (in this case Δx) can be plotted on a log-log graph and the order of convergence can be calculated from the slope. This procedure is illustrated in Figure 3.12. To maximize the effectiveness of the order of convergence analysis, the analytical solution should be as complex as possible to probe all terms of the model. The difficulty with the order of convergence analysis is that it requires an analytical solution. Although analytical solutions can be found for simple geometries and boundary conditions, it is generally difficult to identify the nontrivial analytical solution and this becomes especially true as the equations to be solved become more complex (e.g., for fluid flow problems). The method of manufactured solutions solves this problem by constructing an analytical solution and establishing the source term that is required to balance these equations.

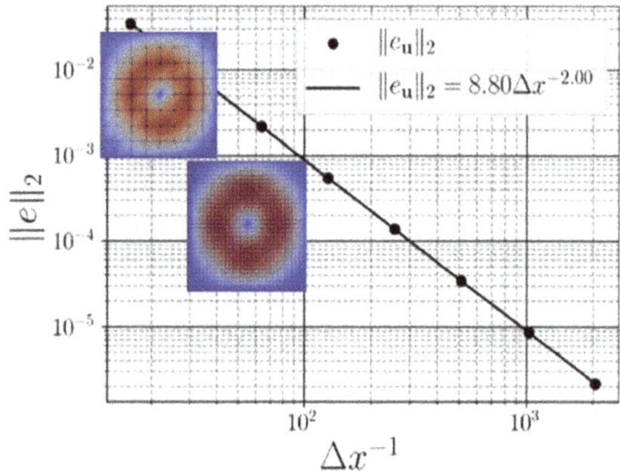

Figure 3.12: Illustration of a manufactured solution and the observed order of convergence.

For example, taking the aforementioned heat transfer equation in 2D, one can define an arbitrary problem with an analytical solution, $T = \sin(\pi x) \cdot \sin(\pi y)$, on a domain that is a unit square $\Omega = [0,1] \times [0,1]$. This solution only satisfies $\nabla^2 T = S$, if S is properly defined. The function that defines S can be calculated by injecting the analytical solution for T in $\nabla^2 T = S$ and, thus, obtaining S. Although this sometimes can be done by hand, it is generally preferable to achieve this using a symbolic manipulator (e.g., Simpy, Maple, Mathematica, Matlab symbolic toolbox). Applying this procedure in our case leads to $S = 2\pi^2 \sin(\pi x)\sin(\pi y)$. With this source term and the known analytical solution, the order of convergence can be carried out directly. The same procedure can be applied to most ODEs or PDEs.

3.9.2 Validation

Although verification is a crucial step, it cannot establish if a model is an accurate representation of the reality of the unit operation. This is the role of validation. Validation is the key that enables the more widespread use of a model for design and scale-up. Validation is conducted by computer code developers (developmental assessment) and/or code users (independent assessment). Generally, validation entails the comparison of the sufficiently accurate simulation results with the measured experimental data. Validation is involved with the identification and quantification of the error and uncertainty in the conceptual and computerized model and its solution, the estimation of the experimental uncertainty, and finally, the comparison between the simulation results and the experimental data from different sources.

Since conducting true validation experiments in many complex systems is infeasible and impracticable, a decomposing-block approach is recommended to use the experimental data from simpler sources [51]. In this approach, the complex system decomposes into some progressively simpler tiers (e.g., subsystem cases, benchmark cases, and unit problems) and the computational results are compared with these reduced scale validation experiments. A systematic validation is conducted through the identification of a validation matrix. A validation matrix is a set of selected experimental data from different tiers [53]. The validation matrix usually includes important local or global measured parameters. Local measurements are generally space-dependent (e.g., flow patterns in an agitated vessel) whereas global measurements are global information (e.g., RTD). Table 3.4 gives some examples of the validation matrix of the local and global measurements for several common unit operations. It should also be noted that the application domain of a model may overlap the validation domain or fall completely outside it. The application domain represents cases of interest for a given application. The validation domain represents all cases where the processes are well understood, and the accuracy of the model has been established through validation. Outside the validation domain, the model may be credible, but there is a lack of confidence in the quantitative capability of the model. In this case, an uncertainty analysis is needed to identify the overall uncertainty of the model.

3.9.3 Uncertainty quantification

As mentioned above, most often both experiments and simulations should be used to solve the scaling. The simulation tool is used to extrapolate from reduced scale validation experiments to the demonstration plant and finally the full industrial scale with sufficient confidence. Simulation of the scaled experiments has a given accuracy defined by the error and it should be determined how this simulation error changes when extrapolating to larger scales. Therefore, scaling applied to a numerical

Table 3.4: Typical validation matrix.

Unit operation	Local measurement	Global measurement
Stirred tank/reactor	Flow patterns measured through PIV	Power consumption obtained through torque measurements
	Dead zones identified through colorimetry	Mixing time obtained through colorimetry and image analysis
		RTD in case of continuous reactor
		Conversion in case of reaction
Fluidized bed reactor	Gas and/or solid flow patterns obtained through radioactive particle tracking	Fluidization surface level
	Solid concentration obtained through tomography	Pressure drop
		Conversion in case of reaction
Rotary kiln	Solid flow patterns obtained through PIV	Relative standard deviation (RSD) obtained through sampling
	Local temperature measurements	
Hopper	Flow patterns measured through PIV	Discharge flow rate and Particles RTD

simulation tool is closely related to the simulation uncertainty evaluation and quantification. UQ is identifying and quantifying the various sources of uncertainty. Different uncertainty sources can be identified and quantified in a well-defined V&V process. Once all sources of the uncertainty have been identified/quantified, sensitivity analysis can show which sources contribute the most [54]. A more detailed list of uncertainty sources can be found in [55, 56, 57]. Figure 3.13 shows different sources of uncertainty within a V&V process, as described below:

- **Parameter uncertainty** is associated with the modeling phase (from reality to conceptual model) and is in close relation to the qualification process. This uncertainty source includes all computer model parameters whose exact values are unknown (e.g., turbulence model parameters).
- **Structural uncertainty** is also related to the qualification process and deals with the model inadequacy, model bias, or model discrepancy due to a lack of knowledge.
- **Algorithmic uncertainty (**also known as numerical uncertainty or discrete uncertainty) arises from the programming phase (from conceptual model to computerized model) and is related to the verification process. This uncertainty source is related to the discretization and solving of the equations and includes temporal and spatial discretization errors, iteration errors, and round-off errors.

- **Parametric variability** is associated with the simulation phase (from the computerized model to reality) and is related to the validation process. This type comes from the variability of the computerized model input variables. Some uncertain input variables do not have a fixed value and most of the time have a distribution (e.g., flow fluctuations).
- **Experimental uncertainty** comes from the variability of experimental measurements. This type of uncertainty is related to the validation process in reduced scale experiments. Measuring equipment used for validation, whether they be for local measurements (such as thermocouple for a temperature or PIV for a flow field) or global measurement (such as the pressure drop through a fluidized bed or the torque generated by the motion of an impeller), all come with their degree of uncertainty. These uncertainties can generally be quantified, either through information from the equipment manufacturer or more often by simply repeating the experiments numerous times and gathering the statistics. Thus, the measurement to which the simulation is compared is not a single measure anymore, but the average of numerous experiments with an appropriate standard deviation. It is important to include this uncertainty when comparing simulation results through the use of a confidence interval. Neglecting to evaluate the degree of uncertainty in the validation data can often lead to not only an erroneous conclusion on the accuracy of those measures, but also on the validity of the simulation themselves.
- **The user/computer/compiler effect** is related to the user's skill of the computer code, hardware, and type of compiler at the time of the simulation.

Moreover, it should be noted that there are two main methodologies for uncertainty quantification: forward propagation of uncertainty and inverse assessment of model uncertainty. In the forward propagation method, the various sources of uncertainty are propagated through modeling, programming, and simulation to predict the overall uncertainty. In the inverse assessment method, however, the model parameters are calibrated using available experimental data (especially at reduced scales). The latter method is in close relation to the parameter uncertainty.

3.10 Conclusion

Conventional scale-up methods are mostly expensive and have various limitations and problems. Simulation, however, can provide valuable quantitative information and minimize the conventional scaling cost by decreasing the size of the pilot or even changing its type to an e-pilot. The objective of this chapter is to explore the possibilities and limitations of using e-pilots in an iterative scale-up methodology.

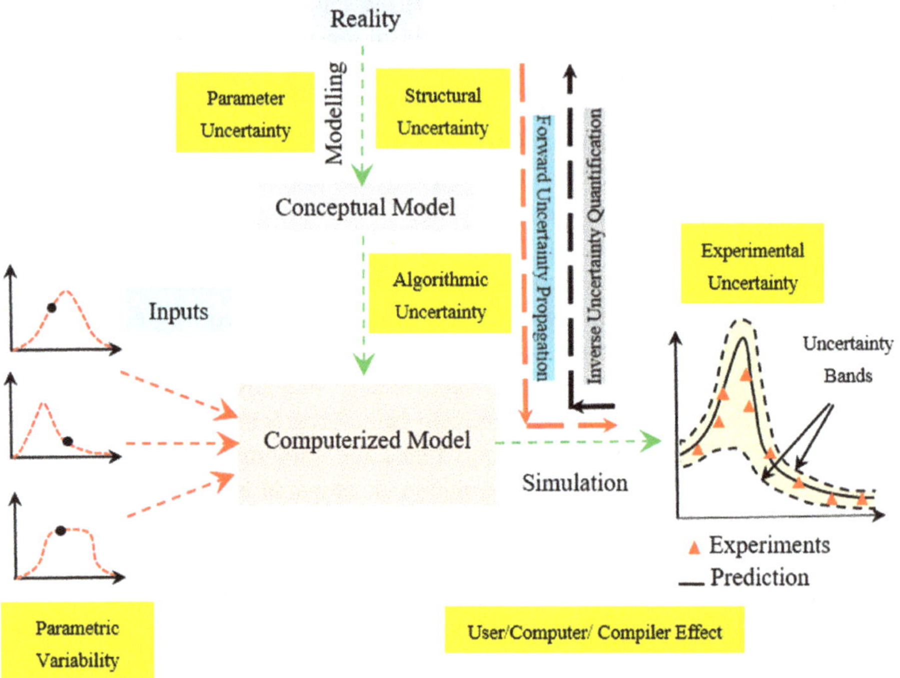

Figure 3.13: Uncertainty quantification.

The role and importance of the simulation in process and unit scales, and consequently in iterative scale-up, are introduced in this chapter. The process simulators are used at the preliminary design stage to predict the behavior of the whole process. However, the unit simulators (like CFD codes) are mostly developed for being used in the detail design phase. The flow of information between process and unit simulations within a multiscale approach is then explored. To overcome the CFD limits and time-consuming issue, the CM and hybrid CFD–CM approaches based on the hydrodynamic similarity are introduced. Model simplification and/or reduction through CMs or even hybrid CFD–CMs can provide a possible simulation of the higher unit scales with relatively good precision and acceptable computation time. Therefore, CMs or hybrid CFD–CMs can be used as e-pilots, which can replace a real pilot. Application of the CFD e-pilot is limited to the unit scale, the needed resolutions, as well as the available computational resources. To have a reliable e-pilot, however, practical V&V methods are needed. Also, the uncertainty quantification methods are required for quantifying the various sources of uncertainty within the e-pilot.

References

[1] J. Harmsen, Industrial process scale-up; a practical innovation guide from idea to commercial Implementation, Second edn, Cambridge: Elsevier, 2019.

[2] D. Staak, T. Grützner, B. Schwegl and D. Roederer, Dividing wall column for industrial multi purpose use, Chemical Engineering and Processing: Process Intensification, 75, 48–57, 2014.

[3] T. Grützner, C. Schnider, D. Zollinge, C. S. Bernhard and N. Künzle, Reducing time to market by innovative development and production strategies, Chemical Engineering and Technology, 39(10), 1835–1844, 2016.

[4] D. Staak and T. Grützner, Process integration by application of an extractive dividing-wall column: an industrial case study, Chemical Engineering Research and Design, 123, 120–129, 2017.

[5] N. Kockmann, M. Gottsponer and D. M. Roberge, Scale-up concept of single-channel microreactors from process development to industrial production, Chemical Engineering Journal, 167(2–3), 718–726, 2011.

[6] G. Heckmann, F. Previdoli, T. Riedel, D. Ruppen, D. Veghini and U. Zacher, Process development and production concepts for the manufacturing of organic fine chemicals at LONZA, CHIMIA, 60(9), 530–533, 2006.

[7] Aspen Technology, Petrofac Improves Process Design Accuracy by Debottlenecking Gas Processes Increasing Capacity by 20%, 2020. [Online]. Available: https://www.aspentech.com/en/-/media/aspentech/home/resources/case-study/pdfs/at-05802-petrofac-cs-2020-0302.pdf. [Accessed August 2020].

[8] Aspen Technology, Inc, Control Column Performance with a Plant Digital Twin Using Aspen HYSYS ®, 2020. [Online]. Available: https://www.aspentech.com/en/-/media/aspentech/home/resources/case-study/pdfs/at-05656-tupras-case-study.pdf. [Accessed August 2020].

[9] Aspen Technology, Inc, Fluor Achieves Significant Time Savings in SRU Simulation, 2020. [Online]. Available: https://www.aspentech.com/en/-/media/aspentech/home/resources/case-study/pdfs/at-05712-fluor-case-study.pdf. [Accessed August 2020].

[10] Aspen Technology, Inc., Production Optimization of Natural Gas Pipelines & Production Facilities Using Performance Engineering, 2017. [Online]. Available: https://www.aspentech.com/en/-/media/aspentech/home/resources/case-study/pdfs/at-05713-cs-ypfb.pdf. [Accessed August 2020].

[11] Aspen Technology, Inc., Saudi Aramco Increases Refinery Capacity by 100,000 Barrels/Day Using Plant Digital Twin, 2020. [Online]. Available: https://www.aspentech.com/en/-/media/aspentech/home/resources/case-study/pdfs/at-05658-cs-saudi-aramco_final.pdf. [Accessed August 2020].

[12] A. Maria, Introduction to modeling and simulation, in WSC97: 1997 Winter Simulation Conference, Atlanta, Georgia, USA, 1997.

[13] S. E. Zitney and M. Syamlal, Integrated process simulation and CFD for improved process engineering, Computer Aided Chemical Engineering, 10(no. -), 397–402, 2002.

[14] L. Vaquerizo and M. J. Cocero, CFD –aspen plus interconnection method. Improving thermodynamic modeling in computational fluid dynamic simulations, Computers & Chemical Engineering, 113(no. -), 152–161, 2018.

[15] O. Levenspiel, Chemical reaction engineering, 3 edn, New York: John Wiley & Sons, 1999.

[16] I. D. G. Chaves, J. R. G. Lopez, J. L. G. Zapata, A. L. Robayo and G. R. Nino, Process analysis and simulation in chemical engineering, New York: Springer, 2016.

[17] M. Marchetti, A. Rao, D. Vickery and t. A. P. D. Team, Mixed mode simulation – Adding equation oriented convergence to a sequential modular simulation tool, Computer Aided Chemical Engineering, 9, 231–236, 2001.

[18] Delacroix, Bastien, Louis Fradette, François Bertrand, and Bruno Blais. "Which impeller should be chosen for efficient solid-liquid mixing in the laminar and transitional regime?." AIChE Journal: e17360.
[19] S. Golshan, N. Mostoufi, R. Zarghami and H. Norouzi, Discrete element method simulation of continuous blenders, in Conference on Modelling Fluid Flow (CMFF2015), Budapest, 2018.
[20] H. R. Norouzi, R. Zarghami, R. Sotudeh-Gharebagh and N. Mostoufi, Coupled CFD -DEM modeling: formulation, implementation and application to multiphase flows, John Wiley and Sons, 2016.
[21] P. V. Danckwerts, Continuous flow systems: distribution of residence times, Chemical Engineering Science, 2(1), 1–13, 1953.
[22] S. Fogler, Elements of chemical reaction engineering, Boston: Pearson, 2016.
[23] A. Egedy, J. A.Oliva, B. Szilágyi and Z. K. Nagy, Experimental analysis and compartmental modeling of the residence time distribution in DN6 and DN15 continuous oscillatory baffled crystallizer (COBC) systems, Chemical Engineering Research and Design, 161, 322–331, 2020.
[24] E. Kougoulos, A. G. Jones and M. W. Wood-Kaczmar, CFD modelling of mixing and heat transfer in batch cooling crystallizers: aiding the development of a hybrid predictive compartmental model, Chemical Engineering Research and Design, 83(1), 30–39, 2005.
[25] M. Öner, S. M. Stocks, J. Abildskov and G. Sin, Scale-up modeling of a pharmaceutical antisolvent crystallization via a hybrid method of computational fluid dynamics and compartmental modeling, Computer Aided Chemical Engineering, 46, 709–714, 2019.
[26] A. Delafosse, F. Delvigne, M.-L. Collignon and M. Crine, Development of a compartment model based on CFD simulations for description of mixing in bioreactors, Biotechnology, Agronomy, Society and Environment, 14, 517–522, 2010.
[27] H. Asadi-Saghandi, A. Sheikhi and R. Sotudeh-Gharebagh, Sequence based process modeling of fluidized bed biomass gasification, ACS Sustainable Chemical Engineering, 3(11), 2640–2651, 2015.
[28] H. Jafari, A. Sheikhi and R. Sotudeh-Gharebagh, Sequential-based process modelling of a circulating fluidized bed reactor, Computer Aided Chemical Engineering, 40, 109–114, 2017.
[29] A. Yousefifar, R. Sotudeh-Gharebagh, N. Mostoufi and S. S. Mohtasebi, Sequential modeling of heavy liquid fuel combustion in a fluidized bed, Chemical Engineering & Technology, 38(10), 1853–1864, 2015.
[30] J. Haag, C. Gentric, C. Lemaitre and J.-P. Leclerc, Modelling of chemical reactors: from systemic approach to compartmental modelling, International Journal of Chemical Reactor Engineering, 16(8), 1–22, 2018.
[31] A. Alvarado, S. Vedantam, P. Goethals and I. Nopens, A compartmental model to describe hydraulics in a full-scale waste stabilization pond, Water Research, 46(2), 521–530, 2012.
[32] A. Delafosse, M.-L. Collignon, S. Calvo, F. Delvigne, M. Crine, P. Thonart and D. Toye, CFD - based compartment model for description of mixing in bioreactors, Chemical Engineering Science, 106, 76–85, 2014.
[33] F. Bbezzo and S. Macchietto, A general methodology for hybrid multizonal/CFD models: Part II. Automatic zoning, Computers & Chemical Engineering, 28(4), 513–525, 2004.
[34] P. Gupta, B. Ong, M. H. Al-Dahhan, M. P. Dudukovic and B. A. Toseland, Hydrodynamics of churn turbulent bubble columns: gas–liquid recirculation and mechanistic modeling, Catalysis Today, 64(3–4), 253–269, 2001.
[35] A. Nørregaard, C. Bach, U. Krühne, U. Borgbjerg and K. V. Gernaey, Hypothesis-driven compartment model for stirred bioreactors utilizing computational fluid dynamics and multiple pH sensors, Chemical Engineering Journal, 356, 161–169, 2019.

[36] E. K. Nauha, Z. Kálal, J. M. Ali and V. Alopaeus, Compartmental modeling of large stirred tank bioreactors with high gas volume fractions, Chemical Engineering Journal, 334, 2319–2334, 2018.

[37] H. Bashiri, M. Heniche, F. Bertrand and J. Chaouki, Compartmental modelling of turbulent fluid flow for the scale-up, The Canadian Journal of Chemical Engineering, 92(no. -), 1070–1081, 2014.

[38] R. Demol, D. Vidal, S. Shu, F. Bertrand and J. Chaouki, Mass transfer in the homogeneous flow regime of a bubble column, Chemical Engineering & Processing: Process Intensification, 144 (no. -), 107647, 2019.

[39] S. Fogarasi, A. Egedy, F. Imre-Lucaci, T. Varga and T. Chován, Hybrid CFD -compartment approach for modelling and optimisation of a leaching reactor, Computer Aided Chemical Engineering, 33, 1255–1260, 2014.

[40] Y. Xue, W. Liu and Z. Zhai, New semi-Lagrangian-based PISO method for fast and accurate indoor environment modeling, Building and Environment, 105(C), 236–244, 2016.

[41] H.-G. Kim, A method of accelerating the convergence of computational fluid dynamics for micro-siting wind mapping, Computation, 7(2), 22 (1–11), 2019.

[42] G. Markou, A parallel algorithm for the embedded reinforcement mesh generation of large-scale reinforced concrete models, in 9th GRACM 2018 International Congress on Computational Mechanics, Chania, 2018.

[43] A. Nikolopoulos, A. Stroh, M. Zeneli, F. Alobaid, N. Nikolopoulos, J. Ströhle, S. Karellas, B. Epple and P. Grammelis, Numerical investigation and comparison of coarse grain CFD – DEM and TFM in the case of a 1 MWth fluidized bed carbonator simulation, Chemical Engineering Science, 163(no. -), 189–205, 2017.

[44] H. Norouzi, R. Zarghami and N. Mostoufi, New hybrid CPU-GPU solver for CFD -DEM simulation of fluidized beds, Powder Technology, 316(no. -), 233–244, 2017.

[45] E. Guerrero, F. Munoz and N. Ratkovich, Comparison between Eulerian and VOF models for two-phase flow assessment in vertical pipes, CT&F – Ciencia, Tecnología & Futuro, 7(1), 73–84, 2017.

[46] B. Blais, D. Vidal, F. Bertrand, G. S. Patience and J. Chaouki, Experimental methods in chemical engineering: discrete element method – DEM, The Canadian Journal of Chemical Engineering, 97(9), 1964–1973, 2019.

[47] AIAA, Guide for the verification and validation, American Institute of Aeronautics and Astronautics, 1998.

[48] H. Coleman, C. Freitas, G. Steele, P. Roache, U. Ghia, B. Blackwell, K. Dowding, R. Hills and R. Logan, Standard for verification and validation in computational fluid dynamics and heat transfer, ASME, New York, 2009.

[49] W. L. Oberkampf and C. J. Roy, Verification and validation in scientific computing, New York: Cambridge University Press, 2010.

[50] P. J. Roache, Verification and Validation in Computational Science and Engineering, -: Hermosa, 1998.

[51] S. Schlesinger, R. E. Crosbie, R. E. Gagne, G. S. Innis, C. S. Lalwani, J. Loch, R. J. Sylvester, R. D. Wright, N. Kheir and D. Bartos, Terminology for model credibility, Simulation, 32(2), 103–104, 1979.

[52] W. L. Oberkampf and T. G. Trucano, Verification and validation in computational fluid dynamics, Sandia National Laboratories, Albuquerque, 2002.

[53] W. L. Oberkampf and T. G. Trucano, Verification and validation in computational fluid dynamics, Progress in Aerospace Sciences, 38(3), 209–272, 2002.

[54] W. L. Oberkampf and M. F. Barone, Measures of agreement between computation and experiment: validation metrics, Journal of Computational Physics, 2017(1), 5–36, 2006.

[55] R. L. Iman and J. C. Helton, An investigation of uncertainty and sensitivity analysis techniques for computer models, Risk Analysis, 8(1), 71–90, 1988.
[56] M. C. Kennedy and A. O'Hagan, Bayesian calibration of computer models, Journal of the Royal Statistical Society: Series B (Statistical Methodology), 63(3), 425–464, 2001.
[57] W. L. Oberkampf, S. M. DeLand, B. M. Rutherford, K. V. Diegert and K. F. Alvin, Estimation of total uncertainty in modeling and simulation, Sandia National laboratorie, Albuquerque, 2000.

Hamed Bashiri, Roshanak Rabiee, Alireza Shams, Shahab Golshan, Bruno Blais, Jamal Chaouki

4 Transition from e-pilot to full commercial scale

Abstract: In this chapter, we first discuss the fundamentals and conventional approaches for the design and scale-up of multiphase mixing systems and highlight the challenges, risks, and consequences of using conventional scale-up. Applications of the iterative scale-up approach are presented as an alternative to conventional scale-up methodologies to find a short and reliable path for large-scale technology implementation. A methodology for developing hybrid multiscale models is introduced as an efficient design tool for the iterative scale-up. We also defined the e-pilot as the largest possible scale to carry out a detailed computational fluid dynamics (CFD) and CFD-discrete element method simulation in a "suitable" time considering all possible governing phenomena. Multiscale gas/liquid flow models for e-pilots to predict the performance of multiphase mixing systems are also introduced, but the transition from e-pilot to an industrial scale remains a real challenge. We present such transition through the development of multiscale models as design tools for the iterative scale-up of gas/liquid stirred tank reactors, and CO_2 capture and acid gas removal using rotating packed beds in which the mass transfer between gas and liquid inside the mixing system is vital for the success of large-scale operations.

Keywords: multiphase flow, mixing systems, multiscale model, iterative scale-up, e-pilot

Hamed Bashiri, CanmetENERGY, Natural Resources Canada, 1615 Lionel-Boulet Boulevard, P.O. Box 4800, Varennes, Québec, J3X 1P7 Canada, e-mail: hamed.bashiri@NRCan-RNCan.gc.ca
Roshanak Rabiee, Department of Chemical Engineering, École Polytechnique de Montréal, P.O. Box 6079, Stn. Centre-Ville, Montréal, Québec, H3C 3A7, Canada, e-mail: roshanak.rabiee@polymtl.ca
Alireza Shams, Department of Chemical Engineering, École Polytechnique de Montréal, P.O. Box 6079, Stn. Centre-Ville, Montréal, Québec, H3C 3A7, Canada, e-mail: alireza.shams@polymtl.ca
Shahab Golshan, Department of Chemical Engineering, École Polytechnique de Montréal, P.O. Box 6079, Stn. Centre-Ville, Montréal, Québec, H3C 3A7, Canada, e-mail: shahab.golshan@polymtl.ca
Bruno Blais, Department of Chemical Engineering, École Polytechnique de Montréal, P.O. Box 6079, Stn. Centre-Ville, Montréal, Québec, H3C 3A7, Canada, e-mail: bruno.blais@polymtl.ca
Jamal Chaouki, Department of Chemical Engineering, École Polytechnique de Montréal, P.O. Box 6079, Stn. Centre-Ville, Montréal, Québec, H3C 3A7, Canada, e-mail: jamal.chaouki@polymtl.ca

https://doi.org/10.1515/9783110713985-004

4.1 Introduction

Industrial processes involve transforming raw materials into added-value products on a large scale. This goal is often achieved by using vessels where chemical and physical transformations of matter take place. These process vessels should be efficiently designed at industrial scale by a process engineer to provide an optimal environment where the raw material can be brought into intimate contact for an adequate length of time (temperature, pressure, etc.).

For an engineer, process or product design often involves a design decision tree whereby the system required to transfer the raw material into the final product is established. The design should guarantee a minimal negative environmental impact and increase the profitability of a given process or, in a nutshell, provide the highest yield for a desired quality product. Therefore, a process engineer has to demystify the relationship between the mixing system (process hardware), operating protocol, and overall performance.

Figure 4.1 shows the steps involved in the design of a mixing system for the transformation journey, which is an iterative procedure. It begins by examining the thermodynamics, chemistry, and reaction mechanisms to identify possible chemical reactions through laboratory experiments or using thermodynamics modeling tools. This step sets favorable operating conditions that maximize the desired products. Moreover, it estimates physical properties (e.g., density, solubility, vapor pressure, heat capacity, and conductivity) and the state of transforming species under the established operating conditions. In the second step, the mixing equipment is designed based on the gained knowledge of chemistry, reaction kinetics, and various required transport processes, such as mixing, heat, and mass transfer. This design can be quite challenging, mainly because chemical reactions are generally related to mass and momentum transfer mechanisms in a complex manner. The efficient design of the mixing equipment maximizes the yield of the desired quality product by selecting and sizing of an optimal hardware.

4.2 Conventional scale-up methods for mixing operations

In spite of advances in engineering design and research tools, the present state-of-the-art regarding the scale-up of mixing operations is still based chiefly on empirical correlations, best practices (know-how routines), and rule of thumb. The rule of thumb method is the most common and the simplest method used for scale-up. For example, the prevalent scale-up criteria used for stirred tank reactors (STRs) are based on keeping one of specific power input per volume, the volumetric mass transfer coefficient, or

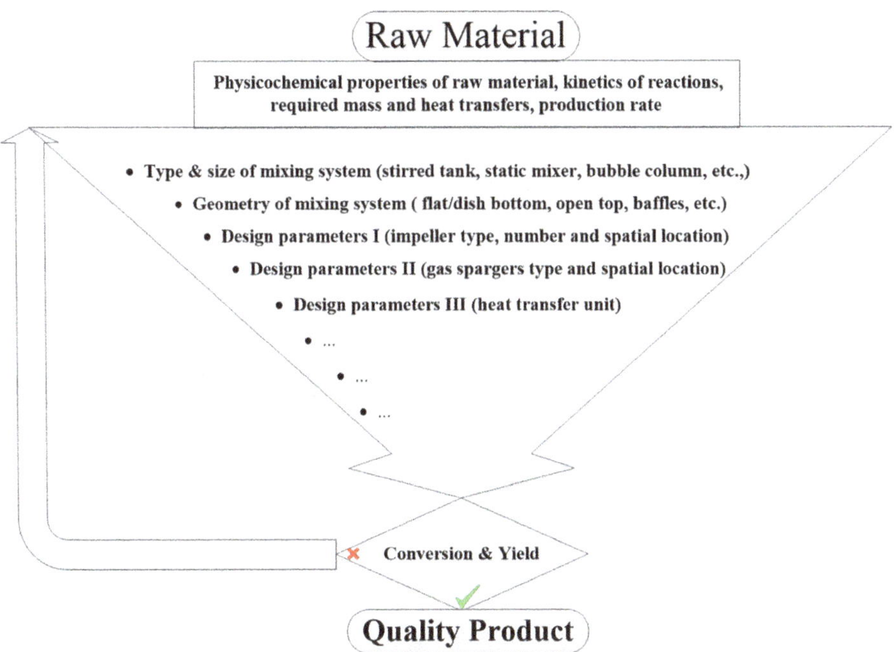

Figure 4.1: Steps in the design of a mixing system.

tip speed of the impeller constant between lab and production scales while maintaining geometrical similarity. However, these criteria result in totally dissimilar process conditions at a production scale compared to lab scale as it is impossible to maintain all the parameters constant simultaneously.

Another common approach that has been used for the scale-up of a mixing system is based on keeping the values of all fundamental dimensionless groups constant during the scale-up. The dimensionless groups are defined as ratios of rates or time constants for the governing mechanisms involved in a process. It is then expected that if all the dimensionless groups are kept constant, the relative importance of the mechanisms or phenomena involved in the process will not change during scale-up. However, this is far from a practical implementation as it is often impossible to keep all the dimensionless groups constant during scale-up. Therefore, an engineer has no choice but to select the most important group and play down the rest.

4.3 Challenges, risks, and consequences of using conventional scale-up in industrial mixing processes

The performance of an industrial-scale mixing system should be similar to that at the laboratory scale, as an unsuccessful scale-up leads to poor product quality, the economic downfall of which can be huge since the product quality is directly linked to its marketability. The estimated cost due to imprecise designs of mixing processes at production scale, in the U.S. chemical industry alone, was about $1 billion to $10 billion in 1989 [1]. It was projected that the pharmaceutical industry could have massive financial losses due to problems in scale-up and process development [2].

The use of conventional methods of scale-up represents a critical risk as they can lead to the poor performance of industrial units. For instance, scale-up based on constant power inputs for two geometrically similar STRs ($V_S = 2$ L and $V_L = 20$ L, a linear scale-up factor of 10) will increase the maximum shear rate by 28% in the large STR, which will dramatically decrease the yield of a bioprocess that contains shear sensitive material (e.g., microorganisms). A process engineer may use the constant impeller tip speed as a scale-up criterion for the advantage of having better similarity in the maximum shear rate at a larger scale STR, yet the power input per unit volume and, consequently, mass transfers are remarkably reduced during scale-up, which leads to poor performance at the production scale. A "well-mixed" hypothesis used in conventional scale-up procedures, in which the values of hydrodynamic parameters are assumed to be constant in the entire system, is another source of concern as such parameters (e.g., the mass transfer coefficient) vary significantly at the production scale.

Figure 4.2 illustrates the dramatic changes in size from bench scale (tank diameter (T) = 0.3 m) to pilot scale (T = 1.5 m) and likewise to larger tanks used for fine chemicals (T = 3 m) and mineral processing (T = 15 m) [3].

Uncertainties and risks associated with extrapolating production scale performance using conventional procedures become increasingly high with the size of the mixing system. Table 4.1 summarizes typical uncertainties and accompanying risks that can arise during the scale-up when following the conventional scale-up methods.

Carrying out experiments on pilot scale is a common strategy that is undertaken by the industry in order to address the scale-up challenges and prevent a painful failure at production scale. Of course, the usefulness of pilot-scale studies relies on how well it can mimic the fluid dynamics and mixing of a large-scale system [4]. Moreover, the required cost and time to perform pilot-scale studies can prevent a new product from ever penetrating the market in a competitive manner. For instance, a recent study estimated that the financial investment to scale-up a microbial process to manufacturing scale is on the order of US $100 million to $1 billion [5]. This investment includes intermediate process validations in pilot scale and demoscales and the construction and start-up of the manufacturing plant that typically takes 3–10

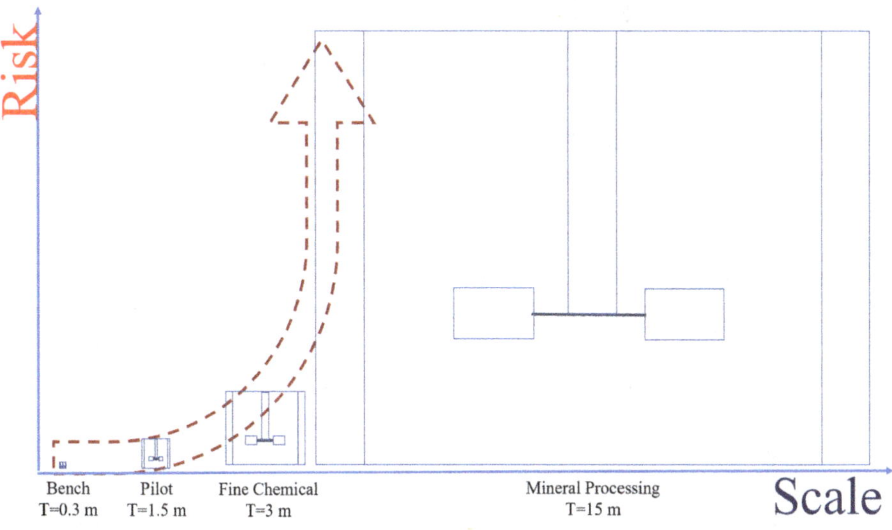

Figure 4.2: Scale of mixing systems versus risk of conventional scale-up procedure failure [3].

Table 4.1: Uncertainties and risks associated with conventional scale-up methods (increase: ↑, decrease: ↓, and variation: ↑↓).

Uncertain factors during scale-up	Potential impacts and hazards at production scale
Flow pattern ↑↓	Heat transfer ↓ and mass transfer ↓
Corrosion, fouling and foaming ↑	Safety concerns (chemical hazard) ↑ and Heat transfer ↓
Heat transfer ↓	Safety concerns (fire hazard and explosion) ↑ and yield and conversion due to runaway reactions ↓
Mass transfer ↓	Concentration gradients ↑ and yield ↓
Variabilities in raw material quality ↑↓	Conversion and yield ↓
Yield and conversion ↓	Process economics and feasibility ↓
Catalyst effectiveness ↓	Process economics and feasibility ↓

years. Even incremental (5–10%) underperformance and/or delays (3–12 months) during scale-up will greatly plunge financial returns to investors and certainly weaken stakeholders as well as customer confidence [5]. A larger decline in performance may actually result in the total failure of the project.

4.4 Applications of iterative scale-up approach for successful industrial mixing operations

In a short path to technology implementation and market introduction, a chief principle is to start development work from the large-scale perspective and then scales it down, not the other way around. This is certainly becoming absolutely vital in a case where global health and security are at stake, such as the production of COVID-19 vaccines to tackle the ongoing pandemic. Obviously, it can be a compromise between the process performance and potential costs with risks of more severe competition in market penetration following a linear scale-up in the hopes of eventually lower final manufacturing costs. Yet, a much higher level of scale-up precision and reliability is required to achieve the feasible economics.

To scale-up a mixing system, an iterative scale-up approach should be used to minimize costs and the risk of operational failures at large scale while reducing the required time for large-scale production. In this approach, the mixing environment that materials experience at large scale should be predicted using reliable models. Next, this condition must be realized at the laboratory scale using experiments and models. Afterward, the design of the system is optimized for identified parameters that govern the efficacy of the system. The reliability of the predictive models can be verified using experiments at this step. Thereafter, the results of the optimized design are applied at production scale. For instance, this could be hardware design adjustments and/or the operating conditions in larger vessels. Certainly, the economic consequences of these optimized designs at large scale can be assessed using reliable models. The potential steps involved in this approach are summarized in a schematic representation of iterative scale-up in Figure 4.3.

One way to identify the critical mechanism in a large-scale mixing system is through the comparative analysis of the characteristic times of governing mechanisms in a process, in terms of orders of magnitude. A fast and slow mechanism has a low and high value for a characteristic time, respectively. This comparative analysis also helps to quantify the transition from the well-mixed conditions at the small scale to the inhomogeneities emerging at the large-scale system. In a bioprocess, for example, seven prevailing mechanisms are: circulation time in mixing system, oxygen transfer from gas to liquid, oxygen consumption in the liquid phase, the growth rate, the substrate consumption, heat transfer, and heat production. The circulation time, oxygen transfer, and consumption are at the same order of magnitude and therefore the oxygen gradient is likely to happen at the larger scale reactor. However, the latter four mechanisms have higher values of the characteristic time compared to the circulation time and therefore it is unlikely that they could affect the performance of the process at the production scale. Iterative scale-up should thus put a spotlight on oxygen mass transfer and eventually oxygen concentration gradients. Accordingly, reliable predictive models are required to serve the iterative

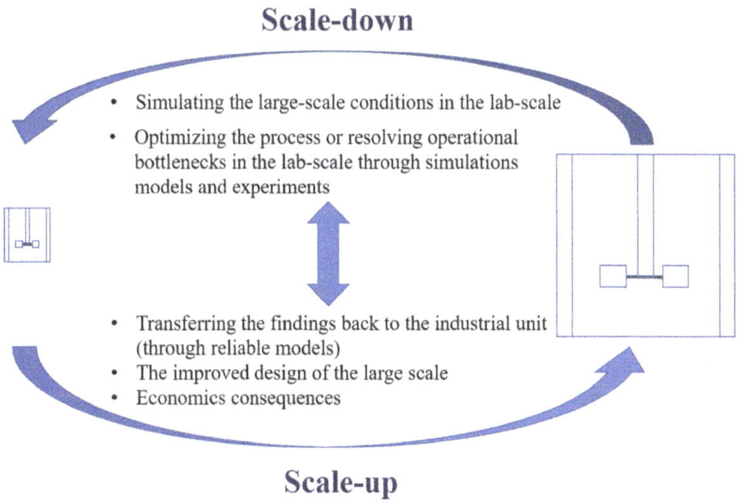

Figure 4.3: Iterative scale-up of mixing systems.

scale-up approach for optimizing oxygen transfer in the production scale of such systems.

4.4.1 Multiscale models as efficient tools for iterative scale-up of fine and bulk chemical manufacturing processes

The design and scale-up of mixing systems have traditionally been carried out using experimental data as a base for empirical correlations. However, this common methodology has faced several challenges, such as the cost of the experimental method used and its scale limitations. Furthermore, industrial processes often involve hazardous conditions (e.g., high temperature, high pressure, and toxic material) for which detailed experimental data acquisition is formidable. The affordability of increasingly powerful computers provides the impetus for using numerical models as a practical design tool.

There are some elements that are occasionally overlooked by scale-up practices in industry, including innovative ideas, mathematical modeling, and fluid dynamics investigations. The mathematical model should be the main tool for synthesizing the innovative ideas for scale-up and improving the performance of an industrial-scale unit. In fact, modeling all pertinent governing mechanisms, at appropriate scale, is one of the predominant challenges of iterative scale-up. The hybrid methods, such as multiscale modeling, are gaining popularity to serve iterative scale-up endeavors [6]. While these models should adequately predict the performance of a mixing system at the local or global scale by transferring data to information, they do not need to be complete. One should note that although a model can fairly describe the reality, its underlying assumptions are not necessarily true, such as an isentropic turbulent viscosity postulate in K–epsilon (k–ε) turbulence models. Therefore, having a fair compromise between complexity and accuracy with convincing simplifications is indeed a smart tactic. Hybrid multiscale models can play a significant role in this regard. Although gaining quantitative know-hows via modeling is very promising, verification and validation with high-quality experimental data remain indispensable [7].

Figure 4.4 illustrates the steps involved in the development of a hybrid multiscale model. The first step is to establish the purpose of developing such a model. Whether this model is used for design, process control and optimization, or scale-up, the required system information and the expected results would be different. For instance, when a process engineer needs to predict the performance of an existing process to improve its design, a model that has an enhanced predictive ability with less uncertainty is desirable. However, when numerous scenarios need to be analyzed for process scale-up in the early stages of the design, model precision can be compromised vigilantly. The system geometry, the process operational parameters, occurring phenomena, and the criteria to strategically compartmentalize the mixing system also need to be defined at this step.

Compartmentalization is frequently done based on the prevailing phenomena and by considering their pertinent time scales. These phenomena are often dependent on the hydrodynamics, which in turn is a function of the system design. These criteria can be simply based on the fluid mechanics (e.g., velocity field, turbulence intensity or dissipation, phase fractions) or they can also consider (bio) chemical reactions (e.g., reactant or product concentration).

In the second step, a wide range of research and engineering tools are used to shed light on the mechanisms governing the system. The computational fluid dynamics (CFD) simulations remain the only tool to obtain detailed local information, yet the model predictions must be extensively validated using experimental data to ensure robust information is provided for compartmentalization. The system scale information can be obtained using engineering calculations, theoretical and phenomenological models and experimental data (e.g., the residence time distribution (RTD), global mass transfer). Each tool can serve various needs while complementing each other to fully characterize the fluid dynamics prevailing in the system.

In the third step, the idea is to determine the number and location of compartments in the system based on the variation in quantities (e.g., reactant concentration) that impact the performance of the system. All obtained local information is compiled using an automated or semiautomated algorithm with given criterion values that divide the system into a compartment network [8]. By integrating the predicted flows by CFD on the surface between two adjacent compartments, the exchanged fluxes between them can be estimated. The average values of the selected parameters in each compartment are then passed along the process-scale model. Finally, the adequacy of

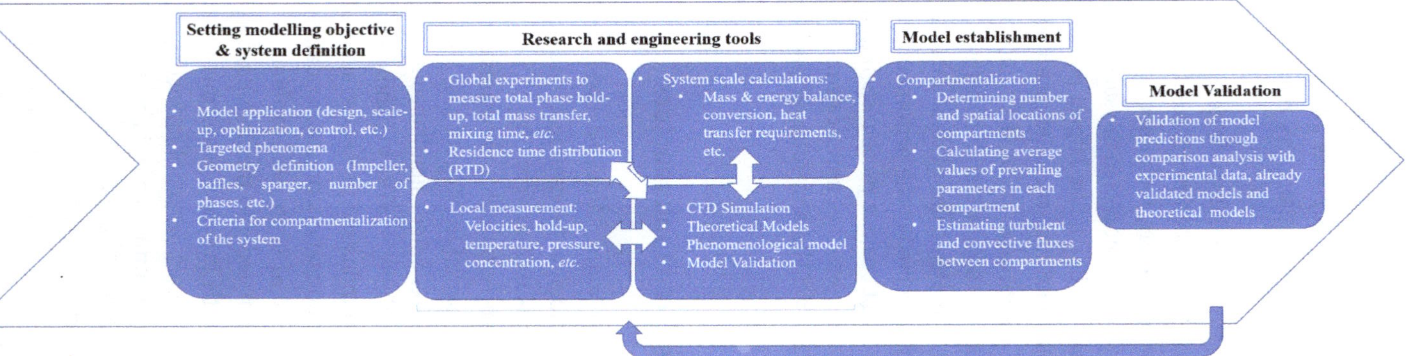

Figure 4.4: Development of the hybrid multiscale models.

model predications is assessed by benchmarking them with experimental works or reliable modeling data obtained in previous investigations dealing with similar systems. If the model does not pass the validation test, the assumptions and physical models used for this development must be revisited until satisfactory results are obtained.

As we have discussed in the previous chapters, the methodology of iterative design requires a process engineer to utilize simulations and experiments results from the laboratory scale to the industrial scale to assess particularly technological risks. In Chapter 3, we discussed the important rule of multiscale models in the successful scale-up of multiphase reactors. We also defined the e-pilot as the largest possible scale to carry out a detailed CFD simulation in a "suitable" time considering all possible governing phenomena. Certainly, the scale of e-pilot depends on the computational resource limits as well as the allocated time to obtain simulation results. Therefore, the vital step in the iterative scale-up is a strategic use of data obtained from an e-pilot in a compartment model (CM) in order to predict the performance of full commercial-scale system. Therefore, in the next sections, we present such transition from e-pilot to industrial scale through the development of multiscale models as design tools for the iterative scale-up of gas/liquid STRs and rotating packed beds (RPBs).

4.4.2 A selected example of gas/liquid stirred tank reactors

Mixing equipment, such as stirred tanks, bubble columns, fluidized beds, and trickle or fixed beds, is commonly being used as reactors in the chemical industry. STRs provide effective mixing for both low-viscosity as well as very high-viscosity and non-Newtonian fluids in which the flow is often turbulent and laminar, respectively. Moreover, they can successfully bring the reactant in different gas, liquid, and solid phases into adequate contact in many processes where simultaneous mixing and phase dispersion are required. The unique flexibility of STRs to handle various process objectives leads to their widespread use in industrial processes, such as aerobic fermentation, hydrogenation, neutralization, chlorination, organic oxidation, polymerization, and gold cyanidation. It is estimated that STRs are used in about half of the chemical processes, by value, that convert raw material to the final product, worldwide [9]. The use of STRs has increased significantly in recent years, due to an increase in the production of added-value materials, such as biopharmaceutical specialty products. The global market of these products only was estimated at US $239 billion in 2015 [10] and it is projected to be worth $395 billion by 2025 [11].

There is not a single optimal design of STRs; however, a few methods have been standardized for particular requirements, such as turbulent mixing with low viscosity fluids as illustrated in Figure 4.5. The tank diameter (T), liquid height (H), impeller diameter (D), impeller blade width (W), impeller off-bottom clearance (C), and baffle width (*B*) are geometrical design parameters that determine the flow pattern of this mixing system. Four baffles usually placed 90° apart are frequently used in this design to break the solid body rotation of the material inside the tank and the central surface vortex.

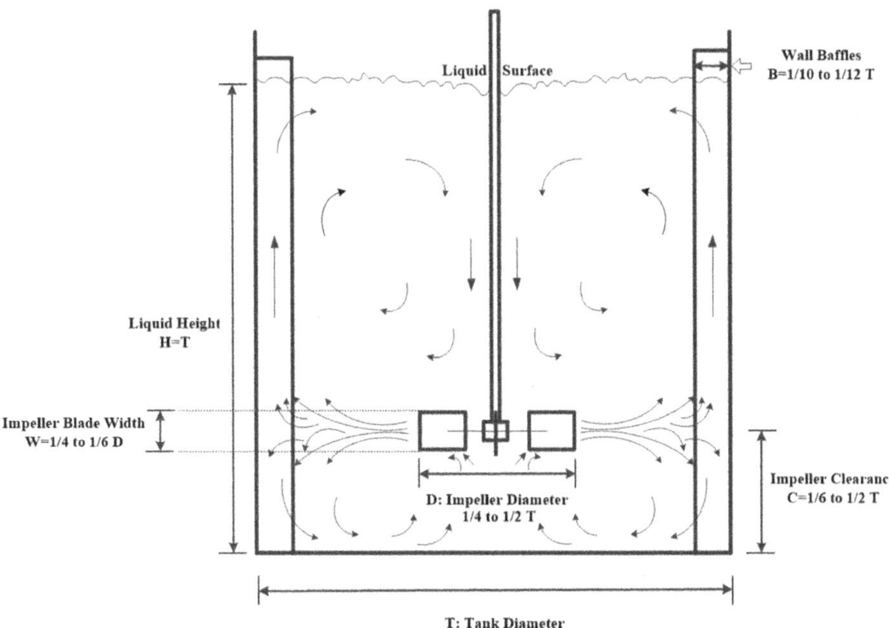

Figure 4.5: Turbulent flow patterns of STRs with the standard tank geometry equipped with a radial impeller (adapted from [12]).

In gas/liquid STRs, the gas is usually injected into the tank through a ring sparger, which is often placed below the impeller. For some processes, such as fermentation, in order to increase the gas residence time, tanks with a higher aspect ratio ($H > T$) are used. For such a tall vessel, multiple impellers are needed to achieve adequate levels of gas dispersion and liquid mixing [13].

4.4.2.1 Development of the multiscale model for iterative scale-up of gas/liquid STRs

Successful production in many processes involving gas/liquid flows is limited by the mass transfer between phases as it can change the reaction rate and, possibly, the selectivity. This becomes particularly critical in the case of processes where low soluble species in the gas phase must be transferred to the liquid phase. Gold cyanidation is one such process where gold is extracted from its ore through a chemical reaction between gold and cyanide in the presence of oxygen. In this three-phase reaction, oxygen is first dissolved in the liquid phase and then along with the cyanide it is transferred to the surface of the gold ore particle where the extraction takes place through a very complicated reaction. This is the case also for many bioprocesses, such as the production of expensive specialty chemicals, including proteins as well as bulk chemicals, such as biofuels, lactic acid, and citric acid, where oxygen transfer is vital for the success of the process [14]. For instance, in the design and scale-up of industrial plants for aerobic fermentations, the most important consideration is often the adequate supply of oxygen to the cells due to its low solubility in the liquid phase.

The mass transfer rate can be quantified as the product of the volumetric mass transfer coefficient (k_La) and the difference between the saturation concentration of gas and its actual concentration in the liquid phase (the driving force, $C^*-C(t)$). The prediction of the volumetric mass transfer coefficient is extremely challenging due to the complexity of the flow structure in STRs as this coefficient is tightly coupled with the hydrodynamics in the reactor. Since the mixing quality and flow characteristics of gas/liquid systems (e.g., bubble dispersion, bubble size distribution, mass transfer resistance in the liquid film around the bubbles, and the mass transfer coefficient) are affected by hydrodynamics, understanding the complex flow structure in STRs is essential for the adequate design and scale-up.

Plentiful empirical correlations have been proposed in the literature that relate k_La to the operating design parameters of STRs, particularly power input per liquid volume (P_g/V_l) and superficial gas velocity (v_{sg}). They were often developed based on experimental data obtained in laboratory-scale reactors and assumed that the STR was "well mixed." More precisely, dynamic methods are often used in which the concentration of dissolved gas over time is measured by a probe that is typically calibrated beforehand. The k_La is then determined from the slope of the natural logarithm of the measured dissolved gas concentrations versus time. Afterward, constants inherent to the empirical correlations are fitted by performing numerous experiments at various operating conditions.

The predictions of these correlations can vary significantly, even without a large difference in the scale of the STR [6]. Therefore, the application of these correlations is limited for the design of large-scale reactors as they are scale-dependent and obtained based on experiments in lab-scale STRs where the volume of a reactor is at maximum about 1 m^3 [6, 13]. Moreover, these empirical correlations cannot bring the concept of imperfect mixing into play as they do not provide any information on the local values of this parameter.

Significant variations in k_La and the driving force inside large STRs may lead to a lower conversion rate and, consequently, longer residence times (such as in cyanidation processes) [6] and higher production costs. In order to improve the performance of industrial-scale STRs through the iterative scale-up approach, it is important to take the local volumetric mass transfer coefficient into account. Thanks to the availability of powerful computers, recently, there has been growing interest in describing the spatial and temporal evolution of the volumetric mass transfer coefficient inside STRs using a simulation model that couples CFD and population balance models (PBM). Although comprehensive multiphase CFD simulations that predict the evolution of bubble sizes and turbulent eddies can help to shed light on the mechanisms governing gas/liquid mixing systems, they still suffer from far-reaching shortcomings, such as the enormous needs for computational resources and a limited understanding of breakage and coalescence processes. The concept of multiscale modeling could partially address the issue related to the computational intensity of such an approach, where the hydrodynamic data obtained by simplified CFD simulations is passed along to a mesoscale PBM. However, the limited understanding of breakage and coalescence phenomena imposes fundamental uncertainties in predictions of such "brute force" simulation models, which makes using them for design and scale-up a formidable challenge. Therefore, for practical uses of local hydrodynamic information in the iterative scale-up of mixing systems, an innovative solution is required.

We developed an original and ground-breaking hybrid multiscale model that brings into play the concepts of multiscale modeling, mass transfer and turbulent theories to predict the local (compartmental) and overall values of the volumetric mass transfer coefficient [6]. It is based on the use of simplified and less computationally intensive CFD simulations (monodispersed bubble size) to strategically compartmentalize the mixing system into a limited number of characteristic zones. The development of this multiscale model follows the steps introduced in the previous sections.

The goal of this development was to introduce a new approach to predict the local values of the volumetric mass transfer coefficient in a typical STR. Figure 4.6 shows the standard design geometry for an STR agitated by a Rushton turbine (RT, the radial flow impeller). The gas was injected into the STR using a ring sparger below the impeller. The STR was fully baffled, including surface baffles to prevent surface aeration. It should be noted that the geometries were prepared with a liquid-free headspace to provide sufficient room for liquid expansion. Therefore, in the dynamic simulation, the liquid surface could freely expand and move while the gas was continuously sparged into the STR. Water and air were used as liquid and gas phases, respectively.

Another important component in the first steps of this development is to map the critical parameters that affect the volumetric mass transfer coefficient. Figure 4.7 is a schematic view of the various factors affecting k_La. As can be seen in this figure, the mass transfer coefficient is enormously affected by hydrodynamics in the reactor and is a complex function of reactor hardware, the physical properties of fluids inside the reactor, and operating conditions. This is probably the reason why simple correlations often fail for scale-up purposes. The turbulent energy dissipation and gas holdup distribution play key roles in the predication of the local values of k_La. Therefore,

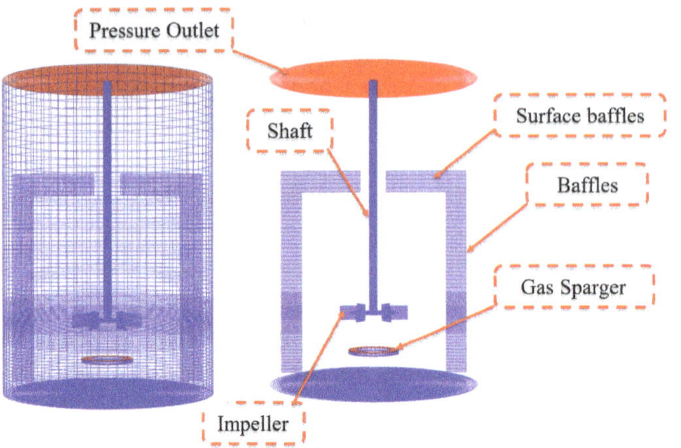

Figure 4.6: The geometry of a mixing system used for the development of a multiscale model [15].

the distributions of the gas hold-up and the liquid-phase turbulent energy dissipation rate should be used as criteria for the compartmentalization of the mixing system.

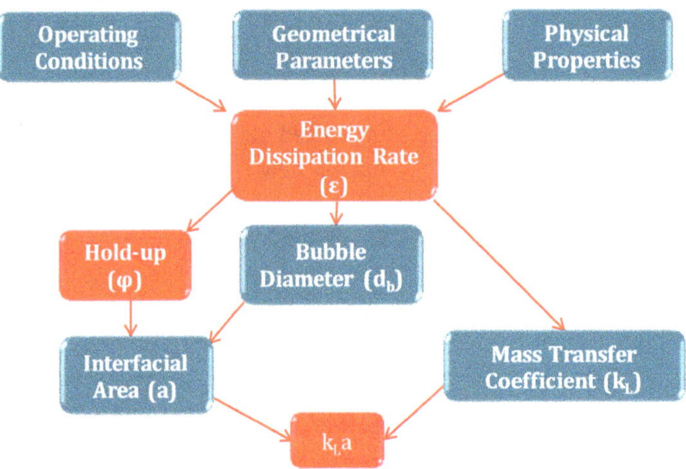

Figure 4.7: Relationships between the volumetric mass transfer coefficient and hydrodynamic parameters (adapted from [3]).

In the second step, we run simplified and less computationally intensive CFD simulations using ANSYS Fluent based on monodispersed bubble size to obtain the turbulent energy dissipation and gas hold-up distribution in this system. The mean bubble size for each simulation case is estimated based on the available empirical correlations in the literature. In the first stage of CFD simulations, solutions of the steady-state turbulent liquid-phase flows were obtained without gas sparging. The adequacy of the numerical models to predict the turbulent single-phase flow was thoroughly assessed using the radioactive particle tracking (RPT) technique. For instance, Figure 4.8 shows the radial profiles of three-dimensional mean velocities in the axial location close to the impeller that are normalized to the impeller tip speed (V_{tip}).

In the second stage, the results of the first stage were used as an initial condition to obtain a solution for the unsteady-state turbulent gas/liquid flow inside the STR. Afterward, the adequacy of the physical models (e.g., turbulent models) and the numerical strategy (e.g., grid size) for the simulations of the turbulent gas/liquid flow inside STRs were assessed by benchmarking their predictions with available experimental data. Good predictions of local and global gas hold-ups and the average value of the liquid turbulent energy dissipation rate inside the system are required to guarantee the suitability of CFD data for a multiscale model.

The spatial variations of the turbulent energy dissipation rate and gas hold-up of turbulent gas/liquid flows generated by the RT inside the STR can be described by five characteristic compartments (Figure 4.9) in order to establish the flow structure. Compartment I has the highest turbulent energy dissipation rate and consequently the bubbles are small. Moreover, gas hold-up is relatively high in this compartment due to gas cavities behind the impeller blades and the continuous flow of the gas toward the wall of the STR through the discharge of the impeller. Most of the bubble coalescence is completed in compartment II and since the turbulent energy dissipation rate is lower and the gas hold-up is high due to gas accumulation near the baffles in this zone, the mean bubble size increases toward its equilibrium value in the bulk of the STR. The bubble formation process at the orifice of the sparger controls the mean bubble size in compartment III, which has a relatively larger mean bubble size and higher gas hold-up. The lowest liquid turbulent energy dissipation rate takes place in compartments IV and V where the bubbles reach their stable size. It should be noted that only when the gas/liquid STR is operating at the complete dispersion flow regime, compartment V will have significant gas hold-up and contribute to the overall gas/liquid mass transfer inside the system.

These spatial locations and boundaries of this compartmentalization are defined based on the local data obtained by CFD for the distribution of the turbulent energy dissipation rate (ε) and gas hold-up (α_g) using a semiautomatic splitting/zoning algorithm. The number and size of these compartments are selected to create an accurate description of the gas/liquid flow structures under various operating flow regimes while the fluctuations of these parameters are sufficiently small within each compartment. More precisely, compartment I radially and axially extended to the edges of the baffles and 1.5 times beyond the impeller blade height above and below the impeller plane, respectively. The boundaries of compartment II are defined from the edges of the baffles to the wall of the STR in the radial direction and to the axial planes where the value of the liquid turbulent energy dissipation rate reaches its average value in the bulk of the STR. The boundaries of compartment III are radially extended to the radius of the ring sparger and axially from the sparger

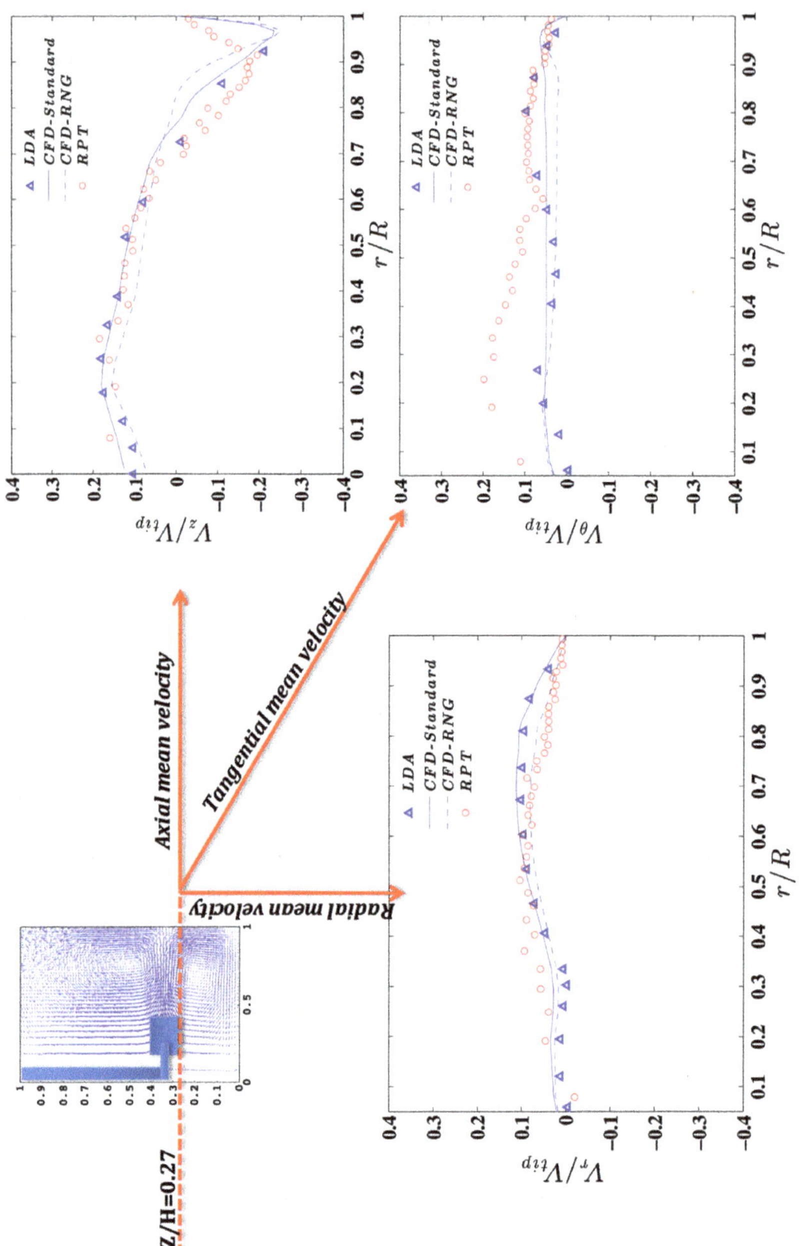

Figure 4.8: Comparison of the simulated and experimental radial profiles of the dimensionless mean velocities (adapted from [16]).

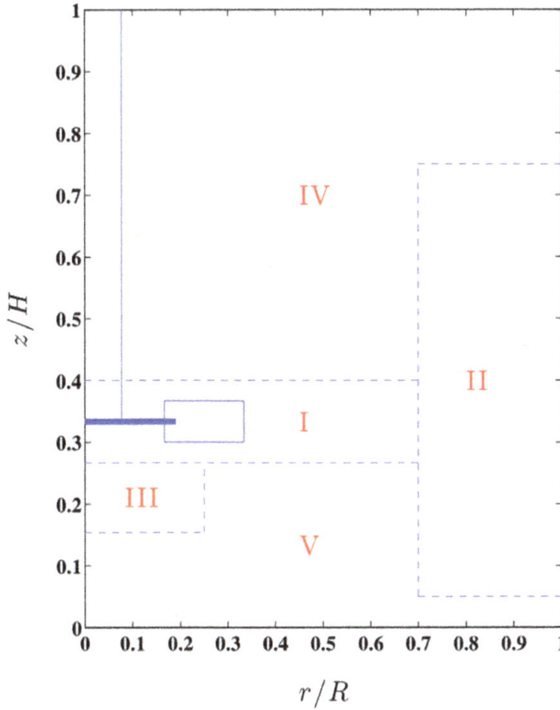

Figure 4.9: Compartments inside an STR agitated by a RT [6].

plane up to the lower horizontal boundary of compartment I. The two remaining zones (compartments IV and V) are those above and below the impeller plane with the average value of the liquid turbulent energy dissipation rate in the bulk of the STR (i.e., the lowest average liquid turbulent energy dissipation rate).

The average values of ε and α_g are calculated in each compartment. These average values are then passed along to semiempirical models and correlations in order to predict the bubble mean Sauter diameter (d_{32}). A volume-weighted average of the compartment-averaged bubble sizes are then compared to the monodispersed bubble size used in the CFD simulation. If there is a good match, the values of the average specific interfacial area (a) in each compartment are predicted based on the value of α_g and d_{32}. Otherwise, another simulation is performed using the updated mono-dispersed bubble size based on the value that was obtained in this step. We demonstrated that only one iteration was required to achieve a convergence. The average value of k_L in each compartment is predicted with local values of ε using the eddy cell model that is based on Kolmogorov's theory of isotropic turbulence and assuming the surface renewal is chiefly dictated by the small-scale eddies of the turbulent flow field. The average value of k_La in each compartment is obtained by multiplying the values of k_L and a. Finally, the overall value of k_La inside the STR is obtained by averaging the values of all the compartments assuming that the mixing time of a system is smaller than the characteristic time for mass transfer (=$1/k_La$) as follows:

$$\overline{k_La} = \frac{\sum_{i=I}^{V} V_i(k_La)_i}{\sum_{i=I}^{V} V_i}$$

Figure 4.10 summarizes the steps of the multiscale model to predict the volumetric mass transfer coefficient in the different zones of STRs. Figure 4.11 shows the prediction of the overall volumetric mass transfer coefficient (k_La) by the multiscale model compared to experimental measurements in a 200-L STR. There is a good correspondence between the mean values of the overall k_La considering the range of uncertainties in the experimental data (~20%). Figure 4.12 shows the effects of increases in the impeller rotational speed at a constant gas flow rate on the contributions of each zone to the overall k_La.

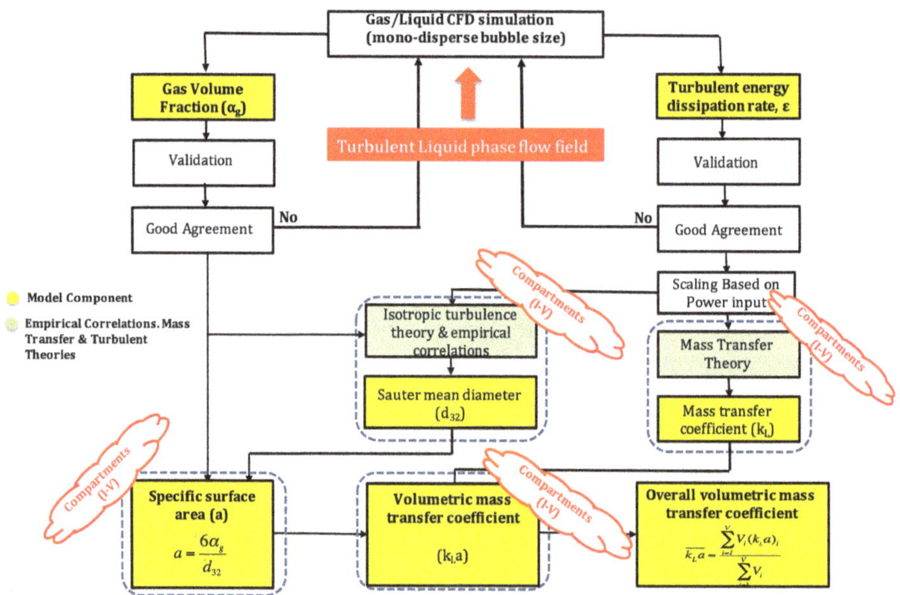

Figure 4.10: Multiscale model [17].

At the relatively low impeller rotational speed (300 rpm), the gas hold-up is low in the zone below the impeller (compartment V) and consequently has a lower contribution to the overall mass transfer inside the STR. However, the contribution of compartments that are or have some part below the impeller plane to the overall mass transfer coefficient is enhanced by increasing the impeller rotational speed to

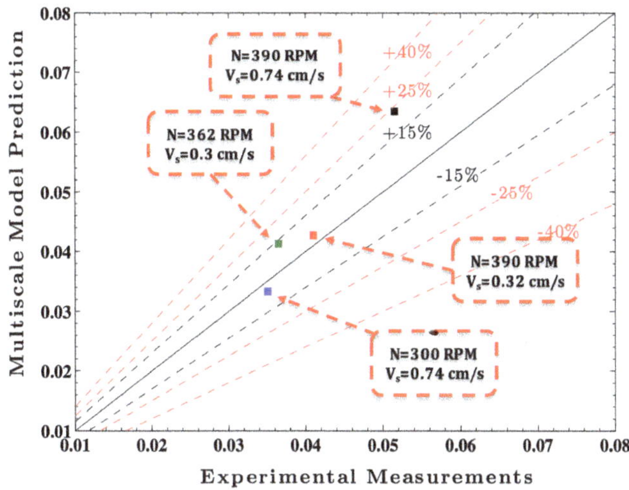

Figure 4.11: Predicted values of the overall mass transfer coefficient by the multiscale versus experimental measurements (adapted from [6]). *N* is impeller rotational speed and V_{sg} is gas superficial velocity.

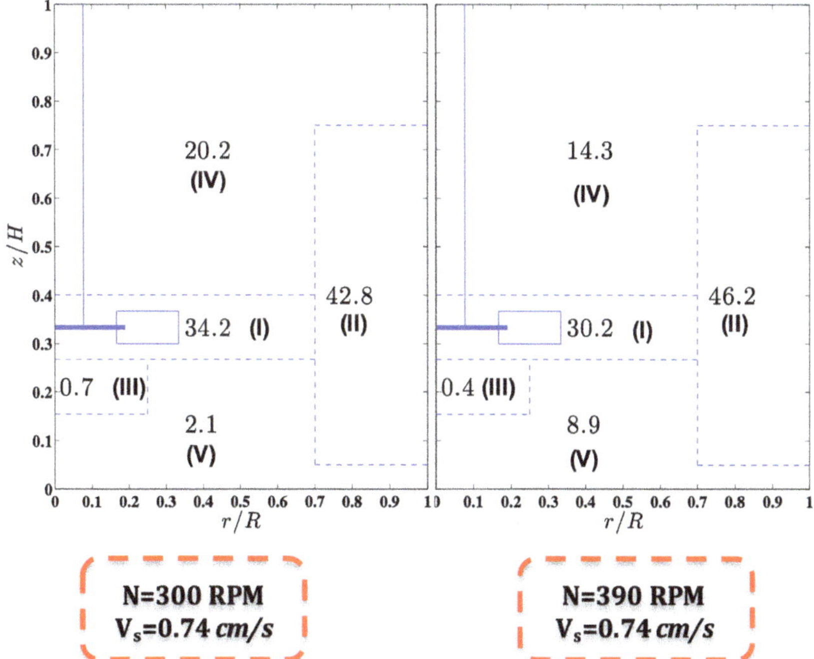

Figure 4.12: Effects of impeller rotational speed on the contributions of the local volumetric mass transfer coefficient to the overall value inside the STR (in %) (adapted from [6]).

390 rpm at a constant gas flow rate as more gas is pushed to the bottom of the STR by the liquid flow (compartment V and the lower part of compartment II).

4.4.2.2 Implications for iterative scale-up of gas/liquid STRs

To maintain the overall value of $k_{\mathrm{L}}a$ constant based on empirical correlations, the values of P_g/V_l and v_{sg} should remain constant. However, the impeller rotational speed is decreased for the scale-up of geometrically similar STRs based on this criterion. Therefore, mixing time is increased as it is inversely proportional to the impeller rotational speed. In fact, as the mixing time increases during the scale-up, the assumption of a well-mixed system will become invalid in industrial-scale STRs and therefore the mixing becomes the limiting factor for mass transfer.

To predict the effects of imperfect mixing in a larger vessel, we performed an iterative scale-up study. More precisely, we first predicted the $k_{\mathrm{L}}a$ distribution and the mixing environment that material experience in a 1,500-L STR (e-pilot) using the validated multiscale model introduced in the previous section. Then using the same model, the environment of an industrial-scale STR was simulated in a 6.3-L tank. To estimate the overall mass transfer coefficient under the effects of imperfect mixing, the RPT was used to determine the RTD of liquid (water) inside the compartments of our lab-scale baffled STR (i.e., scale-down of a large STR). To find the RTD of each compartment, the tracer was followed and the elapsed time in which the tracer crossed the boundaries of the compartment was recorded. The RTD in each compartment was then constructed by repeating this procedure iteratively and assuming ergodic motion for the tracer in the mixing system [6]. We found that the liquid spends more time inside the compartments with lower local mass transfer coefficient values (compartments IV and V), which certainly becomes worse in an industrial scale due to the increase in the volume of low-velocity zones and the reduction of the impeller rotational speed.

When the mean values of residence time in each compartment ($\overline{t_i}$) are non-dimensionalized by the mean circulation time of the mixing system ($\overline{t_c}$), they can be used as a weighting factor to estimate the overall mass transfer coefficient based on the local values of $k_{\mathrm{L}}a$ as follows:

$$\overline{k_{\mathrm{L}}a} = \frac{\sum_{i=I}^{V} \overline{t_i}(k_{\mathrm{L}}a)_i}{\sum_{i=I}^{V} \overline{t_c}}$$

We found that the overall volumetric mass transfer coefficient in a large-scale STR can decrease by at least by 20% due to imperfect mixing.

In this study, we used about 10^6 control volume in the CFD simulation of a 1,500-L STR (e-pilot) as a part of our multiscale model to predict local values of the mass transfer coefficient with good accuracy [6]. However, keeping the same CFD grid density for simulating larger scale STRs ($T > 3$ m) is indeed a formidable challenge since the number of control volumes scales as T^3 (as $H \propto T$ so $V \propto T^3$ where H and V are the height and the volume of an STR). Therefore, keeping the same grid size for larger scales requires a massive number of control volumes and consequently enormous computational power that renders such simulations rather impractical.

Therefore, to assess the adequacy of our multiscale model in transition from e-pilot to a commercial-scale STR, we studied the prediction of contributions of the local volumetric mass transfer coefficient ($k_L a$) to its overall value in a very large industrial-scale STR ($T \approx 5.7$ or $V \approx 1.45 \times 10^5$ L and when one sixth of the geometry and periodic boundary conditions are considered) using a scale-down approach. More precisely, we first calculate the mesh density at the large-scale STR when the same number of control volumes used in this study to simulate the 1,500-L STR (~10^6 control volume) is employed. Then, we run the simulation of the 1,500-L STR again using such a coarse mesh at the large-scale STR, which was represented by using only about 0.05×10^6 control volumes at the small scale. We found that although the values of ε are smaller everywhere in the tank, the impact on the contributions of the local values of $k_L a$ to its overall value are insignificant (below 10% for the compartments with the largest contributions (I, II, and IV), 1% for compartment III and around 15% for compartment V) [6]. It essentially shows that compartment-averaged values of hydrodynamic quantities in the model are not vastly sensitive to local hydrodynamic variations due to using a coarse grid and eventually our multiscale model, even when a coarse grid is utilized, could be adequate for design and scale-up purposes.

The proposed iterative scale-up approach can be indeed used to find an innovative retrofit design for hardware adaptations of an industrial-scale STR by altering the geometrical similarities that serve to improve mixing and enhance the local volumetric mass transfer coefficient inside the identified deficient zones. Therefore, as the monitoring of local variations of hydrodynamic parameters is essential for this purpose, the proposed multiscale model can play a significant role in this regard. Obviously, a good verified computation model should not be the only source of advancement in the scale-up approach and the iterative scale-up and scale-down should be guided by process experts with sound judgement. Nevertheless, such models can replace “know-how” scale-up approaches with “know-why”-based ones and in the long run will pave the way for more “sustainable” processes. Of course, the close collaboration of industrial and academic partners remains an urgent requirement for the iterative scale-up practice to become satisfactorily predictive and precise.

4.4.3 A selected example of rotating packed bed for CO_2 capture and acid gas removal

Chemical absorption-based processes in packed beds (PBs) is a commercially available technology for removing CO_2 and H_2S from process gases and has extensively been used in industry for natural gas sweetening, and post combustion CO_2 capture just to name a few [18]. However, these processes suffer from high capital expenditure (CAPEX) and operating expenditure (OPEX) due to the large size of the equipment and energy-intensive desorption (or solvent recovery) units, respectively [19]. Process intensification is a promising pathway to address these issues by significantly reducing equipment size and energy consumptions of these processes. The rotating packed beds (RPBs) are novel operating units that utilizes the centrifugal force to intensify mass transfer between gas and liquid absorbent [20]. Thanks to this process intensification strategy, the required size of absorption unit and solution circulation flows can decrease considerably which in turn leads to lower CAPEX and OPEX compared to conventional PBs for the same level of acid gas removal. It has been shown that using RPBs can reduce the size of absorption unit up to 90% [20] and can enhance the micromixing by a factor of 2–5 times compared to PBs [21].

The RPB consists of a hollow, rotating tube filled with packings such as wire screens, meshes, or nickel foam (Figure 4.13). In an RPB, a liquid jet, which is generally an amine-based absorbent, is fed into the system as microscopic droplets. More precisely, a jet of liquid splashes through the packing, and due to the strong centrifugal force, the liquid jet breaks into filaments and droplets. A large centrifugal force generated inside RPBs leads to a higher velocity, thinner liquid film, and smaller droplets, thus enhancing interfacial area, which consequently, enhances mass transfer and improves micromixing efficiency.

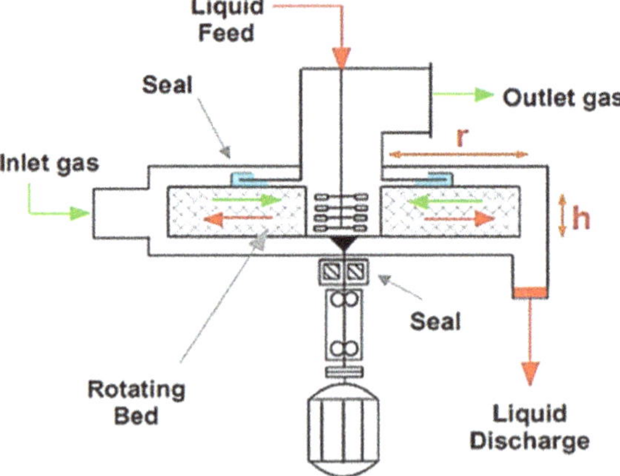

Figure 4.13: Rotating packed bed technology [23]. Green arrows indicate gas flow whereas red arrows indicate liquid flow.

Commercial applications of RPBs encounter considerable resistance by industry due to both real and perceived risks imposed by technical and economic barriers. These issues involve a reliable mechanical design of the sealed rotating equipment with a reliable rotor stability [22]. To the best of the authors' knowledge, except for an industrial-scale RPB unit in China, no other implementation at a commercial scale has been reported yet, in particular for CO_2 capture and acid gas removal applications.

Successful scale-up and commercialization of an RPB require a comprehensive understanding of its performance at a large scale (industrial scale). It is, however, not practical to perform time-consuming experiments on RPBs, as using majority of hydrodynamic characterization techniques is challenging owing to inherent complexities in their structure and operations. For instance, it is not straightforward to measure some parameters, such as droplet size and velocity, in the packing zone, while the PB is rotating. However, the droplet size distribution (DSD) and mass transfer area are among the critical parameters that define the performance of an RPB absorber. Thus, a proper mathematical model could provide more comprehensive and detailed information about the hydrodynamics of an RPB and its overall absorption performance particularly for a scale-up study.

4.4.3.1 Development of the multiscale model for iterative scale-up of RPBs

We developed a multiscale model by a strategic combination of a macro-scale compartmental model (CM), mesoscale two-fluid model (TFM), micro-scale volume of fluid (VOF) model and artificial intelligence (AI)-based neural network method. This model can be used as a sophisticated design tool to simulate commercial-scale RPB absorbers for CO_2 capture and acid gas removal. This multiscale model can simulate RPB configurations at various scales, design configurations, and under different operating conditions within a relatively short computational time. Figure 4.14 summarizes the steps involved in the multiscale model to predict the performance of RPBs.

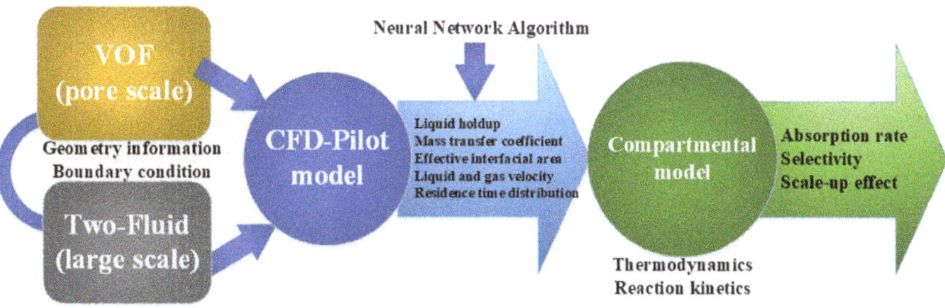

Figure 4.14: The component of the multiscale model for RPBs.

In the first step, we run CFD simulations based on the VOF method using OpenFOAM on the Compute Canada's clusters to obtain the DSD in the RPBs. We assessed the predicted DSDs using literature data [23]. Figure 4.15 shows the simulated droplet sizes at two operating conditions of an RPB where water and air were used as liquid and gas phases, respectively. As can be seen in this figure, by threefold increase in the RPB's rotational speed (ω), the mean droplet size decreases by about 50%. Although, the formation of liquid filaments and droplets or film layers can be tracked during the transient VOF simulations, this simulation method is computationally expensive. We used more than 137 core-years of computational cost for executing 54 VOF simulations and for RPD's diameter up to 18 cm. Therefore, we developed a correlation using these simulation results that relates the mean bubble size to the operating conditions of RPBs. The predicted mean droplet diameter using this correlation is fed as an input into the developed mesoscale TFM.

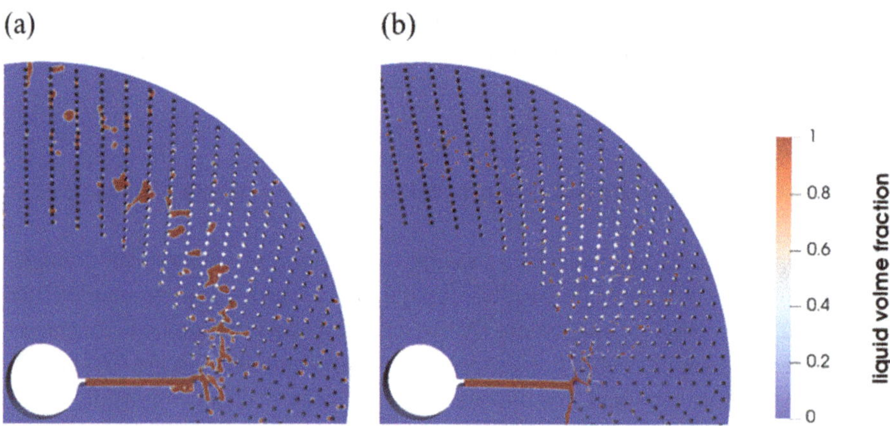

Figure 4.15: Droplet size distributions at (a) $u_l = 1.5$ m s^{-1}, $\omega = 500$ rpm (average droplet diameter = 0.574 mm), and (b) $u_l = 1.5$ m/s, $\omega = 1{,}500$ rpm (average droplet diameter = 0.282 mm). u_l is the liquid jet velocity.

We developed a TFM in OpenFOAM by adapting a coupled porous media two-phase model introduced by Lu et al. [24] to simulate the hydrodynamics of RPBs under various operating conditions and scales. This mesoscale model predicts the two-phase flow hydrodynamics in RPBs such as liquid holdup, pressure drop, and liquid velocity without resolving the gas/liquid interface and with relatively larger mesh. Therefore, it is more suitable for the simulations of large-scale RPBs compared to VOF.

The experimental results for air/water system obtained by Yang et al. [25] and Burns et al. [26] were used to assess our numerical model. Figure 4.16 shows the comparison of the predicted average liquid holdup by TFM in OpenFOAM with the experimental values, predictions by a commercial software (Ansys Fluent) and an empirical correlation [26] for two scales and various operating conditions. We found that the mean relative error between the experimental and numerical results was only around 17% that shows the adequacy of the developed TFM to predict the hydrodynamics of RPBs.

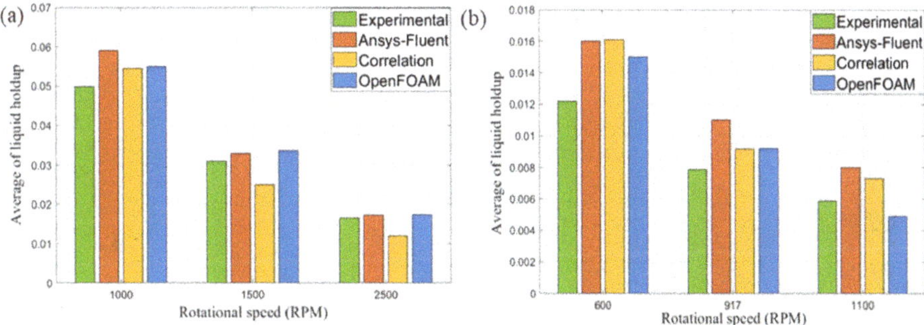

Figure 4.16: Comparison of average liquid holdups obtained from our OpenFOAM model, Ansys-Fluent, an empirical correlation [26] and (a) experimental measurements of Yang et al. [25] experiment in an RPB with diameter = 10 cm (b) experimental measurements of Burn et al. [26] in an RPB with diameter = 32 cm.

Although the TFM is less sensitive to the mesh size compared to VOF model, it still requires a considerable computational expense since many physical models for multiphase flow should be solved simultaneously. For instance, for simulating 5 s of real-time process in an RPB with a diameter of 10 cm using ~200,000 cell numbers, TFM needed about 7 days to ensure that the steady-state condition was established. Therefore, we used these simulations to divide the system into the process-scale compartments in which the reaction models for acid gas absorption into chemical solvents were solved.

We performed the analysis of RTD using the CFD model to divide the system into a compartment network. In a continuous flow system, RTD is defined as the probability distribution of time that fluid parcels stay inside a control volume. Based on RTD data, the flow behavior inside a system can be defined by assembling compartments in order to represent tracer experiments. These compartments can be perfectly mixed reactors, plug flow reactors, axially dispersed plug-flow models, laminar flow reactors, recycles, by-passes or short-circuits, and dead zones and stagnant zones.

We obtained the RTD of an RPB by adding a tracer equation to the liquid phase in CFD models. More precisely, when the CFD model reached to a steady-state solution, the tracer was injected at the inlet and its concentration was measured at the outlet. Figure 4.17 shows the RTD of the liquid tracer in a RPB that can be fitted to an axial dispersion model. However, we further simplified this flow model and estimated the flow behavior in the RPBs with two PFRs, one for the liquid phase and another for the gas phase, as calculated dispersion coefficients ware too small.

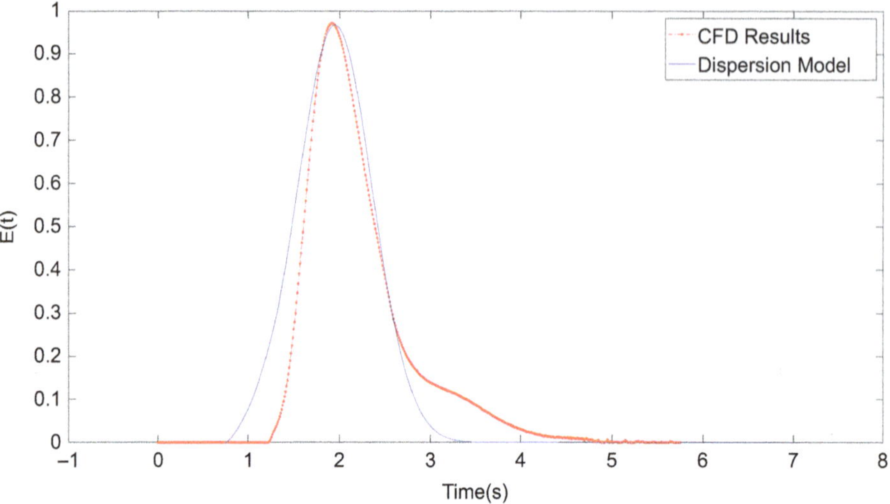

Figure 4.17: The tracer RTD in a large-scale RPB with diameter = 120 cm.

We developed CMs based on chemical reactions and gas–liquid mass transfer in MATLAB for CO_2 absorption into aqueous monoethanolamine solutions and the simultaneous absorption of H_2S and CO_2 from the natural gas stream into methyldiethanolamine (MDEA) solutions. CO_2 (and H_2S for the acid gas removal process) moves from the gas phase to the liquid phase and reacts with the aqueous amine solution. To determine the amount of CO_2 that is absorbed, mass transfer and chemical reactions between the gas and liquid phase are considered under the following assumptions: (1) the absorber operates under steady-state conditions; (2) gas and liquid only flow in the radial direction considering a counter-current configuration (two PFR models); (3) there are uniform conditions in radial directions in both the gas and liquid phases; (4) CO_2 (and H_2S for the acid gas removal process) is the sole component that is transferred from the gas phase to the liquid phase; (5) all reactions take place in the liquid phase; and (6) and the temperature and pressure distributions throughout the bed is uniform.

To solve the model, physicochemical properties, thermodynamics, and hydrodynamic parameters are required. These include density, viscosity, solubility, mass and heat transfer coefficients, and the specific heat capacity of both phases presented in the system. In the first step, we developed a preliminary CM based on the literature data for the mass transfer correlations and hydrodynamic properties. In this model, we assumed that the overall reaction rate of CO_2 can be expressed in terms of reversible reactions. More precisely, for the absorption of CO_2 in MEA aqueous solutions, we considered a fast-reversible reaction of CO_2 and OH− in parallel with the other rapid pseudo-first-order reversible reaction between CO_2 and MEA.

Subsequently, we improved the preliminary CM to a phenomenological CM by considering the reaction rates for all compounds and ions in the system for CO_2 capture and acid gas removal applications (Figure 4.18). In the phenomenological CM, the concentration of each component was calculated in two stages. First, an initial data was obtained using an equilibrium-based model, considering equilibrium reactions and overall mass balance in the liquid phase to determine initial guesses for the liquid inlet compositions. In the second stage, the mass balance equations based on the rate-based reactions and gas–liquid mass transfer, were subsequently used along with the results of the first stage as initial values. The upgraded model was further used to study the acid gas removal process using MDEA as the chemical absorber.

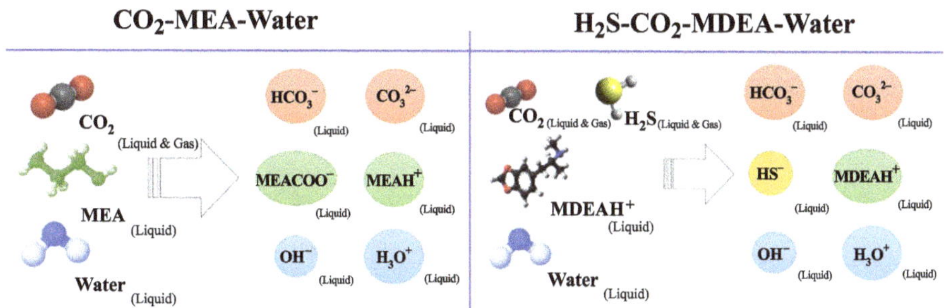

Figure 4.18: Available compounds and ions in gas and liquid phases for CO_2 capture and acid gas removals.

Moreover, in the phenomenological CM, the hydrodynamic properties (liquid and gas velocities, liquid hold-up, pressure drop and the effective mass transfer area) are obtained from TFM and passed along to the CM by using neural networks. More precisely, we developed five neural network (NN) models that were trained by a large database generated by CFD simulation results for small to large-scale RPBs. Finally, a comprehensive lookup tables were generated for the required hydrodynamics data to be used in CMs (Figure 4.19). Other required parameters such as mass transfer coefficient were obtained from the literature.

In order to assess the adequacy of the developed model, we used the experimental data obtained by Thiels et al. [27] and Qian et al. [28] for CO_2-MEA and H_2S-CO_2-MDEA systems, respectively. Figure 4.20 shows effects of gas to liquid ratio on CO_2 recovery of a lab-scale RPB for CO_2 capture. As can be seen in this figure, the CO_2 recovery drops by an increase in gas-to-liquid ratio. Figure 4.20(a) shows that, as we expected, the phenomenological CM better predicts the performance of RPD compared to the preliminary CM. These predictions were further improved when CFD data was fed into the phenomenological CM (Figure 4.20(b)). As can be seen in the Figure 4.20(b), the CFD-based CM predictions are in excellent agreement with the experimental data. These improvements can be better appreciated at a commercial-scale RPB. In addition, we found that the phenomenological CM can predict adequately the H_2S recovery in a lab-scale RPD that uses MDEA as an absorbent (Figure 4.21).

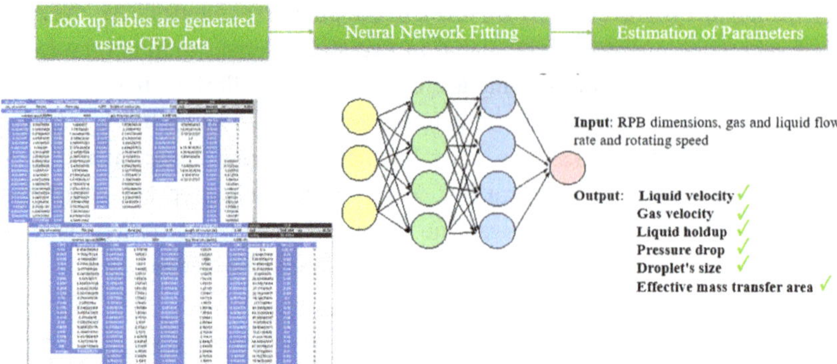

Figure 4.19: Integration of CFD data into the compartment model using trained neural networks.

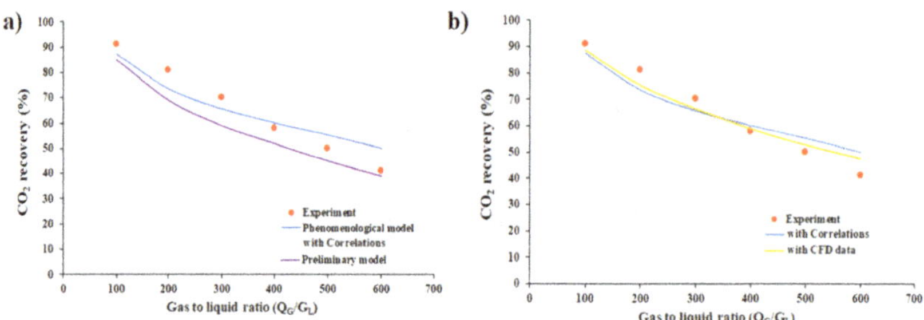

Figure 4.20: Validation of models with experimental data for a CO_2 capture process: (a) preliminary model vs. phenomenological model, (b) the effect of employing CFD data (experimental data from Thiels et al. [27] for an RPB with diameter = 12.5 cm).

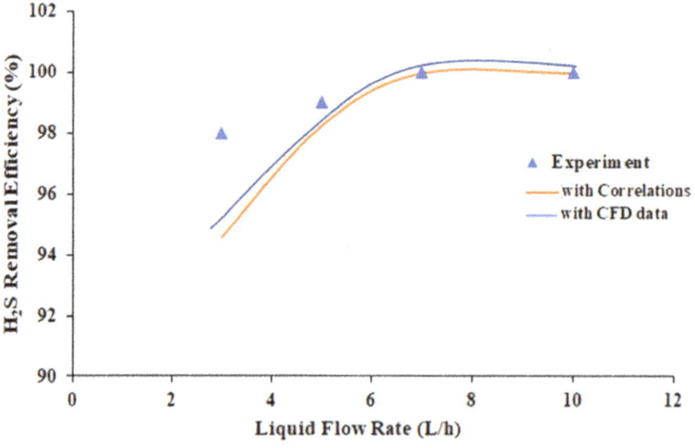

Figure 4.21: Validation of models with experimental data for acid gas removal (experimental data from Qian et al. [28], for an RPB with diameter = 14.6 cm).

4.4.3.2 Implications for iterative scale-up of RPBs

Figure 4.22 summarizes the steps in a transitive design of RPBs from a lab- and e-pilot scale to the industrial scale. We used the validated phenomenological CM as the multiple-size e-pilot to study the effect of the scale on hydrodynamics. Then, we used the CM to design commercial RPB absorbers with pre-defined inputs. The scale-up from e-pilot to commercial scale RPB follows the constant acceleration ($r_0\omega^2$) rule where r_0 is the outer diameter and ω is the rotational speed of the PB. The axial height of the packing (h) was determined based on the loading capacity.

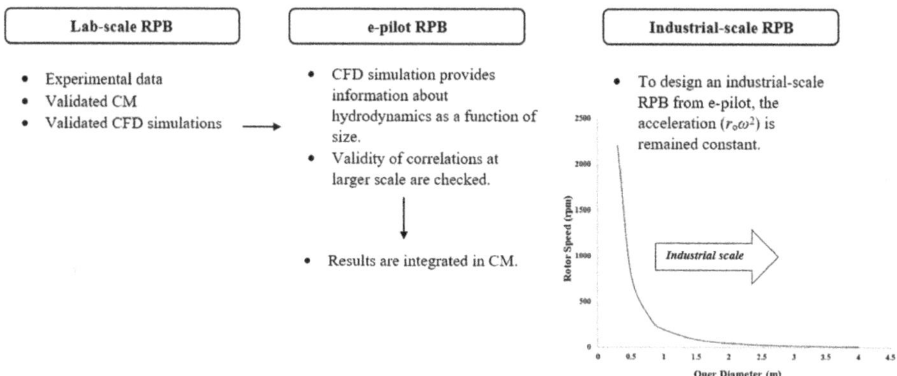

Figure 4.22: The steps in a transitive design of RPBs from the lab-scale and e-pilot to the industrial scale.

In the acid gas removal processes, intensification factor is defined as the volume ratio of a PB to an RPB for the same acid gas recovery. Using the developed model, we predicted the intensification factor by retrofitting the PB to a RPB in three post-combustion CO_2 capture plants that use MEA as absorbent, located in the US (National Carbon Capture Center), Norway (Mongstad CO_2 capture pilot plant), and South Korea (Boryeong power) and a desulfurization unit in China that uses MDEA as absorbent (Figure 4.23). We found that that the acid gas removal can be intensified by replacing PBs with RPBs by an intensification factor ranging from 4 to 9 depending on the rotational speed. The developed model can thus be used as a design tool to predict the performance of large-scale industrial RPBs, particularly, for those operating under extreme conditions where the costs and risks involved in scale-up failure must be minimized.

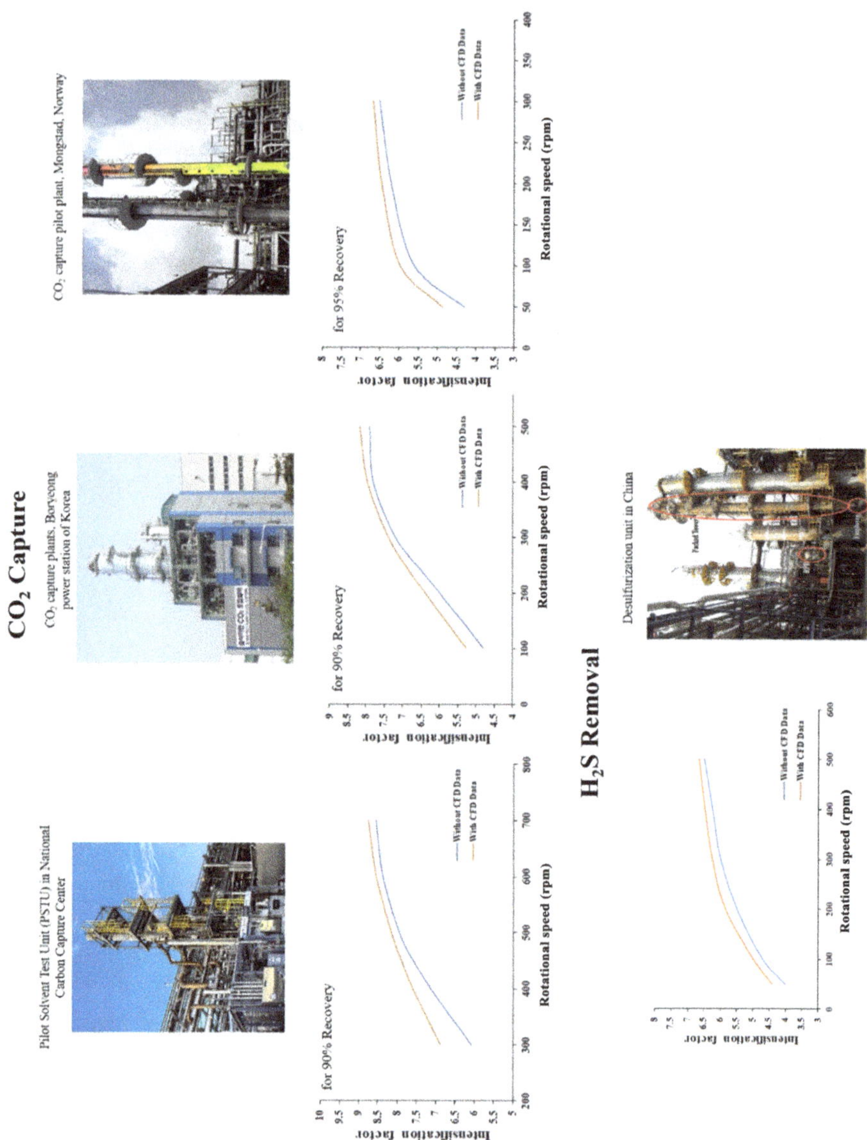

Figure 4.23: The intensification factors for industrial-scale RPB absorbers.

4.5 Conclusion

The conventional methods of scale-up pose a critical risk as they can lead to the poor performance of industrial units, which may lead to catastrophic financial losses. Often, pilot-scale experiments are undertaken by industry as a strategy to mitigate conventional scale-up challenges and painful failure at production scale. However, the required cost and time to perform the pilot-scale studies adversely affect the competitiveness of such processes. The iterative scale-up approach by using e-pilot concept can be a paradigm shift to simultaneously minimize the costs of pilot-scale studies and the risk of operational failures on a large scale. We discuss the development of hybrid multiscale models to predict the performance of a unit operation at e-pilot and industrial scales. Using such a model, we first studied the effects of scale-up on the local and overall volumetric mass transfer coefficient in the STRs. We estimated that the overall volumetric mass transfer coefficient for a larger scale STR (only by scale factor ~5 for the tank diameter) can decrease by at least 20% following the conventional scale-up approach. The proposed model can indeed be used to find an innovative retrofit design for hardware adaptations of an industrial-scale STR, which leads to an improved mixing performance and enhances the overall volumetric mass transfer coefficient as part of the iterative scale-up methodology. We also showed that such models can be indeed an asset to a process engineer in process intensification opportunities in large industrial-scale systems. We used the multiscale model to predict the performance of the acid gas removal in the industrial-scale RPBs compared to PBs and demonstrated that such hybrid multiscale models could be a very efficient design tool for the scale-up studies and the transition from the e-pilot to the production scale.

References

[1] H. Bashiri, F. Bertrand and J. Chaouki, Scale-up effects on turbulence non-homogeneities in stirred tank reactors, presented at the International Gas–Liquid, Gas–Liquid–Solid Congress, Braga, Portugal, Jun. 2011.

[2] E. L. Paul, V. A. Atiemo-Obeng and S. M. Kresta, Handbook of industrial mixing: science and practice, John Wiley & Sons, 2004.

[3] H. Bashiri, Numerical and experimental investigation of liquid and gas/liquid flows in stirred tank reactors, École Polytechnique de Montréal, 2015.

[4] H. Bashiri, N. Mostoufi, R. Radmanesh, G. R. Sotudeh and J. Chaouki, Effect of bed diameter on the hydrodynamics of gas-solid fluidized beds, 2010.

[5] J. S. Crater and J. C. Lievense, Scale-up of industrial microbial processes, FEMS Microbiology Letters, 365(fny138), Jul, 2018, doi: 10.1093/femsle/fny138.

[6] H. Bashiri, F. Bertrand and J. Chaouki, Development of a multiscale model for the design and scale-up of gas/liquid stirred tank reactors, Chemical Engineering Journal, 297, 277–294, 2016.

[7] H. Bashiri, E. Alizadeh, F. Bertrand and J. Chaouki, Investigation of single and multiphase fluid flow behaviour in stirred tanks by means of CFD and particle tracking, presented at the AIChE Annual meeting, San Francisco, Nov. 2013.
[8] H. Bashiri, M. Heniche, F. Bertrand and J. Chaouki, Compartmental modelling of turbulent fluid flow for the scale-up of stirred tanks, The Canadian Journal of Chemical Engineering, 92(6), 1070–1081, 2014.
[9] Butcher, M. and Eagles, W., Fluid mixing re-engineered, Chemical Engineer, 733, 28–29, 2002.
[10] V. G. Pangarkar, Design of multiphase reactors, John Wiley & Sons, 2014.
[11] J. Guo, H. Tu and F. Atouf, Measurement of macro-and micro-heterogeneity of glycosylation in biopharmaceuticals: a pharmacopeia perspective, 2020.
[12] G. B. Tatterson, Fluid mixing and gas dispersion in agitated tanks. McGraw-Hill New York, 1991.
[13] E. K. Nauha, Z. Kálal, J. M. Ali and V. Alopaeus, Compartmental modeling of large stirred tank bioreactors with high gas volume fractions, Chemical Engineering Journal, 334, 2319–2334, 2018.
[14] F. Garcia-Ochoa, E. Gomez, V. E. Santos and J. C. Merchuk, Oxygen uptake rate in microbial processes: an overview, Biochemical Engineering Journal, 49(3), 289–307, 2010.
[15] H. Bashiri, F. Bertrand and J. Chaouki, A multiscale model for the design and scale-up of gas/liquid stirred tank reactors, presented at the 2nd International Conference on Multi-scale Computational Methods for Solids and Fluids, Sarajevo, Bosnia and Herzegovina., Jun. 2015.
[16] H. Bashiri, E. Alizadeh, F. Bertrand and J. Chaouki, Investigation of turbulent fluid flows in stirred tanks using a non-intrusive particle tracking technique, Chemical Engineering Science, 140, 233–251, 2016.
[17] H. Bashiri, F. Bertrand and J. Chaouki, Development of a multiscale model for the design and scale-up of gas/liquid stirred tank reactors, presented at the Mixing XXIV Conference, The Sagamore, Lake George, NY, USA., Jun. 2014.
[18] A. L. Kohl and R. Nielsen, Gas purification, Elsevier, 1997.
[19] H. Bashiri, O. Ashrafi and P. Navarri, Energy-efficient process designs of integrated biomass-derived syngas purification processes via waste heat recovery and integration of ejector technology, International Journal of Greenhouse Gas Control, 106, 103259, 2021.
[20] Z. Qian, Q. Chen and I. E. Grossmann, Optimal synthesis of rotating packed bed reactor, Computers & Chemical Engineering, 105, 152–160, 2017.
[21] G.-W. Chu, L. Sang, X.-K. Du, Y. Luo, H.-K. Zou and J.-F. Chen, Studies of CO2 absorption and effective interfacial area in a two-stage rotating packed bed with nickel foam packing, Chemical Engineering and Processing: Process Intensification, 90, 34–40, 2015.
[22] A. Stankiewicz and J. A. Moulijn, Re-engineering the chemical processing plant: process intensification, CRC Press, 2003.
[23] P. Xie, X. Lu, X. Yang, D. Ingham, L. Ma and M. Pourkashanian, Characteristics of liquid flow in a rotating packed bed for CO2 capture: a CFD analysis, Chemical Engineering Science, 172, 216–229, 2017.
[24] X. Lu, P. Xie, D. Ingham, L. Ma and M. Pourkashanian, A porous media model for CFD simulations of gas-liquid two-phase flow in rotating packed beds, Chemical Engineering Science, 189, 123–134, 2018.
[25] Y. Yang, et al., A noninvasive X-ray technique for determination of liquid holdup in a rotating packed bed, Chemical Engineering Science, 138, 244–255, 2015.
[26] J. Burns, J. Jamil and C. Ramshaw, Process intensification: operating characteristics of rotating packed beds – determination of liquid hold-up for a high-voidage structured packing, Chemical Engineering Science, 55(13), 2401–2415, 2000.

[27] M. Thiels, D. S. Wong, C.-H. Yu, J.-L. Kang, S. S. Jang and C.-S. Tan, Modelling and design of carbon dioxide absorption in rotating packed bed and packed column, IFAC-PapersOnLine, 49(7), 895–900, 2016.
[28] Z. Qian, L.-B. Xu, Z.-H. Li, H. Li and K. Guo, Selective absorption of H2S from a gas mixture with CO2 by aqueous N-methyldiethanolamine in a rotating packed bed, Industrial & Engineering Chemistry Research, 49(13), 6196–6203, 2010.

Guillaume Majeau-Bettez

5 Life-cycle assessment and technology scale-up

Abstract: This chapter addresses the complex and critical task of optimizing the environmental performance of processes under development. A systematic approach is necessary to jointly manage multiple types of environmental impacts and to mitigate the risk of reducing local emissions at the cost of increasing emissions elsewhere in the value chains of a process. A life-cycle assessment (LCA) can yield such a system perspective and inform technology choices. Some obstacles have limited the usefulness of conventional LCAs in guiding process design, however. In order to bring timely and effective guidance, LCAs can benefit from the iterative, uncertainty-focused, model-based approach to process development that is presented throughout this book. This chapter thus explores different strategies to make the best use of available data for environmental assessment throughout the maturation of technologies, from lab-scale proofs of concepts to commercialization, and even all the way to mass deployment. These strategies range from qualitative early technology screening to detailed system modeling and scenario analyses. In this iterative refinement, this chapter argues for a more central role for uncertainty reduction as a guiding principle to focus LCA efforts, and for a closer integration of detailed engineering process modeling and environmental system modeling. Such integration puts new requirements not only on engineers' best practice but also on software and database development efforts.

Keywords: life-cycle assessment, iterative scale-up, sustainable production, uncertainty, ecodesign

5.1 The need for a life-cycle perspective

Confronted with the ongoing climate crisis, the rapid deterioration of natural ecosystems, the increased awareness of emissions' impacts on human health, and the long-term competition for finite resources, society is placing increasing pressure on the industry to help re-establish a balance with the environment [1]. For process and technology development, this constitutes a new and increasingly important objective: not only must a process prove technologically reliable and financially profitable,

Guillaume Majeau-Bettez, Department of Chemical Engineering, Polytechnique Montréal, 2500 Chemin de Polytechnique, Montréal, Québec, H3T 1J4, Canada, e-mail: guillaume.majeau-bettez@polymtl.ca

https://doi.org/10.1515/9783110713985-005

but it should also contribute to reducing unintended stressors on the environment. This has become one of the great imperatives of our time.

As seen throughout this book, anticipating and improving any characteristic of a future technology is highly complex and fraught with uncertainty. In that respect, environmental characteristics are no exception. Improving the environmental performance of a process comes with two extra challenges, however.

First, the environmental performance of a technology cannot be reduced to a single dimension. This stands in contrast to, for example, financial profitability. Although anticipating the capital expenditures, operational costs, and future revenues of a process can prove highly complex and uncertain, all this modeling can at least be reduced to a single objective: a net profit. In contrast, there is no objective way of "adding together" along a single dimension the protection of human health, biodiversity, resources, and the climate. There is always a risk of solving one problem by blindly exacerbating another. For example, the reduction in greenhouse gas emissions brought about by carbon capture and storage (CCS) typically comes at the cost of reduced energy efficiency – and therefore increased fossil fuel depletion – along with increased eutrophication and aquatic ecotoxicity impacts [2]. Sustainable innovation is necessarily a multidimensional balancing act.

Second, the environmental performance of a technology will depend largely on the characteristics of the broader industrial system in which it will be deployed. Indeed, the design of a process will determine not only the direct emissions of this process but also the emissions occurring indirectly in the value chains that will come to sustain this process. For example, the energy efficiency of an aluminum smelter will determine its electricity consumption, which will in turn influence the level of activity of electricity providers and their associated carbon dioxide emissions. The same logic holds for efficiency in material inputs (feedstock, solvents, infrastructure, consumables, etc.) and service requirements (esp. waste treatment services). These inputs are all ultimately linked to resource extractions and substance emissions all over the world through their value chains. In short, sustainable scale-up design must take into account processes that are beyond the immediate control of technology developers. To bring a net reduction in environmental impacts, a holistic system perspective is required.

Such a system-wide assessment of the multiple environmental burdens caused by a process is precisely the aim of a life-cycle assessment (LCA), as briefly explained in Section 5.2. Unfortunately, as detailed in Section 5.2.2, the dominant approach to performing LCA reduces its ability to provide timely and useful guidance to early design and technology scale-up decisions. The challenge is to adapt this method so as to offer increasingly refined assessments at every step of the design and deployment of a process, from conceptual lab-scale exploration (Section 5.3) through scale-up and commercialization (Section 5.4), and sometimes all the way to mass deployment (Section 5.5). These steps present different levels of flexibility, data richness, and uncertainty, but they all can benefit from a life-cycle perspective to ensure an outcome that can prove relevant in an economy increasingly aware of its environmental constraints.

5.2 Life-cycle assessment and its paradox

5.2.1 The basic structure of a standard life-cycle assessment

LCAs offer a standardized, iterative approach to quantify the potential environmental impacts caused directly and indirectly by a human activity [3]. LCAs are typically performed to compare multiple technological systems that yield an equivalent functionality. To this end, the standard LCA procedure follows four mandatory steps.

The first step (Figure 5.1, step I), the *goal and scope*, defines the intended purpose of the study: guide design, communicate environmental performance to customers, and so on. It also specifies the function that is delivered by the system under analysis. Defining an LCA in terms of a unit of functionality allows for flexibility in its application: it can compare any technology or process that successfully delivers a given function (e.g., heating a surface), regardless of their scale or technological differences. For example, the environmental impacts of technologies as different as a hydropower plant and a gas turbine can be compared when normalized relative to the same *functional unit*: 1 kWh of electricity delivered.

This need to compare systems of equivalent functionality largely determines the appropriate boundaries of the system under study. For example, if a study compares the environmental performance of different synthesis pathways yielding the *same* substance (with identical characteristics), these industrial production processes are functionally equivalent. This study can then exclude any activity linked to the subsequent use or end-of-life of this substance, as the environmental impacts of these activities will not be affected by the choice of synthesis pathway. In contrast, if an LCA study compares various car designs with differing characteristics such as material compositions, fuel efficiencies, and expected lifetimes, then the analysis must include not only the activities linked to the production of the vehicles but also the activities linked to their use and end-of-life. This is necessary in order to express the environmental performance of these different prototypes as they deliver a common function: transporting passengers over a distance. A more efficient car will consume less fuel per kilometer during its use phase; a car that lasts longer will "spread" its production impacts over more kilometers driven; and so on.

Having defined the function and boundaries of the system under study, an LCA proceeds to the second step (Figure 5.1, step II): calculating the *life-cycle inventory* (LCI) of all exchanges between this system and the environment. These exchanges include both extractions of natural resources (e.g., use of minerals or fossil fuels) and emissions to the air, water, and soil. The compilation of this LCI starts with the

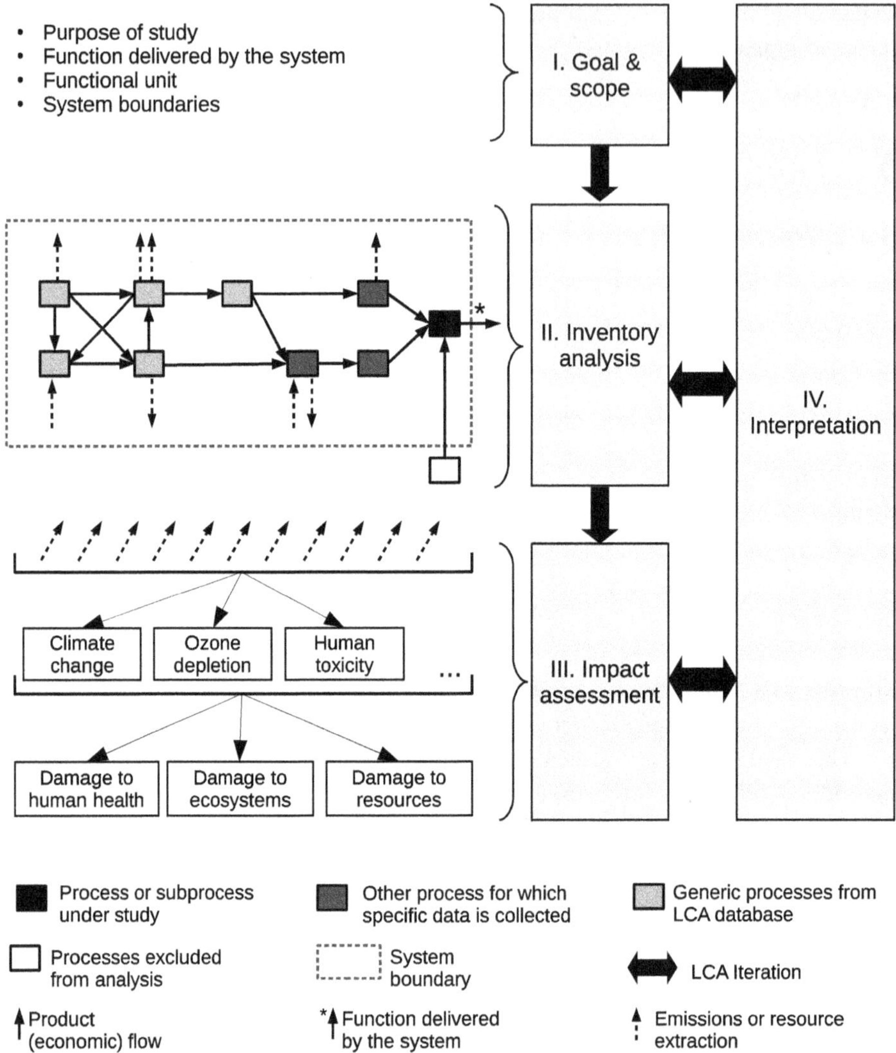

Figure 5.1: The standard, iterative procedure of life-cycle assessment (left), with simplified representations of each step (right).

flows of the process that delivers the system's function (Figure 5.1, dark box). The technology of this first process is described by quantifying its direct exchanges with the environment (dashed arrows), along with the different inputs that it requires from the economy (full arrows). Then the inventory compilation turns to these various inputs from the economy, which must be produced by other processes (gray boxes). These processes have technologies of their own, which cause other exchanges with the environment and require other inputs from the economy, and so on.

The technologies of the key processes at the heart of an LCA study (Figure 5.1, black and dark gray boxes) can typically be described with a high level of confidence and specificity. In contrast, as we move further away along the value chains that supply these key processes (Figure 5.1, pale gray boxes), technology descriptions necessarily become more uncertain and generic. For example, the LCI of a CCS unit may present with a relatively high level of detail the direct inputs and emissions of the process under study [2]: its energy requirements, use of amine-based solvents, and so on. It may also detail the flows and emissions linked with the production of some key inputs: the production of the amine solvent, for example. Such a detailed description of the key processes that are specific to the system under study is dubbed the *foreground data*. However, an LCI necessarily involves a multitude of processes that are *not* specific to the study at hand. The production of a CCS plant certainly requires the production of steel, but there is nothing very specific about this steel production. The same goes for production of the fuel, of concrete, and so on. To describe such processes that are not specific to the study at hand, it is more efficient to re-use precompiled process descriptions from previous LCA studies. Such generic process descriptions – with their inputs and emission flows – are compiled in curated LCA databases. Such data are commonly referred to as *background data*. Thus, an LCA study necessarily combines precise foreground data – to model as accurately as possible a technology's specificity – and more uncertain and generic data from LCA databases [4] in order to complete the description of the entire value chains.

Once all of the system's exchanges with the environment are quantified as precisely as possible, the potential consequences of these flows are estimated in the impact assessment step (Figure 5.1, III). This third step characterizes the potential contribution of each emission or resource extraction to the problems of climate change, ozone depletion, human toxicity, acid rain, fossil fuel depletion, and so on, relying on precompiled characterization methods that reflect the evolving findings of climatology, toxicology, and other natural sciences [5]. The causal chains from emissions to environmental problems can be extended further – though at the cost of greater uncertainty – translating specific environmental problems into potential damages to areas of protection valued by humanity: human health, ecosystem quality, or resource availability. This impact characterization step is typically unaffected by the level of maturity or scale-up of a technology, except in the case where an emerging technology leads to a novel type of emission whose impact has not been sufficiently studied by natural sciences.

In the fourth step (Figure 5.1, step IV), the environmental impact estimates are then interpreted to evaluate the level of confidence in the overall assessment, along with its ability to guide decisions and meet the objectives defined in step I. The processes and value chains contributing the most to the environmental impacts are identified. The interpretation step is also increasingly facilitated by the ability of LCA databases and analysis software to support extensive uncertainty analyses [6, 7].

Crucially, the LCA method is defined as an iterative process. The interpretation step should identify any refinements in the analysis that would be required in order to robustly guide action and satisfy the study's initial objectives, leading to further rounds of targeted data collection, modeling, and analysis. For example, at each iteration, a practitioner can decide where to focus inventory and modeling efforts to "expand the foreground" and replace generic background data with system-specific process models. With such an iterative approach to uncertainty management, LCA proves highly compatible with the general philosophy of this book concerning the treatment of uncertainty throughout the scale-up of a process, as will be discussed extensively in Section 5.4.

The interested reader is referred to complementary documentation on the standard LCA procedure [8, 9], as we turn to the challenges of its application to technologies in different stages of their maturation to commercial use.

5.2.2 The false paradox of LCA

LCA is data intensive. To perform a precise LCA, a detailed understanding is required not only of the inputs to a production process but also of the providers of these inputs, the local energy mix and geographical context, the typical use and end-of-life of the products, the expected yields and efficiencies, the markets for coproducts, and so on. These data requirements typically limit precise LCAs to retrospective analyses of commercially mature technologies [10]. Obviously, it is easier and more precise to assess the flows and value chains of established industrial activities than to anticipate those of emerging ones. Consequently, and rather paradoxically, the detailed understanding of the environmental strengths and weaknesses of an industrial process often *follows* its deployment. At this stage, it is partly too late to significantly alter this established activity based on this detailed LCA understanding. Previous investments in infrastructures and the need to maintain production outputs both limit the range of feasible reconfigurations. Retrospective LCAs then mostly support the communication of environmental performance indicators to the public rather than process development. In LCA, the point of greatest understanding coincides with the point of least flexibility, and vice versa [11, 12].

In response, a growing body of literature is seeking to "streamline" LCAs, so as to simplify its fast and early application in guiding ecodesign efforts. Unfortunately, these streamlined LCAs typically rely on major simplifications [13]: omitting entire life-cycle phases, directly using data from surrogate processes, limiting the range of materials or impacts considered, and so on. Such shortcuts may be perfectly appropriate to provide early guidance for relatively simple product designs, especially if the product's use phase is simple and the material composition of the product determines the majority of its environmental performance (e.g., water bottles or shoes). For complex process development, however, such simplifications risk overlooking the key characteristics that actually determine the environmental performance of design candidates.

Thus, there remains an epistemic tension with the use of LCA in process development and design. If the LCA is applied "early" in the development of an industrial process, the necessary understanding of the characteristics of process may not be available, or it may prove too time-consuming to appropriately assess all design options, leading to an assessment that proves insufficiently specific to guide design decisions. If the LCA is applied "late," key design decisions may have already been made, and it may prove impractical and costly to revise these in order to tackle specific weaknesses identified by the detailed assessment.

The rest of this chapter strives to demonstrate that such an apparently paradoxical situation can be resolved at every stage of a process development, from lab-scale exploration (Section 5.3) to technology scale-up (Section 5.4). The resolution of this paradox resides in a tighter integration of LCA with process modeling, along with a more iterative approach focused on continuous uncertainty evaluation and reduction.

5.3 Early lab-scale life-cycle guidance

Early lab-scale development research and early product design are both characterized by a period of exploration of a large number of possible technological options. This is the development stage with the greatest level of flexibility, and the priority is to "think outside the box" and quickly consider a wide range of material compositions, technological approaches, and so on. Most options considered will not yet have a working model, not even at lab scale, to guide a quantitative assessment. And yet, already at this stage, a life-cycle perspective may help identify potential environmental pitfalls and guide further conception efforts toward the more environmentally promising options.

Semiquantitative environmental screening approaches appeared as early as the 1990s to maintain a life-cycle perspective in early product design [14, 15], and have since been refined and generalized as the iterative LiSET framework [16], which was recently used to guide nanomaterial and battery development research [17]. Rather than obtaining a single, quantitative estimate of the environmental performance of each technology candidate, this approach aims to map the relative strengths and weaknesses of the different candidates along an increasingly detailed range of factors that determine environmental performance.

The basic idea is that the life-cycle impact of any technology can be broken down into a number of contributing factors: its direct emissions, its energy requirements, the environmental intensity of this energy's production, its material requirements, the environmental intensity of these materials' production, and so on. These contributing factors define the columns of a screening matrix, with technology candidates defining the rows (Figure 5.2). Each cell then records whether a given candidate is deemed likely to fall among the better or among the worse candidates in terms of a given factor, or whether there is a complete absence of data. This allows for a first visual, semiquantitative (binary) overview of the state of understanding of the life-cycle profile of a pool of technology candidates.

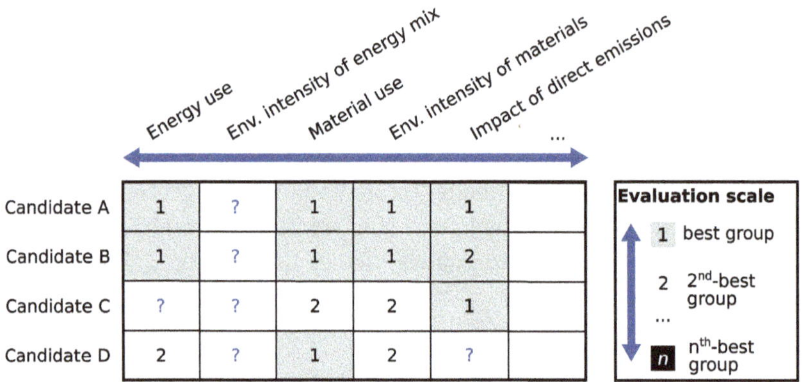

Figure 5.2: Semiquantitative life-cycle screening of lab-scale research candidates (rows) across a range of determinants of environmental impacts (columns) with a semiquantitative grouping (gray-scale, 1 – n) and a mapping of data gaps (question marks). Opportunities for an iterative refinement of the screening are highlighted in blue: filling data gaps (question marks), refining the range of environmental determinants covered (horizontal arrow), and refining of ranking scale (vertical arrow).

This visual map has a certain number of advantages. First, by maintaining a detailed decomposition of the environmental impact contribution (columns of the matrix), it can assist in the early identification of health, safety, security, and environmental (HSSE) issues. Second, it visually highlights data gaps, identifying what cannot be taken into account so early in the development process. Third, by regrouping technology candidates into a small number of groups, it implicitly communicates the high level of uncertainty linked to each environmental characteristic: at this early stage, we do not have enough confidence in our data for an actual ranking of all candidates, but only enough confidence to regroup them into "likely favored" and "likely disfavored" groups.

The most important benefit of this life-cycle screening approach, however, is its compatibility with an iterative reduction in uncertainty and a continuous refinement of the specificity of the life-cycle screening of technology candidates. Indeed, at each iteration, the assessment can be refined in three different ways:

- Important data gaps can be communicated and targeted for data collection.
- The contributing factors to the environmental performance can be disaggregated. For example, the material requirements column can be split into specific metal

requirements, use of bio-based materials, etc. The presence of a greater number of contributing factors reflects an increasing ability to capture the multiple characteristics that determine the overall environmental performance.
- The relative scoring scale can be refined to reflect an increasing confidence in the ability to rank design options, going from two groups (favored/disfavored), to three groups, to ten, potentially all the way to a full ranking of all technological options.

Such a progressively refined semiquantitative mapping of the environmental strengths and weaknesses may thus allow for a first prioritization of research efforts toward the most promising avenues, despite the lack of realistic performance data. It also sets the stage for an agile approach to quantitative LCA guidance of maturation of these technologies, beyond lab-scale candidates toward commercialization.

5.4 Process scale-up with LCA guidance

5.4.1 Issues with process scale-up in LCA

In the scale-up phase, the number of potential design candidates is greatly reduced, and the level of technical understanding is enhanced. A quantitative LCA approach to guide and optimize the environmental performance of the resulting process becomes more feasible. Nevertheless, recent reviews [10, 18, 19] only point to a relatively small number of LCAs that specifically attempt to anticipate or guide technological development as it is scaled-up from lab demonstration to commercialization [20–25].

This relative lack of publicly available scale-up LCAs may be partly explained by questions of industrial secrets and intellectual property. While a lab-scale proof of concepts often leads to an open scientific publication or patent, and is therefore amenable to public LCAs, industries investing massive sums of money in the commercialization of a process will naturally aim to protect the resulting industrial secrets. Any detailed LCA guidance to these crucial steps will likely remain private, for internal use only, and be ignored by literature reviews.

Beyond such considerations, however, a certain number of factors do conspire to complicate the use of LCA as a scale-up design tool. On the one hand, the understandable desire to keep LCA costs and delays to a minimum during this critical and dynamic phase of the process deployment will put pressure to simplify the LCA modeling as much as possible. On the other hand, at this stage of technological maturity, the broad simplifications that were acceptable at lab scale are no longer an option if the LCA is to remain relevant to ongoing decisions.

This tension is complicated by a general lack of capability for complex process modeling by LCA practitioners and software tools (although this is changing, see

Section 5.4.4). This weakness in process modeling may be surprising at first, but it is explained by the typical demands placed upon LCA. Typically, to perform the LCA of an established commercialized process, it is not necessary to describe in great detail the kinetics, thermodynamics, or inner mechanisms of this process; only the ratio of its inputs and outputs under average operating conditions will suffice. With thousands of processes involved in an LCA, this leads to a tendency to treat processes as "black boxes," which are simply defined by their input and output flows, along with their connections to other processes and value chains. In other words, there is a historical tendency in LCA practice to focus on the connections between processes, rather than on their inner mechanisms. This approach obviously hampers the ability of LCA practitioners and tools to engage with the process-focused activities and decisions that accompany technology scale-up.

Such limitations stand in contrast to the engineering modeling employed by the few LCAs that directly aimed to capture scale-up effects [21–24]. These studies can justify their estimates of scaling factors and learning curves with clear thermodynamic principles and chemical engineering simulations. Thus, it is not a fundamental flaw of LCA that prevents it from delivering a detailed guidance during technology scale-up, but rather a lack of engagement with the process development team and detailed process models. Without this engagement, however, an LCA will be limited to two diametrically opposed strategies, both of which will likely limit its contribution to a peripheral role.

One strategy to avoid engaging with detailed process scale-up models is to attempt a rough extrapolation from lab-scale data [26], with minimal adjustments based on data from similar processes that have reached commercialization (cf. [27, 28]). Such an approach will struggle to capture with sufficient specificity the economies of scales, learning curves, and auxiliary processes that necessarily accompany technology scale-ups. Life-cycle impacts per unit of function delivered can sometimes drop by one or two orders of magnitude with industrial scale-up [18, 23]. Consequently, a simple extrapolation risks misidentifying the most important contributors to the overall environmental impact of the future commercialized process, thereby misguiding decision makers. This approach can also be problematic when evaluating the merits of emerging technologies, which might be unfairly placed at a disadvantage when compared with their long-optimized and already commercialized conventional alternatives [19].

The opposite approach to avoid engaging in detailed prospective modeling in LCA is to wait for access to the actual flow data from a pilot or a scaled-down plant. The problem with this approach is that it essentially repeats the "LCA paradox" (Section 5.2.2) on a smaller scale: a more confident understanding of the environmental profile of the future plant is achieved, but the development is then already in a relatively advanced stage [24]. Having invested substantial time and capital in a pilot plant, the range of options that can be cost-effectively explored to minimize the life-cycle impacts of the future process are severely reduced by a partial lock-in effect.

5.4.2 Criteria for an effective LCA guidance of scale-up

From the suboptimal uses of LCA presented in the previous section, we can deduce certain criteria for an effective environmental guidance of process scale-up toward commercialization.

Criterion 1: The ability to capture the consequences of potential design choices prior to their implementation. A preliminary assessment is of little use if it cannot relate to the design questions, or if the assessment arises after substantial real-world investments have been made.

Criterion 2: The ability to robustly rank the environmental performance of design choices. Absolute accuracy and precision are not required across the entire system description in order to guide decision, as long as it is possible to robustly identify the design parameters that improve or deteriorate the environmental performance. In other words, the effectiveness of a scale-up LCA does not depend on the accuracy of the overall assessment of the environmental burdens, but rather on its ability to minimize the risk of conclusion-altering errors in the environmental ranking of design options for key parameters.

Criterion 3: The capacity to keep LCA modeling and data collection costs and delays to a minimum. Beyond the financial consideration, potential delays in analysis are a critical issue, as it is essential to keep pace with process development in order to remain relevant.

As we will argue in the following sections, we believe that failure to meet these criteria is caused by three long-standing issues:
1. A lack of integration between process development and LCA modeling
2. A lack of a genuinely iterative application to the LCA framework
3. An underutilization of uncertainty estimates as a guiding principle to direct LCA modeling efforts

Consequently, the iterative, model-intensive strategy presented throughout this book, with its targeted experimental approach to uncertainty reduction, seems particularly amenable to improve the effectiveness of LCA guidance as technologies mature beyond lab-scale exploration.

5.4.3 Toward a more integrated approach to process development

The different chapters of this book present arguments and case studies in favor of a more important role for process modeling in scale-up planning, with an iterative and targeted approach to uncertainty reduction. This strategy builds more flexibility in the scale-up procedure, and opens the door to the (virtual) exploration of a multiplicity of scale-up scenarios. This increased agility, in turn, should make it easier to explore cost-effective ways to improve multiple performance criteria, including environmental life-cycle performance.

This raises the question: how should standard LCA practice (Figure 5.1) evolve to take advantage of this change in technology scale-up paradigm, in order to better fulfill the criteria for effective scale-up guidance listed above? In other words, given that LCA is already defined as an iterative modeling tool, what synergies can be found with the process development iterations in order to maximize its usefulness and lead to an environmentally optimal final process?

The white boxes in Figure 5.3 synthesize the iterative workflow outlined in Chapter 2, with the numbers in parentheses referring to the specific steps presented in Table 5.6. The gray boxes highlight how this workflow could facilitate, and be guided by, a complementary LCA procedure.

The first steps aim to identify the range of possible design solutions and their key characteristics (Figure 5.3(a)). The scoping reflection in these steps should be completely integrated with the first mandatory step of the LCA method, the definition of the goal and scope. Well-defined objectives and a thorough scoping of possible solution candidates is a necessary foundation for both types of analysis.

The workflow outlined in Chapter 2 then stresses the need for an early identification of HSSE issues (Figure 5.3(b)). Such a screening of potential environmental pitfalls is precisely the aim of the early life-cycle guidance outlined in Section 5.3. This step is thus enhanced and systematized.

The subsequent steps aim to progressively model scale-up scenarios (Figure 5.3 (c)): identify the appropriate scale of a process, perform unit operations design with simple hypotheses, and progressively migrate to detailed simulation (esp., simulation flow diagrams). Crucially, this modeling is iteratively refined (Figure 5.1(d)) with additional data collection, modeling, and experimentation. The successful integration of LCA in the scale-up design depends on the ability to efficiently extract and leverage the environmentally relevant data from process models at every iteration. This requires the following steps (Figure 5.3(I–V)):

I. Extract process inputs, resource use, and emissions directly from the detailed process modeling. To capture the uncertainties and degrees of freedom associated with each design parameter, multiple sets of these flows should be exported to reflect the range of possible development pathways under investigation (fulfilling criterion 1). With the appropriate software investment, this step could become completely automated.
II. Explicitly exclude from the system boundaries (Figure 5.1, dashed box) any input or value chain whose contribution to the environmental impact is unaffected by any of the design decisions. For example, this may include the production of a feedstock whose use efficiency cannot be further improved and remains invariant

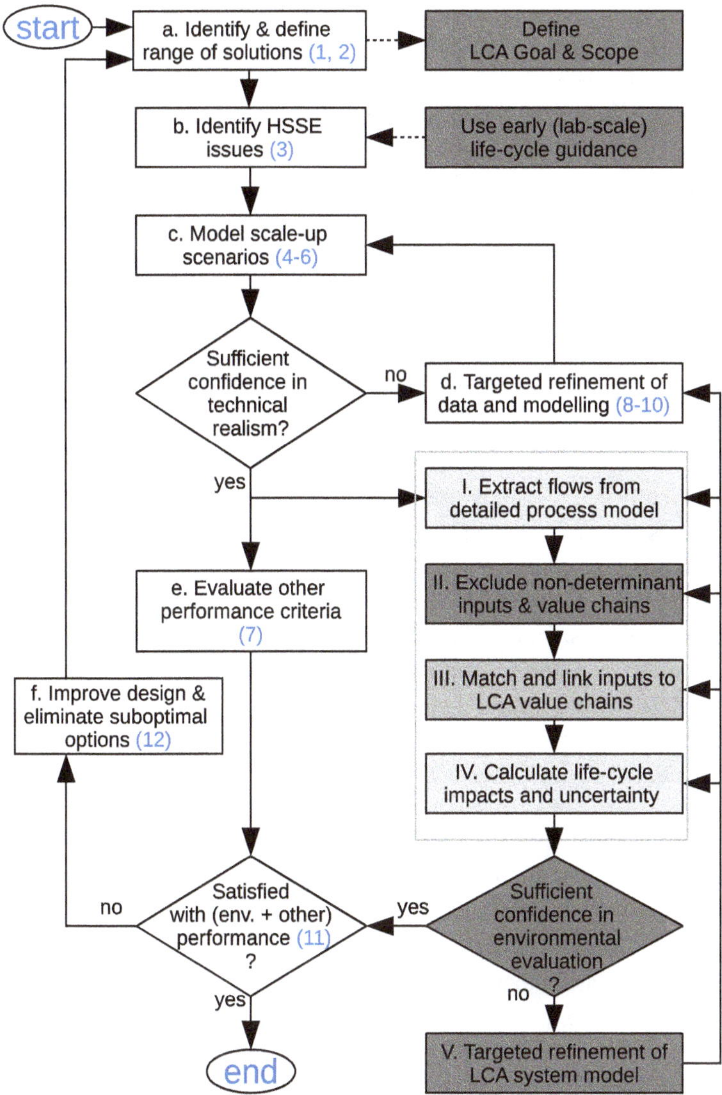

Figure 5.3: Integration of iterative scale-up process modeling (a–e, white boxes) with an iterative life-cycle assessment and guidance (I–V, gray), with potentially automated linkages (light gray), partially automatable steps (medium gray), and manual (dark gray) system restrictions or targeted data collection to reduce uncertainty. Numbers 1–12 (blue) refer to the steps outlined in Table 2.6. HSSE: health, safety, security, and environmental.

across all scale-up scenarios. Such an input certainly contributes to the overall environmental burden of the processes, but as every alternative is equally affected, its description is not required to distinguish between design options (fulfilling criteria 2 and 3).

III. Match the remaining inputs to products within LCA value chain descriptions, preferably from a single, internally consistent LCA database. With the development of concordance tables between process modeling software and LCA databases, this step could be semiautomated. If an emerging technology requires a type of input that cannot be found in LCA databases, the initial strategy should be to select the value chain of a proxy material instead, denoting the added uncertainty caused by such an imperfect match. If the uncertainty surrounding this input proves problematic, then further data collection should be extended to improve the representation of this value chain and thereby reduce this uncertainty in a targeted manner.
IV. The remainder of the LCI calculations and characterization of impacts could then be normally performed in an automated manner. This step should take full advantage of the ability of LCA databases and software to support uncertainty analysis [6, 7]. The uncertainty and sensitivity analysis should focus on the influence of design parameters on the overall performance, and whether the level of precision and confidence is sufficient to distinguish between scale-up scenarios (criterion 2).
V. If the uncertainty is too high to confidently rank the environmental performance of the different scale-up scenarios, then the dominant source of uncertainty is targeted for another round of data collection, refined process modeling, improved matching to LCA databases, and so on (criterion 2). The systematic exploration of uncertainties in life-cycle impacts to prioritize research becomes a core guiding and design principle [29, 30], rather than an afterthought for results communication. This iterative cycle is repeated until we can, with the least effort (criterion 3), provide useful guidance to the design of the technology deployment.

This environmental evaluation will likely be performed in parallel with the evaluation of other indicators, such as financial profitability (Figure 5.3(e)). These evaluations then guide further design selections and improvements (Figure 5.3(f)) until all criteria are robustly satisfied.

This sequence should make maximum use of the technical understanding of the physical process that is brought to commercialization, while minimizing LCA efforts and delays with an iterative treatment of uncertainty. This vision of an LCA process more tightly intertwined with process engineering does not require any fundamental divergence from the principles and standards of the LCA community but hangs on a greater proximity of practice and the integration of modeling tools. Although such linkages between software tools and databases are beginning to emerge [31], a major community effort is still required to ensure an efficient exchange of data, as explored in Section 5.4.4.

Uncertainties linked to the use of novel materials and the release of novel pollutants will continue to constitute a challenge and will continue to justify further investment in open LCA databases and impact characterization methods. We would argue, however, that these issues are not specific to the scale-up activity: if the uncertainties concerning these aspects of the system are not elucidated during scale-up, they will be just as problematic in the LCA of the commercialized process.

Beyond questions of uncertainty, it remains important to keep a scenario-based approach throughout this procedure. These scenarios should allow for the exploration of the key open-ended factors that are deemed most likely to influence the sustainability of the final technology. While some of these factors constitute decisions yet to be made by the process developers (alternative geographical locations for plant deployment, potential industrial symbiosis, etc.), others will simply fall beyond the developers' control and may constitute a source of risk, as discussed in Section 5.5.

5.4.4 Technical considerations

The iterative approach outlined in Figure 5.3 essentially rests on two strategies: (1) a systematic quantification and reduction of uncertainty and (2) an efficient linkage between process simulation data and LCA modeling. The efficiency of these two strategies hinges on recent and ongoing database and software developments.

With respect to the first strategy, the systematic approach to uncertainty reduction proposed in this chapter would have been impossible without the emergence of transparently uncertain databases, where all process requirements and emissions are linked to uncertainty distributions that can be directly inspected by the user [32, 33]. The ecoinvent database stands out as a leader in this respect [4]. Keeping pace with these database improvements, virtually all dedicated LCA analysis software tools allow for statistical uncertainty analyses in the LCI (Figure 5.1(II)) calculations.

Contrary to these advances in LCI uncertainty management, the quantification of the uncertainty associated with the impact characterization step (Figure 5.1(III)) remains an ongoing challenge for multiple types of environmental impact categories. Thus, even though it is now routine to estimate the uncertainty surrounding the different emission flows of a life cycle of a process, it still proves challenging to assess the uncertainty surrounding the potential environmental damage caused by these emissions. This is especially true for ecotoxicity impacts, due to the complexity of the phenomena and the large number of substances and species involved. The integration and refinement of uncertainty factors in impact characterization methods thus remain an active field of research [5], which will help consolidate the minimization of uncertainty as a guiding principle of LCA iteration.

With respect to the second strategy, an efficient integration of specialized chemical engineering process modeling software and LCA tools is still lacking. Recent attempts to leverage detailed process simulations in LCAs involved extensive matching of chemical and product classifications, along with manual data entry (e.g., [34]). The development of an efficient application programming interface for automatic data exchange between chemical engineering and LCA software tools will require further investments. Open-source software, such as OpenLCA [35] and Brightway2 [36], may prove more amenable to such a community-based effort.

The LCA software tool EASETECH presents an alternative approach. Rather than linking to external, specialized design tools, it internally allows for increasingly detailed and parametrized process modeling [37]. Such parameterization notably allows for the exploration of the performance and environmental consequences of changes in feedstock composition, or operation conditions. This type of approach then offers a complementary perspective to guide material flows and circular economy strategies.

In short, despite encouraging advances in terms of inventory uncertainty management, further investment is required in LCA database and software infrastructure to overcome technical inefficiencies. Although this lack of automation in data exchange constitutes an unnecessary hurdle, it is not sufficient to imperil the iterative approach presented in Figure 5.3.

5.5 LCA and scale-up beyond the here and now

The development of novel processes can profoundly transform the physical basis of our economy, with potential indirect repercussions on various markets and resources. At the same time, our societies are already engaged in major transitions, notably with the decarbonization of the energy sector and the rapid industrialization of some parts of the world. Such megatrends may come to completely transform the context of operation of processes, and thereby alter their environmental performance and desirability. In short, whether because of perturbations brought about by the massive commercialization of a new process or because of external trends, the assumption that the interactions of a process with its surrounding economy will remain constant is likely to prove false. Relying on such *ceteris paribus* simplifying assumption increases the risk of a suboptimal environmental performance. This risk can be mitigated through large-scale, prospective scenario analyses.

Needless to say, such analyses do not aim to “predict the future,” or any particular event (war, pandemic, and meteorite). The aim is to investigate the environmental competitiveness of a process along plausible pathways to a range of possible futures, through a limited number of simplified scenarios that are as internally consistent as possible. For example, the International Panel on Climate Change articulates its modeling of climate mitigation efforts in terms of five representative shared socioeconomic pathway scenarios [38]. The range of possible transition trajectories presented within these scenarios translates into different levels of availability of low-carbon energy, which may in turn greatly affect the overall environmental performance and desirability of a scaled-up process. Such transformations of the background economy are thus increasingly included in LCA studies [39].

Beyond a certain scale of mass commercialization, a novel process may also start to interact with the rest of the economy in a different way: through competition for constrained resources. This could be through competition for limited agricultural land, or for limited local hydropower capacity, or for limited emission rights in a cap-and-trade jurisdiction, for example. Such constraints may cause value chains of a process to change, or cause that of a competitor to change, in a nonlinear manner, altering the system's environmental performance. Such dynamics cannot be captured by the usual computational approach to LCA, which relies on a linear, unconstrained technology model [40]. To model a technology deployment beyond a certain scale, LCA can thus be combined with constrained physical and economic models [41–43] to capture the relevant causal links.

The need for such a representation of physical and system constraints is especially critical in the development of recycling and waste management processes, as these processes have limited control on the quantity and quality of their feedstocks. Circular economy initiatives are thus particularly likely to backfire unless guided by a prospective modeling of their integration in their future system. For example, major food companies considered transforming their value chains to produce beverage bottles made with 100% recycled polyethylene terephthalate (rPET), instead of virgin PET (vPET). The production of rPET is less environmentally intensive than that of vPET, so using more of this environmentally friendly material may seem like a sensible approach [44]. In North America, however, the supply of post-consumer PET is limited, notably because of low collection rates. The supply of rPET already proves insufficient to fulfill the demand for low-quality applications, such as sheet, film, and fiber production. Consequently, a system perspective reveals that mandating a 100% rPET content in bottles would not reduce overall vPET consumption; it would only shift its use away from food-grade applications and toward the production of sheets and fibers. Worse, by diverting post-consumer PET away from low-grade applications and toward food-grade use, such a strategy would force additional sorting and decontamination steps, logically causing at least some additional costs, waste generation, and environmental impacts [44]. Fortunately, it is possible to anticipate such dynamics through process and system modeling prior to investment and deployment, so as to avoid well intentioned but unadapted industrial ecodesign strategies.

5.6 Conclusion

In view of the complexity of the environmental challenges, a holistic system perspective is absolutely crucial to guide process development. If a process is conceived in isolation, without consideration for its connections to the broader economic and environmental systems, it is highly likely that suboptimal designs will have unintended consequences upon deployment: underperforming environmentally compared to competitors, tackling one environmental problem while blindly exacerbating another, or even counterintuitively worsening the very problem a technology aimed to resolve.

The emergence of more agile, iterative approaches to both process development and LCA should make it possible to more effectively correct such environmentally problematic designs, and more proactively seek approaches that contribute to a sustainable, decarbonized future. Every stage of process development can benefit from taking into account the broader picture.

For such a system perspective to be useful, however, it must adapt to the level of data richness and complexity of the different stages of process development. Life-cycle environmental performance analysis should therefore iteratively evolve from qualitative screening of lab-scale prototypes to the detailed chemical engineering models of nearly commercialized solutions, and even beyond to guide mass deployment.

At each step, the goal must not be to achieve perfect accuracy in the quantification of all environmental impacts, but to reduce the key sources of uncertainty that prevent a robust environmental ranking of process designs. Consequently, the quantification of uncertainty must go beyond its current use in result communication and become a guiding principle to direct iterative research efforts.

To effectively guide development choices, LCA results cannot afford to lag behind design decisions. Such timely guidance can be achieved by relying more heavily on prospective modeling in LCA, rather than waiting for measurement data from pilot or early commercialization plants. For this to be efficient, environmental assessment work must integrate more closely with process development and simulation. This integration would be facilitated by a more fluid and automatic exchange of data between chemical engineering frameworks and LCA software tools.

Beyond such technical considerations, however, it is the very attitude to environmental performance that must change. Given the urgency of the climate and environmental crises, we can no longer afford to limit the reduction of environmental impacts to a mere afterthought. A life-cycle perspective must become a core design principle, and a constant imperative throughout the conception and maturation of novel processes. Only thus can we maximize the chances of these processes proving relevant in a future that will have no choice but to be sustainable by design.

References

[1] J. Rockström, W. Steffen, K. Noone, A. Persson, F. Stuart Chapin, E. F. Lambin, T. M. Lenton, M. Scheffer, C. Folke, H. J. Schellnhuber, B. Nykvist, C. A. De Wit, T. Hughes, S. Van Der Leeuw, H. Rodhe, S. Sörlin, P. K. Snyder, R. Costanza, U. Svedin, M. Falkenmark, L. Karlberg, R. W. Corell, V. J. Fabry, J. Hansen, B. Walker, D. Liverman, K. Richardson, P. Crutzen and J. A. Foley, A safe operating space for humanity, Nature, 461(7263), 472–475, 2009.

[2] R. M. Cuéllar-Franca and A. Azapagic, Carbon capture, storage and utilisation technologies: a critical analysis and comparison of their life cycle environmental impacts, Journal of CO2 Utilization, 9, 82–102, 2015.
[3] International Organization for Standardization. ISO14040 – Environmental management – life cycle assessment – principles and framework. Technical report, ISO – International Organization for Standardization, Geneva, Switzerland, 1997.
[4] G. Wernet, C. Bauer, B. Steubing, J. Reinhard, E. Moreno-Ruiz and B. Weidema, The ecoinvent database version 3 (part I): overview and methodology, International Journal of Life Cycle Assessment, 21(9), 1218–1230, 2016.
[5] C. Bulle, M. Margni, L. Patouillard, A. M. Boulay, G. Bourgault, V. De Bruille, V. Cao, M. Hauschild, A. Henderson, S. Humbert, S. Kashef-Haghighi, A. Kounina, A. Laurent, A. Levasseur, G. Liard, R. K. Rosenbaum, P. O. Roy, S. Shaked, P. Fantke and O. Jolliet, IMPACT World+: a globally regionalized life cycle impact assessment method, International Journal of Life Cycle Assessment, 24(9), 1653–1674, 2019.
[6] P. Lesage, C. Mutel, U. Schenker and M. Margni, Uncertainty analysis in LCA using precalculated aggregated datasets, International Journal of Life Cycle Assessment, 23(11), 2248–2265, 2018.
[7] S. Muller, C. Mutel, P. Lesage and R. Samson, Effects of distribution choice on the modeling of life cycle inventory uncertainty: an assessment on the ecoinvent v2.2 database, Journal of Industrial Ecology, 22(2), 300–313, 2018.
[8] J. B. Guiné, Handbook on life cycle assessment: operational guide to the ISO standards, Eco-Efficiency in Industry and Science, Kluwer Academic Publishers, 2002.
[9] M. Z. Hauschild, R. K. Rosenbaum and S. I. Olsen, Life cycle assessment: theory and practice, Springer International Publishing, 2017.
[10] S. M. Moni, R. Mahmud, K. High and M. Carbajales-Dale, Life cycle assessment of emerging technologies: a review, Journal of Industrial Ecology, 24(1), 52–63, 2020.
[11] M. Hauschild, H. Wenzel and L. Alting, Life cycle design – a route to the sustainable industrial cultur?, CIRP Annals, 48(1), 393–396, 1999.
[12] W. Haanstra, W.-J. Rensink, A. Martinetti, J. Braaksma and L. Van Dongen, Design for sustainable public transportation: LCA-based tooling for guiding early design priorities, Sustainability, 12(23), 9811, 2020.
[13] J. A. Todd and M. A. Curran, Streamlined life-cycle assessment: a final report from the SETAC North America streamlined LCA workgroup, Environmental Toxicology, 31(July), 1999.
[14] T. E. Graedel, B. R. Allenby and P. R. Comrie, Matrix approaches to abridged life cycle assessment, Environmental Science & Technology, 29(3), 1–6, 1995.
[15] H. Wenzel, Application dependency of LCA methodology: key variables and their mode of influencing the method, International Journal of Life Cycle Assessment, 3(5), 281–288, 1998.
[16] C. R. Hung, L. A.-W. Ellingsen and G. Majeau-Bettez, LiSET: a framework for early-stage life cycle screening of emerging technologies, Journal of Industrial Ecology, 2018.
[17] L. A.-W. Ellingsen, C. R. Hung, G. Majeau-Bettez, B. Singh, Z. Chen, M. S. Whittingham and A. H. Strømman, Nanotechnology for environmentally sustainable electromobility, Nature Nanotechnology, 11(12), 1039–1051, 2016.
[18] S. Maranghi and C. Brondi, Life cycle assessment in the chemical product chain: challenges, methodological approaches and applications, Springer International Publishing, 2020.
[19] J. A. Bergerson, A. Brandt, J. Cresko, M. Carbajales-Dale, H. L. MacLean, H. S. Matthews, S. McCoy, M. McManus, S. A. Miller, W. R. Morrow, I. D. Posen, T. Seager, T. Skone and S. Sleep, Life cycle assessment of emerging technologies: evaluation techniques at different stages of market and technical maturity, Journal of Industrial Ecology, 24(1), 11–25, 2020.

[20] M. Shibasaki, N. Warburg and P. Eyerer. Upscaling effect and life cycle assessment. Proceedings of the 13th CIRP International Conference on Life Cycle Engineering, LCE 2006, pages 61–64, 2006.
[21] M. Caduff, M. A. J. Huijbregts, H. J. Althaus and A. J. Hendriks, Power-law relationships for estimating mass, fuel consumption and costs of energy conversion equipments, Environmental Science and Technology, 45(2), 751–754, 2011.
[22] M. Caduff, A. J. Mark, A. K. Huijbregts, H. J. Althaus and S. Hellweg, Scaling relationships in life cycle assessment: the case of heat production from biomass and heat pumps, Journal of Industrial Ecology, 18(3), 393–406, 2014.
[23] S. Gavankar, S. Suh and A. A. Keller, The role of scale and technology maturity in life cycle assessment of emerging technologies: a case study on carbon nanotubes, Journal of Industrial Ecology, 19(1), 51–60, 2015.
[24] F. Piccinno, R. Hischier, S. Seeger and C. Som, From laboratory to industrial scale: a scale-up framework for chemical processes in life cycle assessment studies, Journal of Cleaner Production, 135, 1085–1097, 2016.
[25] B. Simon, K. Bachtin, A. Kiliç, B. Amor and M. Weil, Proposal of a framework for scale-up life cycle inventory: a case of nanofibers for lithium iron phosphate cathode applications, Integrated Environmental Assessment and Management, 12(3), 465–477, 2016.
[26] A. C. Hetherington, A. L. Borrion, O. G. Griffiths and M. C. McManus, Use of LCA as a development tool within early research: challenges and issues across different sectors, The International Journal of Life Cycle Assessment, 19(1), 130–143, 2014.
[27] L. A.-W. Ellingsen, G. Majeau-Bettez, B. Singh, A. K. Srivastava, L. O. Valøen and A. H. Strømman, Life cycle assessment of a lithium-ion battery vehicle pack, Journal of Industrial Ecology, 18(1), 113–124, 2013.
[28] G. Majeau-Bettez, T. R. Hawkins and A. H. Strømman, Life cycle environmental assessment of lithium-ion and nickel metal hydride batteries for plug-in hybrid and battery electric vehicles, Environmental Science & Technology, 45(10), 4548–4554, 2011.
[29] N. E. Matthews, L. Stamford and P. Shapira, Aligning sustainability assessment with responsible research and innovation: towards a framework for Constructive Sustainability Assessment, Sustainable Production and Consumption, 20, 58–73, 2019.
[30] B. A. Wender, R. W. Foley, V. Prado-Lopez, D. Ravikumar, D. A. Eisenberg, T. A. Hottle, J. Sadowski, W. P. Flanagan, A. Fisher, L. Laurin, M. E. Bates, I. Linkov, T. P. Seager, M. P. Fraser and D. H. Guston, Illustrating anticipatory life cycle assessment for emerging photovoltaic technologies, Environmental Science & Technology, 2014.
[31] F. Rezaei, C. Bulle and P. Lesage, Integrating building information modeling and life cycle assessment in the early and detailed building design stages, Building and Environment, 153 (January), 158–167, 2019.
[32] A. Ciroth, S. Muller, B. P. Weidema and P. Lesage, Empirically based uncertainty factors for the pedigree matrix in ecoinvent, The International Journal of Life Cycle Assessment, 2013.
[33] S. Muller, P. Lesage and R. Samson, Giving a scientific basis for uncertainty factors used in global life cycle inventory databases: an algorithm to update factors using new information, The International Journal of Life Cycle Assessment, 21(8), 1185–1196, 2016.
[34] D. Han, X. Yang, R. Li and Y. Wu, Environmental impact comparison of typical and resource-efficient biomass fast pyrolysis systems based on LCA and Aspen Plus simulation, Journal of Cleaner Production, 231, 254–267, 2019.
[35] A. Ciroth, ICT for environment in life cycle applications openLCA – a new open source software for life cycle assessment, International Journal of Life Cycle Assessment, 12(4), 209–210, 2007.

[36] C. Mutel, Brightway: an open source framework for life cycle assessment, The Journal of Open Source Software, 2(12), 236, 2017.
[37] C. Lodato, D. Tonini, A. Damgaard and T. F. Astrup, A process-oriented life-cycle assessment (LCA) model for environmental and resource-related technologies (EASETECH), The International Journal of Life Cycle Assessment, 25(1), 73–88, 2020.
[38] B. C. O'Neill, E. Kriegler, K. Riahi, K. L. Ebi, S. Hallegatte, T. R. Carter, R. Mathur and D. P. Van Vuuren, A new scenario framework for climate change research: the concept of shared socioeconomic pathways, Climatic Change, 122(3), 387–400, 2014.
[39] A. M. Beltran, B. Cox, C. Mutel, D. P. Vuuren, D. F. Vivanco, S. Deetman, O. Y. Edelenbosch, J. Guinée and A. Tukker, When the background matters: using scenarios from integrated assessment models in prospective life cycle assessment, Journal of Industrial Ecology, 24(1), 64–79, 2020.
[40] A. Azapagic and R. Clift, Linear programming as a tool in life cycle assessment, The International Journal of Life Cycle Assessment, 3(6), 305–316, 1998.
[41] F. Duchin and S. H. Levine, Sectors may use multiple technologies simultaneously: the rectangular choice-of-technology model with binding factor constraints, Economic Systems Research, 23(3), 281–302, 2011.
[42] S. Pauliuk, A. Arvesen, K. Stadler and E. G. Hertwich, Industrial ecology in integrated assessment models, Nature Climate Change, 7(1), 13–20, 2017.
[43] T. Dandres, C. Gaudreault, P. Tirado-Seco and R. Samson, Assessing non-marginal variations with consequential LCA: application to European energy sector, Renewable and Sustainable Energy Reviews, 15(6), 3121–3132, 2011.
[44] G. Lonca, P. Lesage, G. Majeau-Bettez, S. Bernard and M. Margni, Assessing scaling effects of circular economy strategies: a case study on plastic bottle closed-loop recycling in the USA PET market, Resources, Conservation and Recycling, 162(June), 105013, 2020.

Gregory S. Patience

6 Case study I: *n*-Butane partial oxidation to maleic anhydride: VPP manufacture

Abstract: Manufacturing vanadium pyrophosphate catalyst (VPP) comprised synthesizing precursor, micronization, spray drying, calcination, and activation at the laboratory scale. The next step was to conceive a pilot plant that represented the process conditions expected in a commercial reactor. The reactor design requires an estimate of the weight hourly space velocity: the mass flow rate of product per ton of catalyst. A typical value for partial oxidation processes is 150 kg h^{-1} t^{-1}. This is the main contributing factor to calculate reactor volume and investment cost. For exothermic and endothermic reactors (let us say $\Delta H_{rxn} > 100$ kJ mol^{-1}), the next largest factor is available heat transfer surface. Fixed bed reactors have tube banks with up to 30,000 25 mm tubes to extract the energy. Immersed cooling coils maintain the temperature control in fluidized beds. Designing reactors for new chemistry and/or geometries has a considerable amount of uncertainty due to factors like catalyst stability, unknown flow patterns, catalyst manufacturing variability, operational complexity, and pressure to meet financial goals of management. Here, we describe the context of producing vanadium pyrophosphate catalyst and testing it at the pilot scale for DuPont's *n*-butane partial oxidation to maleic anhydride technology. Throughout the process, the R&D team faced uncertainties related to heat transfer, catalyst integrity, low activity/selectivity, and unforeseen operational difficulties such as solid circulation stability, thermal excursions, and steam stripping. The pilot plant was designed to address these uncertainties, and the target production rate was 100 kg h^{-1} (a scale-down factor of 100 versus the commercial process) but only achieved 40 kg h^{-1} and a weight hourly space velocity of 50 kg h^{-1} t^{-1}.

Keywords: vanadium pyrophosphate, attrition, innovation, THF, pilot plant, catalyst, design

Gregory S. Patience, Canada Research Chair in High Temperature, High Pressure Heterogeneous Catalysis, Department of Chemical Engineering, Polytechnique Montréal, 2500 Chemin de Polytechnique, Montréal, Québec, H3T 1J4, Canada, e-mail: gregory-s.patience@polymtl.ca

https://doi.org/10.1515/9783110713985-006

6.1 Introduction

DuPont commercialized the first circulating fluidized bed (CFB) for a specialty chemical maleic anhydride (MAN). Their LYCRA® business was growing and required greater capacity for tetrahydrofuran (THF) and polymethylene ether glycol. Rather than replicating their existing technology which produced THF from formaldehyde and acetylene (Reppe process), they invented a new process. In the first step, vanadium pyrophosphate (VPP) partially oxidized *n*-butane to MAN. In the second step, Pt hydrogenated the maleic acid (MAC) to THF in a reactive distillation column. The novelty of the partial oxidation process was its scale, the attrition resistant catalyst, and the idea of supplying oxygen from the VPP lattice (redox reaction). Selectivity to MAN is higher when the oxygen comes from the VPP lattice rather than from the gas phase like in fixed and fluidized beds. The commercial process shuttled more than 1 t s^{-1} of VPP from the butane oxidation side (reduction) to the regenerator side (oxidation).

This chapter concentrates on VPP synthesis and pilot plant performance. But first, it describes the historical context and mentions the development process. We describe the constraints around developing, testing, and commercializing catalyst, which cost up to $10 million per year for R&D excluding the $10 million pilot plant (author's estimate). It took 16 years to develop the idea in the early 1980s to commercialization in 1996. The catalyst R&D team identified and overcame multiple challenges:

- High VPP attrition rates
- Low spray drying yield
- Safety during modified micronization (VPP dust and toxicity)
- Excessive VPP cost of manufacture
- Registering VPP on the TSCA inventory

The mechanical properties of the VPP catalyst were exceptional and exceeded all expectations based on ASTM tests and pilot plant experience. Catalyst losses due to attrition were 1 kg h^{-1} while some extrapolations from the pilot experience suggested it could reach 20 kg h^{-1}, which translates to a make-up rate of $8 million per year.

6.1.1 Idea to innovation

DuPont's historical success was based on innovation in chemistry and engineering. Jack Welch (GE CEO) wrote: "In my early days of hiring in Plastics, we were always trying to hire people away from DuPont . . ., if you got a DuPont engineer, you were getting the most cutting-edge knowledge of processes and techniques" [45]. In the 1960s, the top money earners for the company were nylon (and intermediates), polyester (films, fiber, and intermediates), and Orlon fibers based on acrylics (Table 6.1). Most of these products were introduced to the market only 10 years earlier. Lycra® –

Table 6.1: DuPont products and sales (cumulative million $) from 1961 to 1970.

Product lines	Year	Sales
Nylon fibers + intermediates	1939	2,314
Polyester fibers, films + intermediates	1950–1953	1,004
Orlon acrylic	1950	441
Auto refinishes + TiO_2	1924, 1931	471
TEL gas additive	1924	420
Neoprene	1932	368
Cellophane	1923	281
Freon	1931	217
Herbicides	1953	151
X-ray products	1932	122
Lycra® spandex	1962	118

a pol yurethane block copolymer – was another fiber that had a tremendous impact as it was invented in 1962 and already contributed over $100 million by the end of the decade.

While in the last couple of decades, we considered that BASF excelled at adding value to their portfolio through synergies and process integration at the Ludwigshafen site, DuPont focused on product development at multiple sites. Throughout the twentieth century, DuPont also had considerable success in engineering and building large-scale commodity plants (e.g., cellophane, TiO_2). Successful commercial launches were based on a pool of thousands of ideas [15]. Besides fibers, DuPont's list of successes include Tyvek®, Corian®, Kevlar®, Nomex®, Silverstone®, Teflon®, Nafion®, and Stainmaster®.

Life cycles of a product last several decades beginning with an embryonic stage when the product is first introduced to the market followed by a growth stage when companies begin to recover the initial investment. During the mature stage of the product, companies make the most money. Without continuous R&D, the product enters a declining stage, in which sales and prices drop. Lycra® invested heavily in product development (end-use research) from the beginning so that the market demand and revenues increased year after year for decades. Hosiery as the initial end-use in the 1960s and in the 1970s they developed the intimate apparel segment, followed by swimwear in the 1980s, ready-to-wear and diapers in the 1990s, and wovens in the 2000s. By 2003, sales had grown to $1.8 billion.

6.1.2 Tetrahydrofuran – THF

DuPont maintained a vertically integrated value chain, including THF, PTMEG, and LYCRA®, and to satisfy the growing demand they required new capacity for the intermediates. In the late 1940s, they synthesized THF as a feedstock (a nylon 6,6 intermediate) based on the Quaker process that converted oats to furfural. They abandoned this route for adiponitrile in the 1960s but continued to manufacture THF for several decades at their Niagara Falls facility as a feedstock for poly(tetramethylene ether) glycol – PTMEG (Terathane®, an intermediate for LYCRA®). They commercialized a second process in LaPorte (TX) based on acetylene and formaldehyde as feedstocks. In the 1970s, BASF proposed supplying THF to DuPont at a price lower than their cost but they refused. Supplying DuPont THF would improve their economies of scale as they could double or triple their capacity and reduce investment according to the six-tenths power law rule, which was developed by C.H. Chilton, a DuPont engineer [9]. D.I. Garnett, another DuPont engineer, described the fundamental basis for the relationship between scale and cost [19, 37].

To maintain competitive, DuPont continued to ream out their LaPorte (TX) facility and increased capacity by 2.5 × the original design capacity in 30 years. The cost advantage of LaPorte was based on a huge acetylene facility next door. To manufacture THF outside of LaPorte would necessarily require prohibitively large investments into an acetylene plant. Furthermore, in the 1980s and 1990s, industry was substituting acetylene and olefins for paraffins at half the price. *n*-Butane oxidation to MAN is an example where a paraffin displaced an aromatic feedstock (benzene) [43]. Monsanto and BASF operated fixed bed reactors (Scientific Design licensed technology) while Alusuisse (PolyNT), Mitsubishi, and BP (Eneos) favored fluidized beds [43]. To compete with these established processes to produce MAN, DuPont developed an attrition-resistant catalyst [3, 4] that could withstand the mechanical stresses in their new reactor concept – a CFB [12, 35]. Major advantages of fluidized bed processes include superior energy integration, higher heat transfer coefficients, and larger capacity, which lower the overall investment [43].

6.2 Initial economic estimation: from laboratory to industrial scale

We demonstrate the feasibility of new chemistry at the bench and then engage cost engineers to estimate capital and operating expenditures based on a flowsheet analysis (e.g., Aspen/HYSIS) [16]. Management reviews the numbers with the researchers to identify research objectives to achieve the economic goals. These goals depend on catalyst performance but CFB technology introduces additional complexity:

- weight hourly space velocity in the riser (*WHSV*), mass hourly production rate divided by the total inventory, which defines reactor dimensions (capital expenditure, CAPEX);
- regeneration rate and thus inventory (CAPEX);
- catalyst cost, which includes the cost benefit of adding promoters and dopants to increase selectivity and conversion (CAPEX);
- catalyst chemical stability (deactivation-OPEX);
- thermal stability (OPEX); and
- mechanical resistance (OPEX).

The economic projections of the initial commercial butane oxidation flowsheet were insufficient to warrant a major development. The evaluation demonstrated that catalyst inventory, and thus reactor volume, contributed most to investment and profitability. Cost engineers are often blamed for being overly pessimistic and for killing programs at their embryonic stages. This was the case in butane oxidation, and one of the early and continuing controversies was catalyst inventory: how much catalyst was required in the fast bed/riser, in the regenerator, stripper, and standpipes. WHSV is the most important design parameter to estimate cost and viability of a catalytic process. It is the ratio of the hourly production rate (in kg h^{-1}) to the mass of catalyst in tons:

$$\mathrm{WHSV} = \frac{\dot{Q}}{M_{\mathrm{cat}}} \quad \mathrm{kg\,h^{-1}t^{-1}} \tag{6.1}$$

The initial price of the precursor was about \$70 kg^{-1}. Typical commercial fixed bed partial oxidation processes operate at WHSV = 150 h^{-1}. (This value may be a design constraint rather than a limit due to catalyst activity. Fixed bed reactors are limited by available heat transfer surface.) To produce 10 t h^{-1} of MAC requires a total active catalyst inventory of

$$M_{\mathrm{cat}} = \frac{\dot{Q}}{\mathrm{WHSV}} = \frac{10{,}000\ \mathrm{kg\,h^{-1}}}{150\ \mathrm{kg\,h^{-1}t^{-1}}} = 67\mathrm{t} \tag{6.2}$$

which corresponds to a cost of \$4.6 million. For strongly exothermic and endothermic reactions, diluting the catalyst with inert at the entrance minimizes hot spots at the expense of increasing reactor volume and thus investment. In the process, only catalyst in the riser/fast bed converts *n*-butane to MAN, while the catalyst in the other vessels and standpipes doubles or triples the total inventory. Minimizing the total catalyst inventory in the regenerator and stripper was one of the main drivers to achieve an economic process.

6.3 Research program

Monsanto was an ideal partner to develop the process as they were the largest MAN producer worldwide, had extensive commercial experience producing MAN, worked on catalyst synthesis for decades, collaborated with top scientists in the world to understand the reaction mechanism [7, 41], and continued to innovate with respect to reactor technology [33], and particularly of feeding *n*-butane with air. However, they pulled out during the design of the pilot plant that was originally to be built at their site in Pensacola (FL). After 3 years, they sold the MAN business to Huntsman who operated 10 sites worldwide producing 249 kt MAN in 2011 [8]. With Monsanto as partner, the plant would dwarf any facility around the world with a design capacity of 120 kt $year^{-1}$, 5× the capacity of standard fixed bed reactors. The final plant design capacity was 80 kt $year^{-1}$, which increased the pressure to minimize operating labor costs and other fixed costs, which are less sensitive to capacity.

When they withdrew from the project, DuPont had a massive task ahead to synthesize precursor (identify a recipe that was independent of anything Monsanto may have shared), micronize the precursor, spray dry it, then calcine and activate it in conditions that were independent of the patent literature to produce active and selective VPP [5]. Together with this daunting task, they had to spend > $1 million for toxicity testing. Yearly costs of catalyst characterization exceeded $1 million per year (excluding personnel) including surface area, particle size distribution, scanning electron microscopy, transmission electron microscopy, oxidation state, X-ray diffraction, and mercury porosimetry.

DuPont engaged professors to assist with the VPP commercialization (Compiegne, Limerick, Bologna) and professors and consultants for reactor design: IGT (standpipe stability), University of Bradford (dipleg design), Monash University (riser pressure drop), Pemmcorp (fluidization and design), University of Houston (kinetics), University College (sparger/reactor design), Polytechnique Montréal (sparger design), City College of New York (reactor studies), and Washington University (solids circulation rate).

The catalyst development program had several objectives:

1. Identify the most economic process to make precursor and negotiate with the vendor to translate the savings into a reduction in the price
2. Minimize losses through the stages of manufacturing (micronization, spray drying, and calcination/activation)
3. Optimize the physicochemical properties (surface area, bulk density, pore size, particle size distribution) to achieve high attrition resistance, activity, and selectivity
4. Identify promoters or dopants that could improve activity, stability, and oxocapacity – the labile oxygen that contributes to the reaction

Besides maximizing the oxygen-carrying capacity of the catalyst (lattice oxygen contribution) [26], activity, and selectivity, the catalyst design had to target a mean particle size of 70 μm with 30% fines (fraction between 20 μm < d_p < 44 μm). The design criteria for attrition resistance was a catalyst loss rate of < 18 mg h^{-1} based on a modified ASTM [2]. The cost of VPP imposed restrictions on the commercial plant design. The directive to the design team was to treat the catalyst gently. Sacrificing standard design practices to accommodate a kinder and gentler reactor configuration would have unforeseen consequences on the final productivity and profitability of the process.

6.4 Synthesis

Commercializing catalyst manufacturing comprised synthesizing VPP at commercial vendors and testing its activity and attrition resistance in the Ponca City pilot plant. DuPont initially worked with a single catalyst vendor to make vanadyl hydrogen phosphate hemihydrate ($VO(HPO_4)\cdot 0.5H_2O$, Figure 6.1) [21], who had many clients also making precursor, so scheduling trials was challenging logistically. What was more challenging was price negotiations. DuPont identified recipes that reduced the time to synthesize precursor with less reagents (isobutanol and benzyl alcohol) but these improvements never translated to lower price. Finally, in 1994 the negotiations turned in favor of DuPont only because they identified an alternative vendor.

Figure 6.1: Scanning electron microscopy micrographs of $(VO)_2P_2O_7$. V_2O_5 was heated, stirred, and refluxed with H_3PO_4 and a mixture of isobutanol and benzyl alcohol. During reaction, the solids form 70 µm rosettes composed of platelets (shown above). Dehydration of the coordinated water from the precursor to form the vanadyl pyrophosphate follows a topotactic transformation (the morphology of the two phases is identical). The width of the image corresponds to 17 µm. Copied from Johnson et al. [21]. Copyright 2010 ACS.

Accepting another catalyst for the commercial plant was somewhat risky since it had not been tested in the pilot plant. The R&D team, commercial team, and business leadership debated how to even make a decision to consider an alternative supplier. Finally, we posed the following question to the group: “At what price would we be willing to risk purchasing precursor, with equivalent performance." The pilot plant had already been mothballed so defining *performance* would have to be based on experimental facilities: *n*-butane conversion, MAN selectivity, oxygen-carrying capacity, and stability measured in the 1/4″ riser reactor, density, and attrition resistance. The consensus of the technical team and management was that the price should be about 1/2 the current price but some were so uncomfortable to make the shift stated that they would not accept it even if it were at no cost. The strategy to talk with this vendor and test their catalyst worked, and the precursor cost dropped by a factor of 2. The final cost for the catalyst including spray drying was \$50 kg^{-1}.

6.4.1 Micronization and spray drying

Precursor synthesis was the first and most expensive step of the manufacturing process. However, to make a viable fluidized bed catalyst requires a narrow particle size distribution around 70 μm with $< 30\%$ fines ($20\ \mu m < d_p < 44\ \mu m$) and attrition resistance [29]. DuPont invented technology to produce spherical micron-sized particles that could resist the mechanical stresses at spargers, in the cyclones, and other internal metal surfaces like cooling coils [3]. In fact, this had to be proven and some engineers characterized the pilot plant in Ponca City, OK, as the world's biggest test mill for attrition.

Typical fluidized bed catalysts – fluid catalytic cracking (FCC) and acrylonitrile, for example – are a composite of an active phase embedded in a silica/alumina binder matrix. The inert binder accounts for up to 50% of the total mass to impart the mechanical strength to withstand particle–particle collisions at high velocity jets and abrasion at cyclone walls [42]. DuPont's innovation was to create a resistant shell with only 10% silica. A slurry of precursor micronized to 2.5 μm with colloidal silica (tetraethyl orthosilicate/polysilicic acid) produced a 10 μm thick shell [38]. One direct consequence of such a low concentration of binder is a reduction of the vessel sizes. Another is that nonselective reactions due to silica–reactant interactions would be reduced (although DuPont never directly attributed low selectivity to the silica).

The critical design parameters to produce the spherical microspheres were the precursor (producing particles <2.5 μm was prohibitively expensive in an air jet mill), mass fraction of solids in the slurry (40% was a target), slurry viscosity (10,000 Pa s), and spray drying geometry. After struggling with a co-current gas flow and wheel atomizer (meaning thousands of kilograms of tests), a counter-current two-fluid nozzle gave the best results at the 3-m scale. The design constraints and difficulties in the 1.4 m Bowen dryer at the experimental station and a commercial 3 m unit disappeared when spray drying in an 8 m unit. Yield losses due to wall build up (bang down) approached 10% in the small units but were much lower in the large unit.

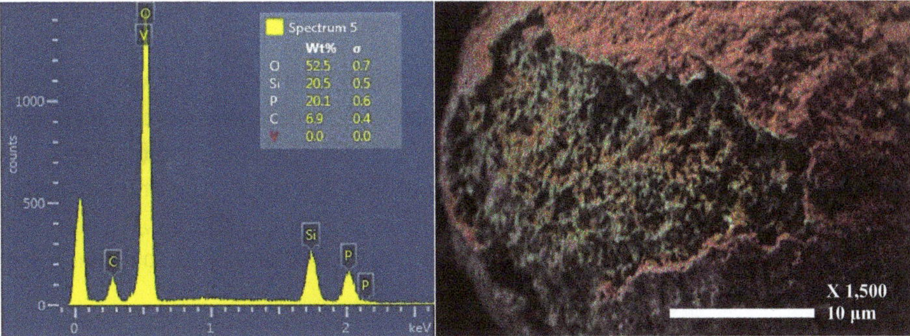

Figure 6.2: Scanning electron microscopic micrographs of the shell and interior of a spherical microsphere of $(VO)_2P_2O_7$ together with an EDX composition profile. The shell region, in red, is predominantly Si, while the interior, in green, is mostly vanadium and phosphorus. Copied from Patience [38]. Copyright 2017 Elsevier.

One of the largest unknowns going forward with the process was the mechanical integrity of the porous microspheres [10]. The ASTM standard to measure attrition rates of powders in jet mills specifies an orifice velocity of 320 m s^{-1} (1,050 ft s^{-1}) [1]. However, particle–particle collisions at this high velocity shattered the VPP catalyst; it cracked the shell producing silica-rich angular shards and releasing the VPP platelets [2, 25]. The rates were much lower at a jet velocity of 231 m s^{-1} (760 ft s^{-1}), which was the velocity chosen for attrition tests and the target loss rate was less than 18 mg h^{-1}. The three charges of pilot plant catalyst were produced in the 3-m diameter spray dryer and the attrition resistance was 9 mg h^{-1}. The engineering team predicted that the attrition rate in the commercial unit would be 10–20 kg h^{-1} based on the pilot plant experience. Hence, the commercial design team minimized orifice velocities to minimize attrition rates. An 8-m diameter spray dryer produced the commercial catalyst and the attrition rate dropped to only 4 mg h^{-1} and the plant attrition rate at steady state was a mere 1 kg h^{-1}.

The decision to treat VPP gently in the design of grid pressure drop may be a contributing factor to poor mass transfer in the fast bed. To distribute gas from the plenum uniformly across the reactor and to prevent nonfluidized regions require that the pressure drop across the bed be greater than 30% of the total bed pressure drop (vertically oriented flow) [22]. For downward nozzles, 10% is acceptable [22]. The higher the better as it directly impacts fluidization quality and bubble characteristics, but less applicable in the turbulent fluidization regime with powder [20, 24, 34]. For unique geometries, such as CFB reactors, a higher pressure drop would be appropriate. The total pressure drop across the riser on top of the fluidized bed was 100 kPa, which would require a pressure drop of 30 kPa but the actual grid operated at a pressure drop below 10 kPa. Oxygen spargers were placed directly over the grid, so the consequences of poor included not only poor gas–solids contacting but the formation of large voids around the oxygen sparger and *n*-butane combustion.

6.4.2 Calcination/activation

Making precursor was somewhat straightforward since DuPont was working with the same manufacturer as other companies that synthesized precursor to manufacture maleic anhydride. However, synthesizing precursor was only the first step. The other steps included dehydration (calcination) and activation. The original procedure to activate precursor required dosing it for 18 h with 1.5% n-C_4H_{10} in air at 460 °C [32]. Other procedures also held catalyst in atmospheres with low concentrations of butane for days [18, 44].

The commercial team refused to implement a protocol requiring an extended activation step or anything above 400 °C as this would double the cost of the reactor. Above 400 °C and 400 kPa requires stainless steel, while carbon steel is acceptable at lower temperatures. Standard catalyst synthesis protocols recommend calcining catalyst at temperatures greater than the operating temperature to impart stability. DuPont had to challenge this notion with an extensive experimental program to establish a relationship between calcining conditions and activity [36]. The research team developed a procedure that operated within the constraints of the commercial reactor. They injected precursor shots of 1–2 t into the regenerator operating at 390 °C. They then transferred and stored the hot catalyst in the fast bed and cooled it in nitrogen to stifle any further oxidation [13].

6.4.3 Reactor facilities

To support the commercialization of n-butane's partial oxidation process (BUOX), the research team had access to multiple reactor types:

A 1/4″ and 5/8″ riser reactor (1 and 10 kg inventory) [35];
One 3″ big bed (fluidized bed with 1 kg capacity to calcine and activate precursor);
One 40 mm ID fluidized bed reactor (90 g capacity);
One micropulse fixed bed reactor [23];
One TAP reactor [31];
Two Sohio pipe reactor (one 1/2″ stainless steel pressure fluidized bed) [36];
Six multiple automated reactor systems (MARS – fixed beds) [30]; and,
One fixed bed coupled with XPS (X-ray photoelectron spectroscopy) [14]; and
One TGA

Each of these reactors contributed to understanding the chemistry, preparing the catalyst, and qualifying the commercial catalyst production. The Big Bed calcined and activated catalyst for the two riser reactors but also tested VPP stability for 1,500 h. The MARS reactors helped develop the calcination/activation protocol and assessed the catalyst activity and selectivity of over 300 samples in about 6 months. The Sohio pipe reactor calcined and activated these catalysts at up to 5 bar and

420 °C with varying concentrations of air and water vapor to simulate commercial conditions. The TAP reactor and fixed bed XPS system concentrated on identifying the fundamental reaction processes on the catalyst. The micropulse fixed bed attempted to simulate riser operation but alternatively dosing the reactor with pulses of oxygen and butane. A thermoconductivity detector of a gas chromatography followed the changes in composition.

The bottleneck for designing new catalysts was the 1/4″ riser which took 2–3 days to complete an analysis. This was the only reactor system in which DuPont qualified commercial catalyst or new compositions. However, from the pilot plant trials to the making commercial catalyst, seldom did researchers have the chance to test new promoters and dopants. The researchers were more creative in identifying novel structures, supports, and compositions then DuPont had capacity to test in this reactor [23]. For several years from 1990 to 1996, it was almost exclusively dedicated to qualifying pilot plant catalyst and commercial production.

These facilities were insufficient to address key uncertainties necessary to successfully start up a commercial facility:

- attrition due to particle–particle collisions at oxygen spargers, recycle gas grid, and regenerator and stripper grids, attrition due to particle–wall cyclones, slide valves, elbows, and exits;
- the effect of pressure on reaction rates;
- purity profile (acetic acid, acrylic acid, and other carboxylic acids)
- calcination and activation protocols;
- thermal and catalytic stability due to cycling and interactions with pure oxygen at the spargers;
- productivity (WHSV); and
- intangibles (unknowns) that become evident at a larger scale.

6.5 Pilot plant – Ponca City, OK

The pilot plant was based on a scale-down factor of 100 versus the commercial process. The pilot plant cost $10 million to build and $0.5 million per month to operate (Figure 6.3). Butane feed was initiated in the summer of 1990 and was shut off on February 29, 1992. During this time, DuPont loaded three charges of catalyst (2,000 kg each). Each charge demonstrated the catalyst synthesis protocol and was instrumental in confirming and refining the commercial design parameters. One of the major contributions was identifying how much acetic acid and acrylic acid would be produced at higher pressure and under reducing conditions, which was only seen once in the experimental reactors.

Figure 6.3: Circulating fluidized bed pilot plant. The 40 m tall unit was built at Conoco's Ponca City refinery. The reduction side of the process had a 0.30 m diameter fast bed and a 0.15 m diameter riser. The oxidation side had a 0.53 m diameter regenerator, a 0.44 m diameter riser stripper to separate interstitial gas from catalyst, and a regenerator stripper to remove nonselective oxygen (the pilot plant demonstrated the latter function to be superfluous: all oxygen is good and selective oxygen) [35].

6.5.1 Circulation stability

The first challenge faced by the pilot plant team was how to circulate VPP catalyst. They had completed a series of tests with FCC catalyst and the plant operated well. However, they were incapable of turning butane on because of periodic fluctuations in the system with VPP even though the particle characteristics were similar (Table 6.2) [28, 29]. The biggest differences were U_{mf} and d_{10} but standard tests were incapable of identifying major differences that would account for the poor flowability. The source of these fluctuations was uncertain but the candidates were a poor particle size distribution (poor fluidizability of the VPP), flooding in the strippers, and instability in the 12 m standpipe from the reactor stripper to the regenerator (Figure 6.3). Long-narrow standpipes are notoriously unstable and require a rigorous aeration strategy to ensure that the solids remain fluidized. Catalyst with higher fines content circulates more easily. Although the VPP particle size distribution was close to that of FCC, its flowability was dissimilar.

The pressure at the bottom of the reactor stripper standpipe was close to 1.5 bar higher than at the bottom of the cone. To ensure a continuous circulation requires the standpipe to operate above the minimum fluidization velocity. Aeration must provide

Table 6.2: VPP commercial catalyst physicochemical properties compared to fluid catalytic cracking catalyst (FCC).

Property	FCC	Precursor	Calcined	Equilibrated	Comment
U_{mf}, mm s^{-1}	2.2	4.7	5	3.9	–
ρ_b, kg ms^{-3}	874(4)	905(9)	899(33)	1,098(3)	Scott density
	853(19)	913(8)	913(17)	1,114(9)	Poured density
	965(10)	997(16)	1,033(18)	1,259(18)	Tapped density
ρ_{sk}, kg ms^{-3}	2360(2)	2496(3)	2,739(1)	2,774(6)	Gas
ε	0.44	0.42	0.56	0.46	–
H_r	1.13(3)	1.09(2)	1.13(3)	1.13(3)	Hausner's ratio
θ_{angle}, °	24(1)	26(2)	24(2)	23(1)	Angle of repose
d_{10}, µm	39.4(4)	50(1)	28(1)	32.3(2)	Laser diffraction
d_{50}, µm	64.0(6)	80(2)	59(3)	55.7(1)	Laser diffraction
d_{90}, µm	104(1)	137(5)	116(3)	91.3(3)	Laser diffraction
ϕ	0.99	0.45	0.3	0.43	Ergun equation
S_A, m^2 g^{-1}	93	22	23	12	BET
	35	34	28	15	BJH[1]
V_{pore}, mL g^{-1}					
d_{pore}, nm	38	14 [17]	–	15 [17]	BJH[1]

[1]BJH: desorption branch.
The uncertainties represent standard deviation of up to five measurements. Precursor – $VOHPO_4{\cdot}0.5H_2O$, calcined – $(VO)_2P_2O_7$, equilibrated after several years in the commercial plant – $(VO)_2P_2O_7$ + β-$VOPO_4$ [29].

sufficient gas to compensate for the change in gas density going down the standpipe. The aeration tap design had a common manifold for all taps. Since the pressure increased going down of the standpipe, more aeration gas would go to taps higher up at a lower pressure. Therefore, the solids at the bottom, near the regenerator, defluidized. This was the origin of the circulation fluctuations of solids.

The solution was to meter gas to each aeration tap independently. We modified the manifold to do this and then IGT personnel came to help supervise the pilot plant test, which succeeded. Circulation stability was still problematic when heating up the reactor to 330 °C. Curiously, as soon as operators began to feed butane, the pressure fluctuations dropped considerably. Butane or adsorbed species must have changed the interparticle forces.

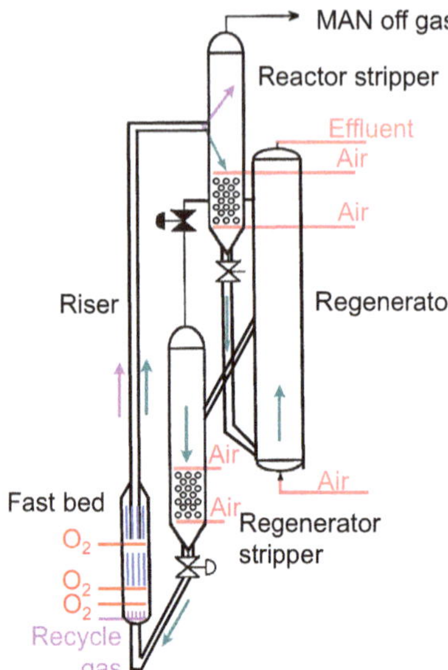

Figure 6.4: Circulating fluidized bed pilot plant. The 40 m tall pilot plant was built at Conoco's Ponca City refinery. The reduction side of the process had a 0.30 m fast bed and a 0.15 m riser. The oxidation side had a 0.53 m regenerator, and seven sets of horizontal tubes in the 0.44 m strippers. The teal-colored arrows demonstrate the solids circulation path while the magenta color represents the gas on the reduction side (mostly recycle gas) [35].

6.5.2 Catalyst reaction performance

In the 1/4″ riser, MAN selectivity approached 80% at 360 °C with an exit O_2 concentration of 4% [35]. Selectivity dropped linearly to 60% when it was below 2%. During the design of the pilot plant, by-product acid concentrations were assumed to be negligible, but just 1 month before the pilot plant was to start up, the 5/8″ reactor produced up to 5% acetic acid + acrylic acid when the reactor conditions stressed the catalyst – highly reducing atmosphere in which the butane concentration at the exit was high (10% and more) and the oxygen concentration was below 4% [27]. Together with demonstrating the mechanical integrity of the plant, one of the objectives was to maximize the productivity (WHSV), which entailed identifying the maximum concentration permissible at the reactor stripper (after catalyst was removed from the riser gas). Based on the catalyst performance in the 1/4″ and 5/8″ risers, DuPont designed the pilot plant to produce 100 kg h^{-1} of MAC but the demonstration run only produced 40 kg h^{-1} at an exit oxygen concentration of 4% and above 15% butane. Not only was MAC yield lower, but the contribution from lattice oxygen was half the design rate (even at the lower production rate). To compensate for the low performance required feeding more oxygen to the fast bed.

The first charge of catalyst produced closer to 45 kg h^{-1}. In the first plant, turnaround operations doubled the length of the fast bed from 3 to 6 m in an attempt to increase productivity. At the same time, they removed a shower cap recycle gas sparger that had completely eroded with a five-point candelabra-type sparger. They included a set of cooling coils in the extended fast bed and an additional set of oxygen spargers that were pointed downward at an angle of about 24° so as not to impinge on the wall or any internal metal surfaces.

Despite the longer residence time in the fast bed, MAN yield remained the same. The factors that had the greatest impact on yield were temperature and oxygen concentration, then butane and solid circulation rate and pressure. The regenerator residence time seemed to have only a marginal impact on it. To improve WHSV, they again modified the reactor configuration (in the second turnaround) and eliminated the regenerator and fed air to the regenerator stripper (rather than nitrogen). The regenerator was still in the loop but operations only fed nitrogen to the vessel. This was a huge change in philosophy because originally the thinking was that loosely bound oxygen from the regenerator was nonselective and should be stripped. Incorporating oxygen into the lattice was thought to follow the reaction sequence [11]:

$$O_2(gas) \rightarrow O_{2(ads)} \rightarrow O_{2^-(ads)} \rightarrow 2O^-_{(ads)} \rightarrow 2O^{2-}_{(lattice)} \tag{6.3}$$

where some of these species were considered strong oxidation agents that could combust butane [6].

6.5.3 Attrition

The catalyst did not shatter as might have been expected from the ASTM D5757 jet mill test. The attrition rate measured in the jet mills for catalyst manufactured with commercial facilities (including a 3 m diameter spray dryer) was twice as good as that based on the experimental station of 1.4 m Bowen spray dryer: it dropped from the target of 18 to 9 mg h^{-1}. The question remained what would be the attrition rate in the commercial facility. However, a priori deciding the cost of attrition versus the risk of poor distribution was problematic. Other contributing factors to attrition include pressure drop across, solids impinging at the riser exit, and the transition from the fast bed to the riser. Erosion and attrition of the catalyst at the cooling coil bends was eliminated as a factor (although the catalyst accumulated iron with time on stream – the source of the iron was never identified). The question was whether or not the design criteria should depend on the solids circulation rate, catalyst inventory, sparger/grid velocities, or a combination of these factors. Only after DuPont commercialized the process were these contributions addressed more systematically [42].

Clearly, if the plant were to be designed minimizing jet velocities at the grid, spargers, and cyclones, catalyst losses would be manageable – economically. The pilot plant personnel were pessimistic with respect to catalyst mechanical integrity and overestimated attrition rates for the commercial reactor. They recommended assuming that the attrition losses could realistically reach 20 kg h^{-1}, which translates to a cost of $8 million per year. A rule of thumb for catalyst makeup is 2% of gross revenues, so this recommendation was 5 times the standard operation. The final design basis for the economic analysis was 10 kg h^{-1}. Whereas the pilot plant catalyst was spray dried in 3 diameter unit, the commercial VPP was spray dried in an 8 diameter unit. Coincidentally, the attrition rate improved by a factor of 2 and reached 4 g h^{-1} This the larger unit produced 800 kg h^{-1} at a cost of \$800 h^{-1}. The superior mechanical resistance of the VPP together with the gentle design translated to an attrition rate in the plant of just 1 kg h^{-1} [35]. This is another example where the lack of scale-up criteria introduces differences of opinion that impact the go–no go decision to commercialize.

To further mitigate catalyst costs, DuPont continued to negotiate with suppliers but also had a program to recover attrited catalyst, micronize it, and respray dry it. Although the Si content each time the catalyst is reconstituted this way it reduces makeup costs by more than 80% [40]. This idea was successfully demonstrated in the commercial CFB.

Manufacturing catalyst was a huge success – precursor synthesis, micronization, spray drying, and calcination/activation – so much so that some engineers would have preferred that the catalyst attrited more so that they could add virgin catalyst that was more active.

6.5.3.1 Thermal excursions

Together with extending the fast bed, changing the recycle sparger geometry, adding oxygen spargers further up the bed, and loading a new charge of catalyst, plant operations switched to low sulfur butane. Within days of starting operations with the new configuration, thermocouple readings in the vicinity of the oxygen spargers increased suddenly. Simultaneously, the oxygen concentration dropped and the carbon dioxide concentration increased. (The CO concentration remained constant.) Together with these changes, some catalyst particles turned black. As the percentage of black particles increased with time on stream, the catalyst surface area dropped as did the catalytic activity in the ¼″ riser (Figure 6.5). These were the only two analytical measures that correlated with pilot plant activity. Initial surface area was a poor indicator of catalytic activity but the change in BET surface area with time was a good indicator.

Since less oxygen was coming from the catalyst lattice, the pilot-scale program focused on how to maximize the oxygen feed rate to the fast bed while minimizing. The implications for the commercial plant was nozzle design and number – what would be the maximum oxygen feed rate per sparger? The other design factors were the nozzle velocity and angle. Programs were initiated to examine various alternatives [39] but the plant design team forged ahead and assumed that if it did not work, they could redesign the spargers (a sparger assembly costs about \$300,000 ignoring lost production to install the new set or the cost of damaging the catalyst).

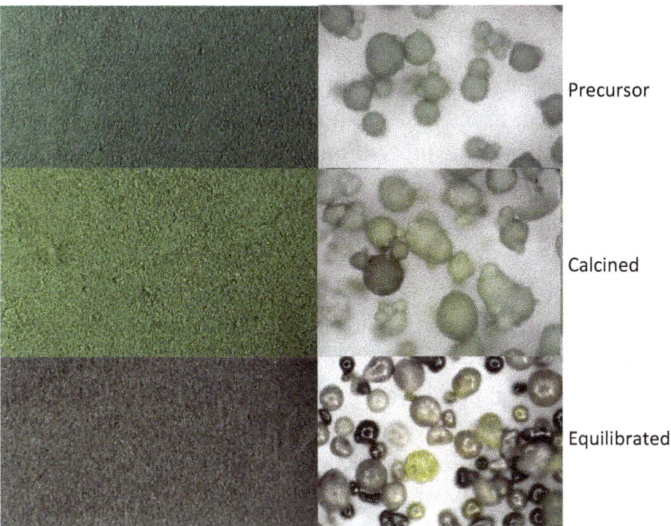

Figure 6.5: Precursor, calcined precursor, and equilibrated catalyst (from Asturias). The color changes from a teal green after spray drying precursor to a dark lime green after calcination, to brown after a couple of years operation in the commercial reactor. The commercial catalyst is very shiny as the lights from the optical camera reflect of its surface. Most of the particles are brown and black with only a few green. The calcined catalyst is mostly green but some microspheres are dark brown and even black. The precursor and calcined catalyst have more 100 μm particles than the equilibrated catalyst, which indicates that the mechanical resistance of the larger particles is poorer than the smaller particles.

6.6 Commercial experience

The catalyst manufacture was a resounding success. We produced several hundred tons of catalyst (approximately $20 million in a 2-year period) on schedule without any safety incidents. The logistics included testing precursor activity, measuring the particle size distribution after milling, and assessing catalyst activity and attrition rate of each of the 2 t commercial batches after spray drying. The precursor was delivered to the Asturias commercial reactor and calcined and activated according to the protocol based on experimental facilities [36]. The by-product acid profile in the commercial reactor was essentially identical to the pilot plant. The distillation columns produced THF that met all of the Lycra criteria and also seemed to make better fiber. The mechanical integrity of the VPP was a tremendous surprise as the attrition resistance was an order of magnitude better than projections from the pilot. Rather than 10–20 kg h^{-1} losses, the demonstrated attrition rate was 1 kg h^{-1}. This fantastic attrition rate decreased the operating costs by several million dollars per year.

6.7 Conclusions

DuPont's R&D team faced substantial uncertainties throughout the design and manufacture of the VPP catalyst. The rigor in the scale-up from experimental to commercial and then addressing the uncertainties in the process was successful. The original focus had been on attrition resistance and although the catalyst did not fall apart in the pilot plant, some engineers still recommended to assume high attrition rates for the commercial economic calculations. Unfortunately, this had a huge impact on the final commercial design, which compromised its viability. The original calcination activation protocol required a temperature hold of 18 h at 460 °C. The commercial team rejected this protocol as it would require stainless steel and thus double the reactor cost. The research team was able to respond to this constraint and after completing several hundred tests in a pressure fluidized bed, they developed a procedure that produced a better catalyst. The thermal and chemical stability of a catalyst was excellent. The VPP operated at a pressure of 350 kPa (absolute) and 400 °C with 9% n-C_4H_{10} and 3–4% O_2 at the exit. The commercial facility operated for almost a decade mostly with the same catalyst from the start-up period. The R&D team developed a recipe to recycle catalyst that further reduced operating costs and attested to their ingenuity.

References

[1] American Standard Testing Method. D5757, standard test methods for determination of attrition of fcc catalysts by air jets, 2011.

[2] P. Asiedu-Boateng, R. Legros and G. S. Patience, Attrition resistance of calcium oxide–copper oxide–cement sorbents for post-combustion carbon dioxide capture, Advanced Powder Technology, 27(2), 786–795, 2016.

[3] H. E. Bergna. Attrition resistant catalysts, catalyst precursors and catalyst supports and process for preparing same, US 4677084/B2, 1987.

[4] T. A. Bither. Vapor phase oxidation of n-butane to maleic anhydride, US 4371701/B2,1983.

[5] E. Bordes-Rchards, A. Shekari and G. S. Patience, Vanadium-Phosphorus Oxide Catalyst for n-butane Selective Oxidation: From Catalyst Synthesis to the Industrial Process, in Handbook of advanced methods and processes in oxidation catalysis, D. Duprez and F. Cavani, editors, Chapter 20, London: Imperial College Press, 549–585, 2011.

[6] G. K. Boreskov, Catalytic Activation of Dioxygen, in Catalysis – science and technology, J. R. Anderson and M. Boudart, editors, Vol. 3, Berlin: Springer, 39–137, 1982.

[7] G. Centi, F. Trifirò, G. Busca, J. R. Ebner and J. Gleaves, Nature of active species of (VO)2P2O7 for selective oxidation of n-butane to maleic-anhydride, Faraday Discussions, 87, 215–225, 1989.

[8] -Z.-Z. Chen, L.-F. Zhang, Y. Li, F. Yang, X.-H. Wang, Y.-X. Qi and Y.-F. Li., Progress in huntsman process of maleic anhydride by n-butane oxidation, Chemical Engineering (China), 39(11), 92–98, 2011.

[9] C. H. Chilton, Six-tenths factor applies to complete plant costs, Chemical Engineering, 57(4), 112–114, 1950.

[10] R. M. Contractor, H. E. Bergna, U. Chowdhry and A. W. Sleight, Attrition Resistant Catalysts for Fluidized Bed Systems, in Fluidization VI, J. R. Grace, L. W. Shemilt and M. A. Bergougnou, editors, New York: Engineering Foundation, 589–596, 1989.
[11] R. M. Contractor and J. Chaouki, Circulating Fluidized Bed as a Catalytic Reactor, in Circulating fluidized bed technology III, P. Basu, M. Horio and M. Hasatani, editors, Oxford: Pergamon Press, 39–48, 1991.
[12] R. M. Contractor, D. I. Garnett, H. S. Horowitz, H. E. Bergna, G. S. Patience, J. T. Schwartz and G. M. Sisler, A New Commercial Scale Process for n-butane Oxidation to Maleic Anhydride using a Circulating Fluidized Bed Reactor, in New developments in selective oxidation II, volume 82 of studies in surface science and catalysis, V. Cortés Corberán and S. Vic Bellón, editors, Elsevier, 233–242, 1994.
[13] R. M. Contractor, H. S. Horowitz, G. S. Patience and J. D. Sullivan. Process for the calcination/activation of V/P/O catalyst, US Patent 5895821/B2, 1999.
[14] G. W. Coulston, E. A. Thompson and N. Herron, Characterization of VPO catalysts by x-ray photoelectron spectroscopy, Journal of Catalysis, 163(1), 122–129, 1996.
[15] E. L. Cussler and G. D. Moggridge, Chemical product design, 2 edn, Cambridge: Cambridge University Press, 2011.
[16] J. DeTommaso, F. Rossi, N. Moradi, C. Pirola, G. S. Patience and F. Galli, Experimental methods in chemical engineering: process simulation, The Canadian Journal of Chemical Engineering, 98(12), 2000–2100, 2020.
[17] N. F. Dummer, W. Weng, C. Kiely, A. F. Carley, J. K. Bartley, C. J. Kiely and G. J. Hutchings, Structural evolution and catalytic performance of DuPont v-p-o/sio2 materials designed for fluidized bed applications, Applied Catalysis A: General, 376(1), 47–55, 2010.
[18] J. R. Ebner and W. J. Andrews. Process for the transformation of vanadium/phosphorus mixed oxide catalyst precursors into active catalysts for the production of maleic anhydride method for preparing maleic anhydride, US 5137860/B2, 1992.
[19] D. I. Garnett and G. S. Patience, Why do scale-up power laws work, Chemical Engineering Progress, 89(9), 76–78, 1993.
[20] D. Geldart, Types of gas fluidization, Powder Technology, 7, 285–292, 1973.
[21] J. W. Johnson, D. C. Johnston, A. J. Jacobson and J. F. Brody, Preparation and characterization of vanadyl hydrogen phosphate hemihydrate and its topotactic transformation to vanadyl pyrophosphate, Journal of the American Chemical Society, 106(26), 8123–8128, 1984.
[22] R. Karri and J. Werther, Gas Distributor and Plenum Design in Fluidized Beds, in Handbook of fluidization and fluid-particle systems, W.-C. Yang, editor, Chapter 6, New York: Marcel Dekker, 155–170, 2003.
[23] K. Kourtakis and G. C. Sonnichsen. Catalyst and method for vapor phase oxidation of alkane hydrocarbons, US Patent 5543532, 1996.
[24] D. Kunii and O. Levenspiel, Fluidization engineering, 2 edition edn, Elsevier: Amsterdam, 2013.
[25] F. Li, C. Briens, F. Berruti and J. McMillan, Particle attrition with supersonic nozzles in a fluidized bed at high temperature, Powder Technology, 228, 385–394, 2012.
[26] G. S. María-Jesus Lorences, R. C. Patience, F. Díez and J. Coca, VPO transient lattice oxygen contribution, Catalysis Today, 112(1–4), 45–48, 2006.
[27] M.-J. Lorences, G. S. Patience, F. V. Díez and J. Coca, Butane oxidation to maleic anhydride: Kinetic modeling and byproducts, Industrial & Engineering Chemistry Research, 42(26), 6730–6742, 2003.
[28] M.-J. Lorences, G. S. Patience, F. V. Díez and J. Coca, Fines effects on collapsing fluidized beds, Powder Technology, 131(2), 234–240, 2003.

[29] M. Menendez, J. Herguido, A. Bérard and G. S. Patience, Experimental methods in chemical engineering: Reactors–fluidized beds, The Canadian Journal of Chemical Engineering, 97(9), 2383–2394, 2019.
[30] P. L. Mills and J. F. Nicole, Multiple Automated Reactor Systems (MARS). 1. A novel reactor system for detailed testing of gas-phase heterogeneous oxidation catalysts, Industrial & Engineering Chemistry Research, 44(16), 6435–6452, 2005.
[31] P. L. Mills, H. T. Randall and J. S. McCracken, Redox kinetics of VOPO4 with butane and oxygen using the TAP reactor system, Chemical Engineering Science, 54(15–16), 3709–3721, 1999.
[32] R. A. Mount and H. Raffelson. Method for preparing maleic anhydride, US Patent 4111963/B2, 1978.
[33] M. J. Mummey. Process for the production of maleic anhydride, US Patent 4855459/B2, 1989.
[34] J. M. Paiva, C. Pinho and R. Figueiredo, Influence of the distributor plate and operating conditions on the fluidization quality of a gas fluidized bed, Chemical Engineering Communications, 196(3), 342–361, 2008.
[35] G. S. Patience and R. E. Bockrath, Butane oxidation process development in a circulating fluidized bed, Applied Catalysis A: General, 376(1–2), 4–12, 2010.
[36] G. S. Patience, R. E. Bockrath, J. D. Sullivan and H. S. Horowitz, Pressure calcination of VPO catalyst, Industrial & Engineering Chemistry Research, 46(13), 4374–4381, 2007.
[37] G. S. Patience and D. C. Boffito, Distributed production: Scale-up vs experience, Journal of Advanced Manufacturing and Processing, 2(2), e10039, 2020.
[38] G. S. Patience, M. G. Rigamonti and H. Li, Analysis of Powders and Solids, in Experimental methods and instrumentation for chemical engineers, 2 edition edn, G. S. Patience, editor, Chapter 10, B. V., Amsterdam, Netherlands: Elsevier, 293–337, 2017.
[39] P. Sauriol, H. Cui and J. Chaouki, Gas jet penetration lengths from upward and downward nozzles in dense gas-solid fluidized beds, Powder Technology, 235, 42–54, 2013.
[40] J.-A. T. Schwartz and D. T. Cline. Process for the manufacture of an attrition resistant catalyst, US 6878668/B1, 2005.
[41] M. R. Thompson, A. C. Hess, J. B. Nicholas, J. C. White, J. Anchell and J. R. Ebner, A Concise Description of the Bulk Structure of Vanadyl Pyrophosphate and Implications for n-butane Oxidation, in New developments in selective oxidation II, volume 82 of studies in surface science and catalysis, V. C. Corberan and S. V. Bellon, editors, 167–181, 1994.
[42] A. Thon, A. Puettmanna, E.-U. Hartge, S. Heinrich, J. Werther, G. S. Patience and Bockrath, Simulation of catalyst loss from an industrial fluidized bed reactor on the basis of labscale attrition tests, Powder Technology, 214(1), 21–30, 2011.
[43] F. Trifirò and R. K. Grasselli, How the yield of maleic anhydride in n butane oxidation, using VPO catalysts, was improved over the years, Topics in Catalysis, 57(14), 1188–1195, 2014.
[44] K. C. Waugh and Y.-H. Taufiq-Yap, The effect of varying the duration of the butane/air pretreatment on the morphology and reactivity of $(VO)_2P_2O_7$ catalysts, Catalysis Today, 81(2), 215–225, 2003.
[45] J. Welch, Winning, New York: Harper Collins Publishers, 2005.

Gregory S. Patience, Jamal Chaouki

7 Case study II: *n*-Butane partial oxidation to maleic anhydride: commercial design

Abstract: DuPont applied an iterative design procedure to build a process to partially oxidize *n*-butane to maleic anhydride in a circulating fluidized bed reactor (CFB). The first step, after generating experimental data in the laboratory, was to develop a flowsheet for a full-scale plant. To address the uncertainties raised in the design of the industrial unit, a pilot plant was built at a scale 100 times smaller than the full scale. In parallel to the detailed design of the commercial unit, the third step dedicated resources to address knowledge gaps remaining after the pilot demonstration. The final step was to support plant operations during the commissioning stages. R&D costs prior to the pilot plant ramped up from 0.5 to 4 million$ year^{-1} and then jumped to 10 million$ year^{-1} through pilot plant operation into design and support during commissioning. Many challenges at the pilot plant were resolved in the commercial design (attrition resistance, heat transfer, solids circulation instability, and catalyst cost), while some remained to be resolved (thermal excursions at the oxygen spargers, stripping efficiency, lattice oxygen contribution, kinetics, and hydrodynamics), and others emerged (agglomeration, solids entry from the side, and combustion downstream of the cyclone). Moreover, the engineering uncertainties were compounded by constraints imposed by business team to minimize catalyst inventory that was a major impediment to achieving the economic goals. Due to new uncertainties discovered during the industrial-scale operations and the constraints imposed by business objectives, this process was unsuccessful and after almost 10 years of operation, it was shut down and the metal sold for scrap value.

Keywords: butane oxidation, pilot plant, technical uncertainties, iterative design, industrial plant

---

Gregory S. Patience, Canada Research Chair in High Temperature, High Pressure Heterogeneous Catalysis, Department of Chemical Engineering, Polytechnique Montréal, 2500 Chemin de Polytechnique, Montréal, Québec H3T 1J4, Canada, e-mail: gregory-s.patience@polymtl.ca
Jamal Chaouki, Department of Chemical Engineering, Polytechnique Montréal, 2500 Chemin de Polytechnique, Montréal, Québec H3T 1J4, Canada, e-mail: jamal.chaouki@polymtl.ca

https://doi.org/10.1515/9783110713985-007

7.1 Introduction

DuPont's aspiration was to build facilities in Europe to serve the growing LYCRA® business and become a technology leader for circulating fluidized beds (CFB). The enormous scale of the n-butane oxidation (BUOX) would provide economies of scale versus fixed bed processes. The overlying concept was analogous to fluid catalytic cracking (FCC) shuttling solids from a regenerator with a long residence time to a riser with a short residence time. In FCC risers, hot catalyst cracks oil but coke forms within seconds, which deactivates it. In BUOX, the reaction kinetics of the vanadium pyrophosphate catalyst (VPP) are more complicated. The VPP has multiple active sites and coke forms more slowly.

To accompany the commercial design process, DuPont implemented the Process and Cycle Time Excellence (PACE) methodology. Although they had successfully commercialized hundreds of products, they continued to strive to identify and adopt better ways of doing research and development. The basis of PACE for new product development (including manufacturing) was to identify systematically:

What must be done?
When it should be done?
Who should do it?
How all elements are related?

It had a series of go-no-go gates and redirects to ensure that projects remained on target or to take a decision to terminate a project. In fact, this program was designed to accelerate the development process but also to kill programs that would just not die. PACE required cross-functional teams to work together from the beginning and included manufacturing (commercial design team), technology (R&D), engineering (hired technology providers like Fluor Daniels), and marketing (business leadership). The idea was to treat development as a business and not a service or function. The five steps after the concept include evaluate, optimize, pilot (develop), design, and start up (Figure 7.1). DuPont began to implement the PACE methodology already at the pilot plant stage. According to a smooth development, then entire process should have taken less than 4y. If companies were to maintain the PACE schedule it would accelerate the introduction of new technology three times faster but perhaps it ignores the complexity of commercializing innovation.

This chapter details DuPont's scale-up methodology for their process from interpreting laboratory experimental data, to developing initial flowsheets, designing a pilot plant, and then identifying shortcomings experienced in the commercial operation. We divide scale-up uncertainties into five classes related to commercial catalyst manufacture, reactor performance, sizing vessels, operating the process, and mechanical integrity. Pilot plants must address all of these uncertainties to ensure a successful start-up. However, business objectives (and thus interventions) can impact. Engineers must have a detached attitude when faced with external interjections and

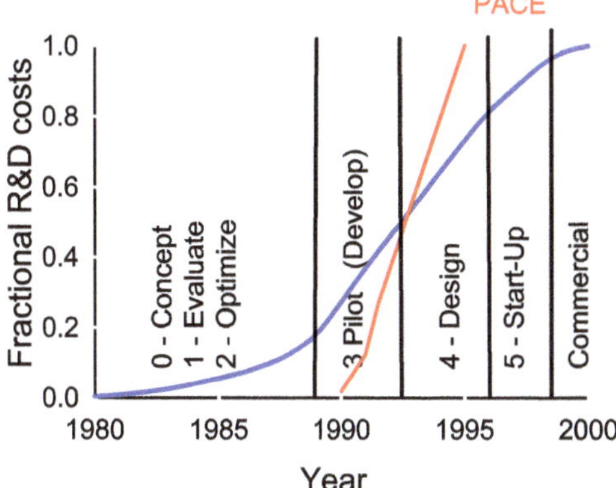

Figure 7.1: PACE project schedule versus actual development schedule.

reject any that risk operability. New technology development requires a substantial financial commitment from management beyond the initial investment cost. Recovering costs from a single plant puts excessive pressure on the technology team that promotes taking unwarranted risks.

7.2 Laboratory to commercial projection

The experimental program in the early 1980s concentrated on developing catalyst manufacturing and testing protocols. Fixed bed tests explored the reaction kinetics, a laboratory-scale CFB qualified catalyst (1 kg inventory), and a larger reactor (10 kg) identified performance parameters for the initial commercial design – regenerator residence time, solids circulation rate, conversion and selectivity as a function of temperature, and butane and oxygen partial pressures. DuPont adapted ASTM tests to characterize the abrasion and mechanical stresses of the novel core-shell catalyst that were expected in the commercial unit. The standard test was designed for FCC catalyst and the high gas velocity shattered the VPP shell. Chemical stability was tested by cycling a TGA between oxygen and reduction cycles for over 1000 times.

In parallel to the catalyst work, DuPont collaborated with institutes to measure the pressure profile in risers (O.E. Potter at Monash University), flammability limits in bomb calorimetry tests, and engaged consultants for the design of fluidized bed systems in general and circulating solids in particular.

7.3 Preliminary commercial plant design

Based on laboratory reactor performance data, DuPont generated a commercial flowsheet that corresponded to the dimensions and performance of an economic process: vessel geometry, heights, flow rates, pressure, temperature and catalyst performance – butane conversion, selectivity, attrition, and inventory – and process parameters, like heat transfer, stripping efficiency, and recycle rates.

To achieve high circulation rates requires tall reactors to generate a sufficient head differential between the riser side and regeneration side to compensate for the pressure drop across slide valves, horizontal pipes, and the riser and fast bed ΔP (Figure 7.1) [26]. The original design had strippers after riser the and regenerator (but the regenerator stripper was removed due to problems with steam stripping in the pilot plant). Provisions were made to inject pure oxygen into the fast bed. Over 2 km of cooling coils were mounted throughout to maintain iso-thermal conditions and remove the 35 MW of energy produced by the highly exothermic reaction.

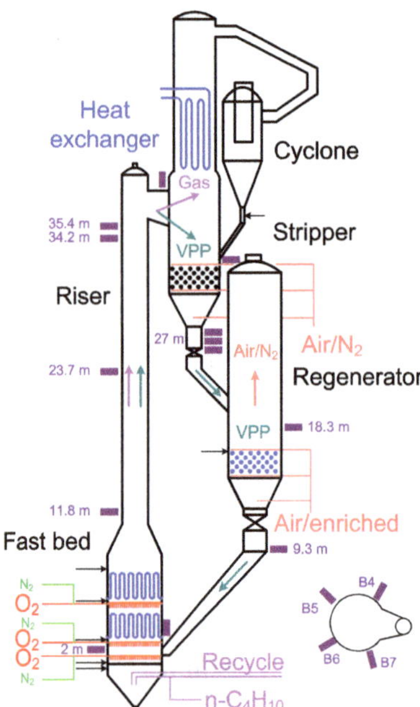

Figure 7.2: Asturias reactor schema (to scale). The fast bed was 4.2 m in diameter, the riser was 1.8 m in diameter, and the regenerator and stripper were 4 m in diameter. To minimize thermal excursions, N_2 was co-fed to the O_2 spargers. Provisions were added to feed enriched air to the regenerator to accelerate the reoxidation rate but also to increase the feed of oxygen to the fast bed and reduce the gas recycle rate. The brandy-wine colored rectangles show the placement of NaI scintillator detectors for the solids radioactive tracer study.

7.3.1 Uncertainties

The pilot plant was required to address five classes of uncertainties: VPP catalyst manufacture, reactor performance and kinetics, fluidization design and geometry, operation, particularly safety, and mechanical (Table 7.1). The experimental reactors operated at low pressure and conversion. Achieving an economic process required high pressure, recycling butane, and higher productivity (WHSV).

DuPont designed the pilot plant to respond to over 60 design uncertainties. We group these uncertainties (Table 7.1) with respect to catalyst manufacture, reactor performance, reactor design, operations, and mechanical design.

Table 7.1: Commercial design uncertainties.

VPP	Kinetics	Design	Operation	Mechanical
Synthesis	Pressure	Pressure loop	Control	Materials
Promoters	Recycle gas	Cooling coils	Safety	Corrosion
Activation	By-products	Riser exit	Operability	Erosion
PSD	Inventory	Grids	Start-up	Filters
% fines	Riser	Regen	Shutdown	Blowback ΔP
% Si	Regen	Standpipe	VPP circulation	Coil support
Chemical stability	Strippers	Cyclones:	Stable	Aeration
Thermal stability	Conversion	Combustion	Failure	Compressor
Attrition resistance	Selectivity	Backflow	Agglomeration	Silos
Oxo-capacity	Circulation rate	Diplegs	n-C_4H_{10}, % S	MAN stripper
Density	WHSV	Flapper valves	CO converter	
Surface area	% butane	Fast bed entry	Sampling	
	% O_2	O_2 spargers	T	
	Temperature	Geometry	ΔP	
		Number	VPP	
		Levels	Analytical	
		Strippers	MS	
		Steam	GC	
		Coils	HPLC	
		Flow rate	MAC for THF	

Design questions in green were addressed in the pilot plantpilot plant and successfully demonstrated at Asturias. Design issues in magenta, which we thought might be problematic in the plant but in the end worked fine. New problems in blue were unexpected in the commercial unit but were successfully resolved (with investment). The text in red relates to continuing problems in the plant that limited production.

7.4 Pilot plant

The research program to develop catalyst and test it in the pilot plant answered most of the design questions that came about from the conceptual commercial design. The catalyst manufacturing process was an unqualified success. The pilot plant operated safely with a complete control strategy and system of interlocks. VPP agglomerated in the pilot plant when we attempted steam stripping. It agglomerated into football-sized boulders following emergency shutdowns during the commercial demonstration phase of the project but this was solved with procedures and eliminating horizontal surfaces (over cyclones, for example). All mechanical components, including metallurgy, thermal expansion allowances, coil support, and blowbacks for the filters, functioned flawlessly.

The pilot plant experience answered as many questions as it uncovered: it operated at a weight hourly space velocity of 40 h^{-1} (based on the product), the turndown ratio was greater than 10, catalyst attrition resistance was an order of magnitude better than that predicted by the ASTM standard, the plant operated safely despite feeding pure oxygen through spargers into a stream with as much as 20% *n*-butane operating at 400 °C and 4 bar (absolute).

The remaining design issues depended on VPP performance and oxygen management: increasing selectivity and activity, maximizing the oxygen carrying capacity of the catalyst, reduce bypassing in the regenerator, improving stripping efficiency to reduce carbon carryover, enhancing the mixing of solids and gas at the grid, and eliminating thermal excursions at the oxygen spargers, which turned catalyst black and reduced surface area and activity. To better design the reactor required a kinetic model that could account for the change in selectivity and productivity as a function of regenerator residence time (something that is still missing). The sparger design was based on reducing the oxygen feed rates per nozzle, but the relationship between velocity and momentum on mixing intensity and the induction time as the concentration passes into the flammability zone remains unknown.

The research team developed power law regression models for butane conversion ($X_{C_2H_{10}}$), maleic anhydride selectivity (S_{MA}), and yield (Y) as a function of the main factors of fast bed temperature (T), gas flow rate (Q), entrance mole fraction ($y_{C_4H_{10}}$), entrance mole fraction (y_{O_2}), ($\dot{M}_s$), regenerator inventory (M_{rg}), and pressure (P):

$$X_{C_4H_{10}}, S_{MA}, Y = \beta_0 \exp\left\{\frac{E_a}{R}\left[\frac{1}{T} - \frac{1}{T_o}\right]\right\} Q^{\beta_1} y_{C_4H_{10}}^{\beta_2} y_{O_2}^{\beta_3} \dot{M}_s^{\beta_4} M_{rg}^{\beta_5} P^{\beta_6} \tag{7.1}$$

We applied these models to the best data point in the set to then identify conditions that met the economic goals. However, the final conditions for the commercial design was somewhat outside the range of data (within all the range of each of the factors, but not the combination). The power law model afforded the team some latitude to reduce the operating temperature to achieve an acceptable selectivity even though very few experiments were run below 375 °C. However, the pseudo activation energy (E_a) was low compared to the activation energy from the experimental reactors and thus afforded too much latitude in the design to operate at lower temperatures while maintaining high selectivity.

Within DuPont, some senior engineers had severe reservations about the viability of the and interpretation of the data. So, the director of engineering R&D (a research division outside the business that was in charge of running the program) assembled a group of engineers with experience in fluidization, FCC, reactor modeling and kinetics to review the design basis and comment on the risks and what steps were available to mitigate these risks. They recommended that a detailed kinetic model coupled with a hydrodynamic model be created to substantiate the regression model. Furthermore, they identified the change in the entrance geometry of the solids to the fast bed from the regenerator as an unnecessary risk and recommended an experimental program to measure the hydrodynamics of the fast bed.

7.5 Shortcomings

DuPont constructed the Asturias plant on-time, within budget, and with zero lost work cases – recordable injuries for which an individual would miss at least one full day of work. However, like the Ponca City Pilot Plant experience, the plant had problems that went unrecognized in the design phase or the data were inconclusive and so no action had been taken.

To respond to the assessment of an independent group of engineers, the Butane Oxidation Reaction Engineering team (BORE) was created This team initiated cold flow modeling studies at PEMMCORP to assess possible scenarios related to the side entry. They engaged Polytechnique to study gas injection into fluidized beds. During the start-up phase, the team created other programs to address shortcomings that emerged [36].

Figure 7.3: Circulating fluidized bed model of the Asturias reactor in plexiglass (1999). Large bubbles (and standing bubbles) were present throughout the bed. At the entrance to the riser, sheets of powder would adhere to the outer wall and the sporadically tear off. The powder in the Asturias reactor made a characteristic sound of three short bursts in 2 s followed by a whooshing sound of about the same duration. Early tests at PEMMCORP were conducted simulating the side entry with shots of colored water into clear water and later of colored FCC into FCC. (Photo from J. Zenz at PEMMCORP.)

The uncertainties in the design phase resulted design shortcomings related to:
1. butane back flow up the regenerator;
2. combustion in the freeboard and cyclone;
3. thermal excursions at spargers;
4. insufficient lattice oxygen (that required too much cofeed of oxygen to spargers);
5. gas–solids contacting in the fast-bed riser;
6. poor stripping efficiency resulting in excessive carryover of carbon to the regenerator;
7. overestimating the cost of catalyst (which forced the design team to make too many concessions with respect to catalyst inventory);
8. poor fluidization characteristics (a higher concentration of fines with a lower average U_{mf} would have decreased thermal excursions at the sparger);
9. scale-up based on an optimistic regression analysis;
10. inadequate distribution of solids to the regenerator that resulted in severe bypassing; and
11. too low a pressure drop across the grid, which probably resulted in maldistribution of the gas.

A more thorough understanding of the reaction mechanism would have increased the chances of success of the program as it would have differentiated the shortcomings related to gas–solids hydrodynamics versus reaction kinetics – lattice oxygen, temperature, pressure, and species concentrations.

7.5.1 Butane backflow

Within 12 h of turning butane feeds on, as much as 10% of the VPP turned from green to black. The thermal excursions in Ponca City turned catalyst black over several months. Researchers at the Experimental Station attempted to reproduce this phenomenon by heating VPP and found that it would turn black at temperatures above 750 °C. The reason the VPP turned black so quickly at Asturias was due to design flaws and poor operation practice. Rather than ramping up butane and oxygen concentrations gradually, within hours of turning butane on, the lead engineer ramped up butane concentration to the design conditions (9%).

The design issue was the side entry: A slide valve below the regenerator controlled the solids circulation rate. Immediately below the valve was a straight 1.3 m section that led to an inclined pipe that changed shape to resemble a squashed oval (Figure 7.4). The oval had the same cross-sectional area of the 1.3 m pipe but it was only 0.6 m high. This geometry minimized the vertical height of the entrance pipe and thus the distance between the oxygen sparger below the cooling coils and the grid (2 m). The solids slid down the standpipe and entered the bed at 3–6 m s^{-1}. They only slid on the bottom, and even though the pipe was just 0.6 m, it was insufficient

Figure 7.4: Fast bed side entry standpipe. The six aeration taps were completely unnecessary.

to form a plug. Rather, the solids shooting in the bed created a vacuum that sucked gas (butane) from the fast bed up the standpipe. Air coming down from the regenerator combusted the butane coming up the standpipe. The temperature became hot enough to deform the slide valve.

This problem was solved by placing an insert in the bed that created a pressure drop (and sort of plug) by directing the flow upwards by about 30°. The plant continued to generate black particles but not at such a disastrously fast rate. The possibility of back flow had been recognized in the PEMMCORP cold flow experiments but the construction phase had advanced too far to do something about it and it was unclear that it would be a problem.

The possibility of backflow up the standpipe was a huge issue for the pilot plant but was somewhat disregarded for the commercial unit design. The decision to introduce solids from the side was driven by the directive to reduce catalyst inventory and thus cost. The huge cross flow of solids coming from the side was thought to improve radial mixing. However, some with experience recommended a symmetrical geometry for flows in fluidized beds.

7.5.2 Combustion in the stripper/cyclone

DuPont had conducted an enormous amount of research to ensure that the plant would operate safely. The principal preoccupation was the possibility of combustion at the nozzle tips but also at the exit of the stripper and downstream of the stripper cyclone. They conducted hundreds of experiments in bomb calorimeters to determine the lower explosive limit (LEL – highest volume fraction of a fuel for which a flame propagates) and upper explosive limit (UEL – lowest volume fraction of a fuel for which a flame fails to propagate). A flame requires a fuel, oxidant, and energy source. Autoignition refers to the minimum temperature at which an external source of energy is unnecessary. Above 400 °C, *n*-butane combusts in air without an external energy source. The design consideration when operating above the temperature is the induction time: how much time at the specified conditions does it take to ignite.

The bomb calorimetry experiments demonstrated that a 4% volume fraction of oxygen in butane was close to the safe limit. (The pilot plant target was 3% O_2 but they were forced to drive toward 4% to increase maleic anhydride yield.) The tests also showed that increasing the diameter of the vessel decreased the oxygen concentration at which a flame propagated. Only once in the pilot plant had they detected combustion downstream of the cyclone. On the last day of testing, the downstream piping was hotter than it had ever gotten and as a result the oxygen in the stream reacted with butane (not maleic anhydride) until all the oxygen was consumed. As a consequence, the temperature in the downstream pipe increased even further. The oxygen concentration would build up and the butane would combust when the oxygen concentration reached the LEL. The cycle time was on the order of tens of minutes and the sequence – combustion, oxygen concentration building up – lasted for 4 h.

In Asturias, after addressing the problem of combustion in the standpipe, a high temperature interlock downstream of the stripper shut the plant down. The temperature at the exit was 399 °C with 3.7% O_2 and 10% C_4H_{10}. The thermocouples were unable to detect the extent of the temperature rise while as much as 1% of the VPP microspheres turned black.

Due to the intervention of the butane oxidation reaction engineering team, the BORE team designed and built a 128 mm diameter fluidized bed reactor that expanded to 305 mm at the top. It was equipped with spargers, many thermocouples, pressure transducers, and analyzers to monitor the oxygen concentration and carbon species. It could only operate from 23h30 to 00h30 because of Delaware butane emissions restrictions for the site that allowed them to operate just 30 min per day. Introducing moderate pressure spikes to the fluidized bed caused thermal excursions and they increased from less than 5 °C when the bed was at 320 °C to almost 100 °C when it was at 370 °C [20]. These studies combined with an elementary step level free-radical kinetic model helped identify temperature, pressure, residence time, and solids density as factors that contributed to the off gas combustion.

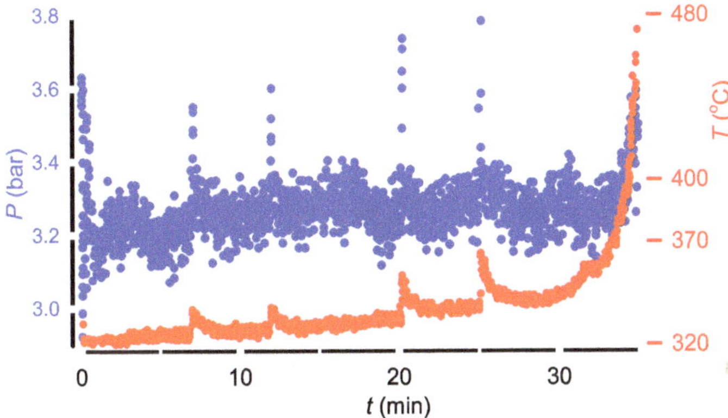

Figure 7.5: Pressure and temperature spikes in the 127 mm fluidized bed. Introducing a pressure spike to the bed initiated combustion that was recorded by thermocouples in the bed. The temperature spikes increased as the bed temperature increased and approached 100 °C at 370 °C. Adapted from Hutchenson et al. (2010) [20]. Copyright 2010 Elsevier.

The Asturias plant had few options to eliminate thermal excursions above the stripper. Either reduce the in the area (to below the induction time) or reduce the temperature. The plant was only operating at 35% capacity at the time, so we needed a definitive solution. Some ideas to reduce temperature included removing insulation along the pipe (but MAN would condense), spray water inside the pipe downstream the cyclone, or maleic anhydride from the scrubber. Initially, we considered reducing the induction time but finally decided to extend the stripper vessel and install a heat exchanger in it to reduce the temperature. The largest crane in the world was contracted to make the 5 t lift. One of its previous commissions was to pour cement over the Chernobyl nuclear reactor. The Spanish population is quite anti-nuclear energy and protested bringing in the crane for fear of radiation.

This design worked perfectly and the plant was able to ramp up to 65% capacity, but only temporarily because the next bottle neck was the spargers.

7.5.3 Thermal excursions at the sparger

Two sets of spargers were originally installed in the fast bed each with 926 nozzles pointing at a 30° angle from the vertical. One set was below the cooling coils at 1.9 and 5.5 m above the grid [40]. (The pilot plant nozzles were pointing at 24° angle from the vertical and pointing away from the wall at 9°.) The design was conservative in that the oxygen flow rate through each nozzle was approximately 50% of the demonstrated rate in the pilot plant. The rationale for sparging oxygen so high in the fast bed was to allow butane to reduce the catalyst and thereby increase reaction rates higher in the bed and hopefully increase the reaction rate. Neither of these

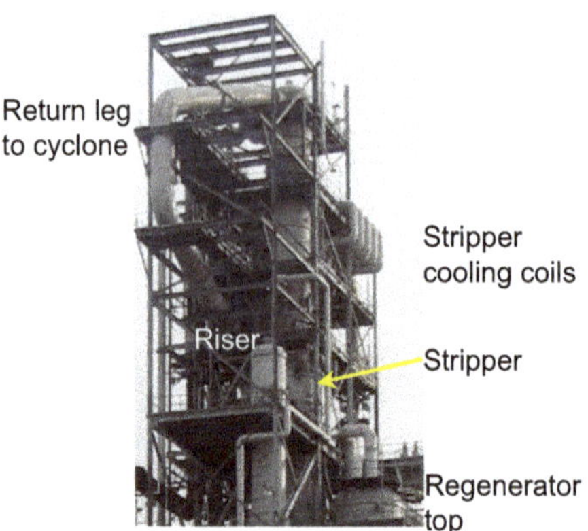

7-6_hxtop

Figure 7.6: Reactor stripper extension to house the heat exchanger to drop the temperature below 300 °C and eliminate off gas combustion.

hypotheses were validated with subsequent experimental data. Catalyst is more selective when it is more oxidized rather than reduced.

To improve oxygen feed capability in the riser, a third sparger was installed such that the tips were only 0.45 m above the grid. Acrylonitrile experience would have recommended a separation of only 0.1 m with a sparger nozzle directly above every grid hole. The third (lower) sparger was the most effective and the top sparger was the least effective: thermocouples at the top detected thermal excursions with as little as 1,500 kg h^{-1} while the design basis was 8,000 kg h^{-1}. Substituting the with sintered metal frits allowed operators to increase the feed rate but at one point they stopped feeding gas to this sparger and MAN condensed in the pipe and on the frits blocking it entirely.

7.5.4 Insufficient lattice oxygen

Sparging oxygen into the fast bed was the next immediate bottleneck to increasing MAN production rates. The plant was able to operate at 65% capacity but it was unsustainable because thermal excursions at the oxygen nozzles turned the catalyst black. If the catalyst were able to supply more oxygen, the feed rates to spargers would decrease (at the same capacity) because more oxygen would come from the regenerator and the selectivity would rise. In the 1/4" riser-reactor, as much as 50 % of the oxygen was supplied from the regenerator via the catalyst lattice, the pilot plant regenerator supplied about 20%, but the contribution of oxygen from the Asturias regenerator was less than 10%, while the two oxygen spargers with 926 nozzles was to feed 8 000 kg h^{-1} to the fast bed.

Enriching oxygen to the regenerator and the stripper were two options to increase the oxygen concentration at the entrance, which was demonstrated to improve selectivity and productivity [40]. However, the higher concentration might increase the risk of thermal events when sparging butane to the recycle gas upstream of the plenum.

FCC regenerators have elaborate solids feed entrance configurations to ensure that they are evenly distributed across the reactor. DuPont's regenerator design assumed that they would provide a sufficient amount of staging and avoid severe bypassing. To test the hypothesis, the research staff added sample ports in different areas of the regenerator (all at the same height) to measure the gas concentrations (Figure 7.7). The lowest oxygen concentrations in the regenerator corresponded to the highest carbon dioxide concentrations and vice versa. The O_2 concentration was highest in the center and lowest at the walls, which indicates that most of the reaction was happening along a straight path from where the solids enter the regenerator at the side. These plots suggest that distributing the VPP would improve the overall effectiveness of the regeneration step.

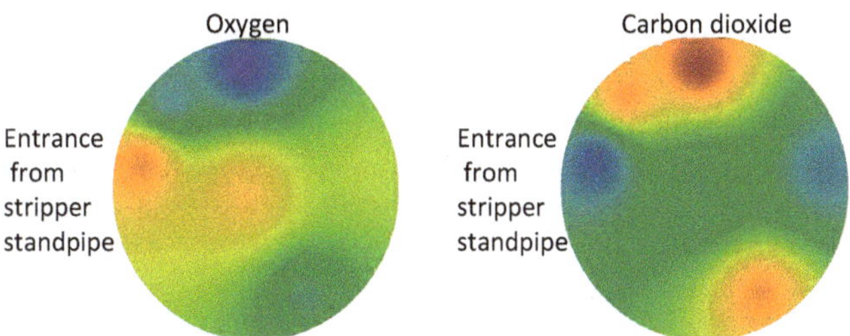

Figure 7.7: Oxygen and carbon dioxide concentration contour plots in the 4 diameter regenerator. The areas of high concentration are in blue and dark green and low concentrations are orange tending toward red. The volume fraction of oxygen ranges from 15% to 19%. The volume fraction of carbon dioxide ranges from 0.8% to 1.8%. Only seven sample points make up the data set – a regression model generated the color distribution (SigmaPlot12). Gregory S. Patience, Experimental Methods and Instrumentation for Chemical Engineers, Elsevier, 2nd Edition, 2017.

7.5.5 Gas–solids contacting/hydrodynamics

One of the most contentious unknowns with respect to the commercial operation was the solids circulation rate. DuPont engineers relied on this parameter to calculate regenerator residence time to size vessels, design the stripper coils to avoid flooding, and determine the ΔP across the riser and fast bed to calculate compressor capacity. The 1/4″ reactor and pilot plant applied the same simple procedure to measure the solids circulation rate accurately: When the slide valve underneath the reactor stripper is closed, catalyst accumulates in the bed and the change in is directly proportional to mass:

$$\frac{dM}{dt} = \frac{X_A}{g}\frac{d\Delta P}{dt} \tag{7.2}$$

Concurrently, the drop in ΔP in the regenerator is also proportional to a loss in mass.

This technique was unworkable with 1.2 m standpipes in Asturias. Rather they relied on the slip factor correlation as well as the correlations the slide valve manufacturers provided [22, 30]. However, to confirm these two measures were valid, DuPont executed the first open-source radioactive test ever to be conducted in Spain. They produced a VPO tracer doped with La^{140} that was activated in a nuclear reactor in Pavia, Italy. Tru-Tec placed 16 lead collimated NaI scintillator detectors throughout the reactor: 5 in the fast bed at the bottom, 4 below the stripper and regenerator slide valves, 4 along the riser standpipe, another in the cross-over pipe and the final one was placed at the top of the stripper (Figure 7.2).

A clean injection pulse facilitates interpreting radioactive tracer data. DuPont assumed that injecting at the top of the riser would be best as the solids loading was low and they could consider the pulse a Dirac δ-function. Injecting just at the cross-over would make it easy to detect the first cycle and thus calculate the solids circulation rate: The total solids inventory divided by the difference in time between the second peak t_2 and the first, t_1. They also anticipated measuring the solids circulation rate, M_s by subtracting the residence time derived from the detector below the stripper ($t_{det,2}$) from signal of the detector at the cross-over between the riser and stripper ($t_{det,1}$):

$$M_s = \frac{M_{stripper}}{t_{det,2} - t_{det,1}} \quad \text{or} \tag{7.3}$$

$$M_s = \frac{M_{total}}{t_2 - t_1} \tag{7.4}$$

A third way to assess the solids circulation rate was in the standpipe below the stripper, but here the signals were too weak and positioning the detectors and housing among the piping was cumbersome. Unfortunately, the injection quill broke in the first trial. Then the Tru-Tec staff had problems injecting and one of the three trials had a double peak at the injection. The background radiation continued to increase with each test, which complicates interpreting the signal. A worse problem was that a detector just above the entrance of the riser (11.8 m above the tan line) detected tracer injected at the top of the riser (35.4 m above the tan line). This spread the peak so it was not a clean sharp pulse.

The backmixing and bypassing were so high in the stripper, regenerator, and fast bed that assuming of the tracer (eq. (7.3)) would give solids circulation rates 5× higher than those predicted by the slide valve vendors and the slip factor correlation. Short regenerator residence time was a prime factor in low selectivity and productivity. Although the regression model assigned a low coefficient to regenerator, subsequent experiments in a pressurized fluidized bed demonstrated the importance of dosing the catalyst in oxygen for extended periods of time (like in the 1/4" riser reactor) [38, 39]. The short residence times were not due to high circulation rates but rather rat-holing, which is consistent with the gas sampling data (Figure 7.7).

To characterize the riser hydrodynamics, DuPont developed a two-zone model with solids upflow through the core ($M_{s,c}$) and downflow through the annulus ($M_{s,a}$). The fitted parameters to the model include the void fraction in the core and annulus, $\varepsilon_c, \varepsilon_a$ and the radial flow rate of solids between the two zones, $\dot{m}_r$ (the radial distribution of solids follows a power-law type distribution but dividing the riser in two zones is a simplification to help understand gross flow patterns [15, 24, 28]):

$$M_c \frac{\partial C_c}{\partial t} + \dot{m}_z \frac{\partial C_c}{\partial z} = \frac{\dot{m}_r}{R\sqrt{\phi}}(C_c - C_a) \tag{7.5}$$

$$M_a \frac{\partial C_a}{\partial t} + \dot{m}_r \frac{\partial C_a}{\partial z} = \frac{\dot{m}_r R\sqrt{\phi}}{R(1-\sqrt{\phi})}(C_a - C_c) \tag{7.6}$$

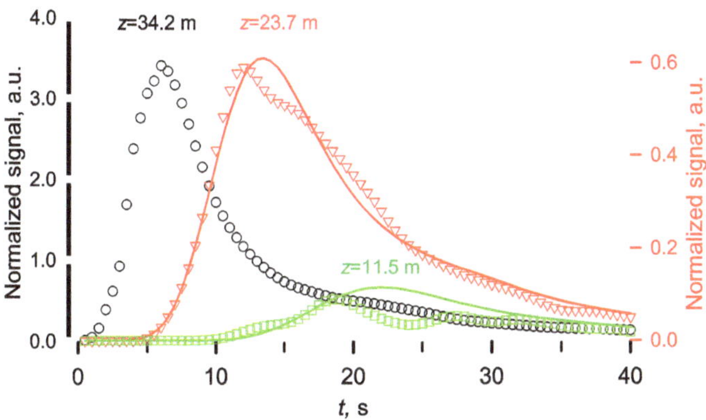

Figure 7.8: Riser solids RTD injecting La^{140} doped VPP to the top of the riser at 35.4 m. The gas velocity was 6 m s^{-1} and the nominal solids circulation rate reported by the slide valve correlation was 2,500 t h^{-1}. A core-annular model fits the data ($R^2 = 0.98$) assuming the solids fall along the wall at 1.5 m s^{-1}, they rise in the core at 2.7 m s^{-1} and the flux downward is equal to 25% of the flux upward. The axis for the green curve is on the right.

The parameters that best fit the data assumed the solids dropped along the wall at 1.5 m s^{-1} and that the density in the vicinity of the wall (annulus) was twice that of the core. To account for the large peak mid-way down the column and at the bottom of the column required that the downward mass flux was 25% of the upward mass flux – assuming the standpipe circulation rate was correct at 2,500 t h^{-1}.

On the regenerator side, the cross-over peak was sharp versus the other peaks, but it was long considering that the injection pulse at the quill was about 1 s – it spread to 15 s. The leading edge at the top of the stripper was sharp, like the cross-over, but the trailing edge did not quite reach the baseline (Figure 7.9). It reached a minimum at 125 s and then began to rise due to the tracer completing the cycle. The tracer peak at the standpipe was also relatively sharp but only returned to within 50% of its baseline. The signal from the detector at the top of the stripper returned to within 15% of the baseline (200 counts). The stripper VPP inventory was 40 t, so the solids circulation rate, considering the time to reach the mid-point counts of the leading edge ($\Delta t = 12$ s), would be 12,000 t h^{-1}. This rate is impossible if the flooding equations through horizontal tubes are correct. Furthermore, since the signal was only within 50% of the baseline, the data demonstrate that a fraction of the VPP was bypassing the stripper coils but another fraction was being held-up. Carry-over of carbon to the regenerator was much higher than the demonstration rates, which substantiates the hypothesis that the stripper was operating poorly. Evidently, the solids entered in a stream following the contours of the vessel and would shoot through the coils, which is very surprising and inconsistent with pilot plant data and cold flow studies that were conducted to design the stripper.

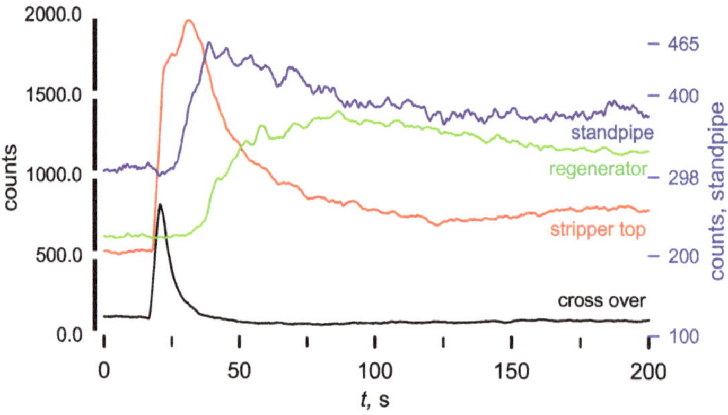

Figure 7.9: Solids RTD on the regenerator/stripper side. The inlet boundary condition is the detector at the cross-over, which is sharp with little tailing. The peak height is much shorter than the stripper and regenerator signals because of the 76 mm thick pipe that attenuates the signal. The metal thickness at the top of the stripper was 32 mm and it was 22 mm at the top of the regenerator. The standpipe wall thicknesses were 38 mm.

The signals at the bottom and the top of the regenerator standpipes were weak due to the pipe wall thickness (50 mm): The maximum number of counts was 115 at the bottom and 70 at the top. The wall thickness in the fast bed was 22 mm and the number of counts approached exceeded 1,100. To compare the curves quantitatively, we set their baseline to 580 and fit the data to a logistic dose response.

$$I = I_0 + \frac{\Delta I}{1 + \exp - \frac{t - t_{1/2}}{\beta}} \tag{7.7}$$

The best fit parameter $t_{1/2}$ is a measure of the inflection point between the entrance background radiation and the point at which the signal reaches a maximum. Based on these values, the catalyst arrives at the back wall first (to within 1 s), then 5 s later it arrives at the detector near the right wall entrance point, and finally 7 s later at the left nearest the entrance (Figure 7.10). Clearly, the catalyst flowed preferentially to the right of the bed and eventually got to the left. This flow maldistribution was non-ideal for gas injection and probably indicated that there were dead zones in the bed.

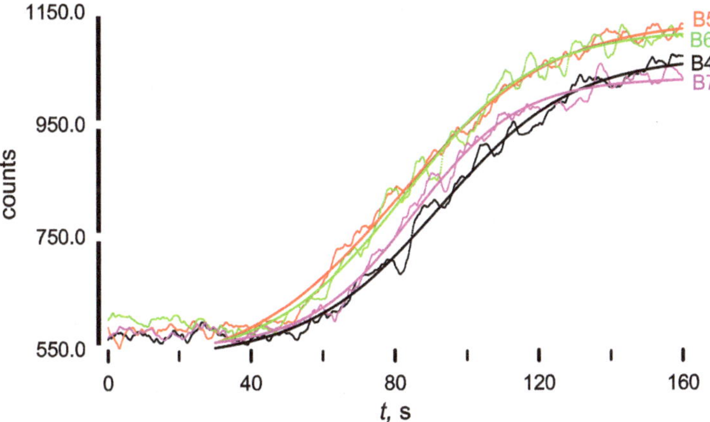

Figure 7.10: Fast bed solids RTD. B4 is at the right side of the solids entrance, B5 is at the opposite wall on the right side, B6 is at the opposite wall on the left side, and B7 is at the left side of the solids entrance (Figure 7.2). Based on a logistic dose response regression model, the tracer hits the opposite side of the bed 81 and 82 s, then arrives at the entrance to the left side at 86 s, and finally arrives at the right side of the wall near the entrance at 93 s.

Fitting the radioactive signal from the exit of the fluidized bed with the logistic dose response gives $t = 100$ s and $t = 115$ s at the top of the riser. The reactor was operating at 205 kg m^{-3} so applying eq. (7.3) gives a solids circulation rate of 2,850 t h^{-1}, which agrees well with the slip factor correlation and the slide valve that reported about 2,500 t h^{-1}.

The side-entry pipe was a squashed oval that was just 0.6 m high and 1.9 m wide (Figure 7.4). Solids came shooting down the standpipe from the slide valve at 3 m s^{-1} or more and made it to the other side of the fast bed in just 12 s. This high velocity demonstrates good mixing but not in all directions. The fact that it took another 7 s longer to get to the right side of the entrance versus the left side is evidence of non-uniformity. No cold-flow modeling or simulation at the time would have been credible enough for the engineers designing the unit to modify the plant to accommodate changes.

7.5.5.1 Stripping

Stripping the interstitial gas of n-butane was one way to minimize carbon carry-over to the regenerator and thus minimize yield losses. DuPont conducted stripping experiments and installed horizontal coils to stage the gas and increase the efficiency. The open area was 50%. The gas velocity was maintained such that the difference between the velocity of solids going down and solids going up was at the minimum fluidization velocity. Stripping efficiency improved with increased gas velocity but air as a stripping gas introduces nitrogen to the recycle stream. Consequently, the purge rate must increase in proportion to the gas introduced. The ideal stripping gas is steam as it condenses with maleic anhydride in the scrubber so the recycle stream concentration of inerts does not build up. The two trials on the first and second charge of catalyst to feed steam to strip carbonaceous species failed miserably. Even though the steam was superheated, immediately when it was injected into the reactor, the catalyst formed a cake and the regenerator formed an annular ring only leaving at 0.1–0.2 m cylindrical hole down the center. Steam stripping was unworkable and so the commercial plant would have to live with either increased carbon carry over to the regenerator due to poor stripping efficiency or increased purge losses due to the higher nitrogen concentration in the stripping gas. (An alternative design was to enrich the air with oxygen, which increases lattice oxygen and gas-phase oxygen to the fast bed via the gas recycle. It would also increase the risk of combustion in the piping downstream of the stripper and cyclone.)

The early estimation of the cost engineer for heat transfer was unfounded and delayed the project unnecessarily. Cost engineers must be accompanied by experts in the field in order to estimate equipment sizes more accurately. DuPont, in general, engages experts and consultants but maybe not at the early stages.

7.6 Piloting deficiencies

The business management required that the pilot plant shut down on February 29, 1992. During the last 6 weeks, the team was completing an extended run on the third charge of VPP to evaluate catalyst stability. The conditions of this run were supposed to represent the desired conditions of the commercial operation but the conditions were unable to meet the business team's economic objectives. The R&D team was tasked to identify operating conditions that would meet the goals. Commercial design and pilot plant operation/data reduction are two activities that proceed in parallel. New data or interpretation of existing data is often too late to input to the commercial schedule as it delays the construction process and introduces excessive costs.

Commercializing new technology requires deep pockets and patience. Programs like PACE introduce structure to the methodology but is sometimes insufficient when the data generated is imprecise or open to interpretation. Six Sigma, which DuPont adopted in the early 2000s, focuses on collecting actionable statistical data and reduces the variance that accompanies the diversity of engineering personnel or corporate culture.

The reason for the change in fast bed entry geometry was to satisfy the business' economic objectives: the projected catalyst cost was too high (>70 $ kg) and the attrition rates were too high (the design basis was 10 kg h^{-1}). Reducing catalyst inventory would reduce the initial capital investment (but not the make-up charges if attrition rates depended on operating conditions). One idea that emerged was to feed solids from the side instead of from below through a J-bend. The 10-foot J-bend beneath the riser in the pilot plant ensured that backflow of butane to the regenerator was virtually impossible. However, since the J-bend and standpipe operated near the minimum fluidization velocity, the suspension density was high, with an inventory that equaled about 25% of the regenerator inventory.

Just as some researchers would not accept a catalyst that had not been tested in the pilot plant, some engineers applied the same reasoning to the commercial design: Building a reactor with a modified geometry would be like running catalyst from an alternative vendor without testing it. The pilot plant introduced catalyst from the bottom and so catalyst should come from the bottom in the commercial reactor. Conventional design practice respects symmetry and straight lines. Hurtling VPP across the fast bed at 1 t s^{-1} is far from symmetrical. Moreover, the entrance point was just 0.6 m above the grid plate. The low pressure drop grid coupled with high velocity solids could induce instability of the jets. On the other hand, solids coming in from the side would contribute to increasing radial mass transfer.

The major failure was to ignore kinetic modeling that characterizes quantitatively regenerator residence time, pressure, temperature, and solids hydrodynamics. Regression models are appropriate when interpolating within a given data set, but when the design point is at the edge of the data cluster, its reliability is questionable. The attitude of management had been to understand the risk and develop contingencies around the risk. This is appropriate for established processes but when the phenomenology is unknown applying a safety factor of two is a guess.

Tackling three new technologies simultaneously in a single project was a colossal task. It took at least 6 months to get the purification train running, another 1.5 years to figure out how to make THF from maleic acid. (The stumbling block was deactivation of the rhenium catalyst and in the end commercial catalyst based on Pt from Englehardt worked fine.) Although the plant engineers were improving the CFB technology, the business management had lost enthusiasm. (Perhaps if the team would have agreed on a unified strategy, they might have convinced management to continue.)

7.7 Business strategy

In the 1970s, BASF offered to sell THF to DuPont for cheaper than they could make it, but DuPont refused. In 2001, DuPont tested their PTMEG (polyTHF) to make Lycra® and it passed all the protocols. Terathane® would no longer be the only supplier for LYCRA®. The technology hiccups throughout the 20 years of development must have been a factor in management's decision to shutter the business and seek higher margins in other growth businesses. The early costs in the 1980s were a couple of principle investigators and technicians working at the Experimental Station (Figure 7.11). In the late 1980s, the strategic business unit began to contribute much more to development costs (including the pilot plant). Engineering research (at the Experimental Station) and plant research at Asturias overtook the research from the pilot plant in the mid-90s and most of these programs ramped down toward the beginning of 2000.

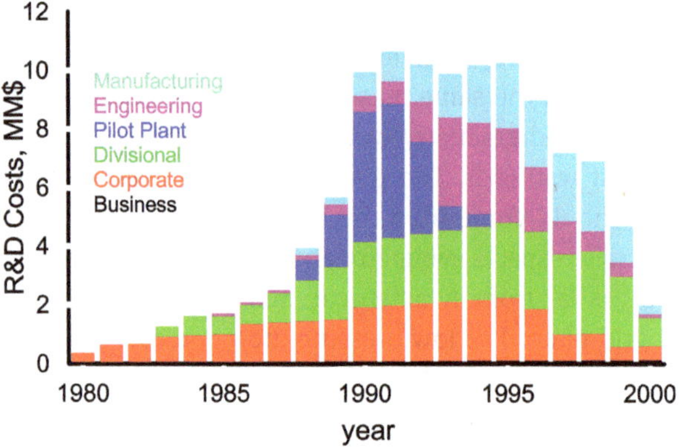

Figure 7.11: Estimated R&D costs through the 20-year development process. The chart includes estimated costs for analytical, pilot plant modifications, catalyst, and personnel to operate the plant (not the cost to build and design), modifications to the commercial reactor (insert, heat exchanger, additional spargers), and personnel. We assumed that each PI + technician incurred a cost of 400 k\$ year^{-1}, which includes management, building, and other overhead.

The market pressures were as disconcerting as the technology challenges. Sisas, a privately owned chemical company that operated the Pantochim maleic anhydride fixed bed reactors in Feluy, Belgium, announced they would expand capacity from 65 to 300 kt year^{-1} [42]. The objective was to make more MAN but also butanediol, and THF. They attempted to attract customers to buy their THF by reducing prices. BASF would match any price they offered and so it dropped almost 1 \$ kg^{-1}, which was less than the cost to make THF in Asturias. Finally, in 2004, DuPont sold its apparel textile business that included LYCRA® and Tactel® to Koch. Soon after, the plant was closed.

The original vision for the CFB technology was to build them big and gain economies of scale through size and plant integration. The Asturias plant was not to be the only one; they were to build others as LYCRA® was continuing to grow and the market was growing even faster. The appetite for larger plants is being challenged by the philosophy of process intensification (PI) that is driving toward smaller plants and distributed production [12, 27]. PI has a future but it will have to address OPEX related to labor costs and transportation. CFB technology provides economies of scale as production rates are several fold larger than conventional technology (the opposite trend of distributed production and PI) and applies to processes limited by heat transfer, coking, catalyst deactivation, and highly explosive mixtures. Operational uncertainty and complexity are greater because CFB structures are taller, compressors are larger and it comprises multiple vessels – fast bed riser, strippers, regenerator (CAPEX). Many development processes have applied CFB technology in the past and many more have been pursued in the last couple of decades [4]:

- chemical looping combustion [1],
- calcium looping [19],
- propane/propylene to acrylic acid (Arkema) [14, 43],
- propane/propylene/glycerol to acrolein (Arkema) [10, 29],
- propane dehydrogenation (Dow) [32],
- isobutane dehydrogenation (Snamprogetti-Yarsintez) [34],
- methanol-to-olefins MTO (ExxonMobil, Total, Haldor-Topsoe) [8, 16, 37]
- polypropylene (LyondellBasel) [25, 44],
- HCN (DuPont, Evonik) [41],
- propane/propylene ammoxidation to acrylonitrile (DuPont) [11, 43],
- transesterification/cracking of oils to biodiesel/biofuel [5, 6],
- fructose/glucose to carboxylic acids, furfural [7, 13],
- methanol to formaldehyde (Haldor-Topsoe) [14],
- dimethyldichlorosilane (Dow Corning) [21],
- ε-caprolactam (Sumitomo) [3],
- syngas/water splitting [9, 23, 31],
- depolymerization of polymeric waste (Recycling Technologies, Mitsubishi Rayon) [17, 35],
- biomass pyrolysis/gasification (Repotec) [2].

The major question going forward with CFB technology is whether or not moving catalyst is cost-effective. For slower reactions switching gas feeds – the huff and puff mode of the Houdry process – may be more appropriate [18, 33]. DuPont broke new ground in developing CFBs but required more time to perfect the technology. Development costs are substantial (Figure 7.11) but, optimistically, if a company can weather price swings in the market and the design flaws, the upside potential is tremendous.

7.8 Conclusions

DuPont commercialized a process that was to be the next step in building massive single-train reactors to produce specialty chemicals. Coupling the partial oxidation of butane with the hydrogenation of maleic acid to THF in an intensified reactive distillation unit was certainly more economical than the technology practiced at the time. They applied an iterative design process to go from laboratory to design a commercial scale facility then built a pilot plant and cold flow facilities to resolve engineering uncertainties posed by this scale-up. The plant operated for almost 10 years. It was shutdown due to design flaws, market pressures, and a change in corporate vision. Continuing to modify the plant configuration represented millions of dollars of investment that could only be justified if management believed that the market would continue to grow and that they had a low-cost process.

This example shows the importance of raising all technical uncertainties during the industrial design stage to better address them. A failure to recognize and to assess these uncertainties due to a lack of experience or lack of time can be catastrophic. It also shows the importance of business leadership allowing the technical team the sufficient time to resolve the problems encountered during the piloting period and commercial start-up.

References

[1] J. Adanez, A. Abad, F. García-Labiano, P. Gayan and L. F. De Diego, Progress in chemical-looping combustion and reforming technologies, Progress in Energy and Combustion Science, 38(2), 215–282, 2012.

[2] M. Aghabararnejad, G. S. Patience and J. Chaouki, Techno-economic comparison of a 7-MWth biomass chemical looping gasification unit with conventional systems, Chemical Engineering & Technology, 38(5), 867–878, 2015.

[3] R. Beerthuis, G. Rothenberg and N. Raveendran Shiju, Catalytic routes towards acrylic acid, adipic acid and -caprolactam starting from biorenewables, Green Catalysis, 17, 1341–1361, 2015.

[4] F. Berruti, T. S. Pugsley, L. Godfroy, J. Chaouki and G. S. Pati- Ence, Hydrodynamics of circulating fluidized bed risers: a review, The Canadian Journal of Chemical Engineering, 73(5), 579–602, 1995.

[5] D. C. Boffito, F. Galli, C. Pirola and G. Patience, Cao and isopro- panol transesterify and crack triglycerides to isopropyl esters and green diesel, Energy Conversion and Management, 139, 71–78, 2017.

[6] D. C. Boffito, C. Neagoe, M. Edake, B. Pastor-Ramirez and G. S. Patience, Biofuel synthesis in a capillary fluidized bed, Catalysis Today, 237, 13–17, 2014.

[7] D. Carnevali, O. Guévremont, M. G. Rigamonti, M. Stucchi, F. Cavani and G. S. Patience, Gas-phase fructose conversion to furfural in a microfluidized bed reactor, ACS Sustainable Chemistry & Engineering, 6(4), 5580–5587, 2018.

[8] C. D. Chang, C. T. W. Chu and R. F. Socha, Methanol conversion to olefins over zsm- 5. 1. effect of temperature and zeolite SiO_2 /Al_2 O_3, Journal of Catalysis, 86(2), 289–296, 1984.
[9] F.-X. Chiron and G. S. Patience, Kinetics of mixed copper–iron based oxygen carriers for hydrogen production by chemical looping water splitting, International Journal of Hydrogen Energy, 37(14), 10526–10538, 2012.
[10] M. Dalil, D. Carnevali, M. Edake, A. Auroux, J.-L. Dubois and G. S. Patience, Gas phase dehydration of glycerol to acrolein: coke on WO_3 /TiO_2 reduces by-products, Journal of Molecular Catalysis A: Chemical, 421, 146–155, 2016.
[11] A. H. Fakeeha, M. A. Soliman and A. A. Ibrahim, Modeling of a circulating fluidized bed for ammoxidation of propane to acrylonitrile, Chemical Engineering and Processing- Process Intensification, 39(2), 161–170, 2000.
[12] D. F. Rivas, D. C. Boffito, J. Faria-Albanese, J. Glassey, N. Afraz, H. Akse, K. V. K. Boodhoo, R. Bos, J. Cantin, Y. W. Chiang, J.-M. Commenge, J.-L. Dubois, F. Galli, J. P. Gueneau De Mussy, J. Harmsen, S. Kalra, F. J. Keil, R. Morales-Menendez, F. J. Navarro-Brull, T. Noël, K. Ogden, G. S. Patience, D. Reay, R. M. Santos, A. Smith-Schoettker, A. I. Stankiewicz, H. Van Den Berg, T. Van Gerven, J. Van Gestel, M. Van Der Stelt, M. Van De Ven and R. S. Weber, Process intensification education contributes to sustainable development goals. part 1, Education for Chemical Engineers, 32, 1–14, 2020.
[13] T. Ghaznavi, C. Neagoe and G. S. Patience, Partial oxidation of d- xylose to maleic anhydride and acrylic acid over vanadyl pyrophosphate, Biomass and Bioenergy, 71, 285–293, 2014.
[14] A. Godefroy, G. S. Patience, T. Tzakova, D. Garrait and J. L. Dubois, Reactor technologies for propane partial oxidation to acrylic acid, Chemical Engineering & Technology, 32(3), 373–379, 2009.
[15] L. Godfroy, G. S. Patience and J. Chaouki, Radial hydrodynamics in risers, Industrial & Engineering Chemistry Research, 38(1), 81–89, 1999.
[16] M. R. Gogate, Methanol-to-olefins process technology: current status and future prospects, Petroleum Science and Technology, 37(5), 559–565, 2019.
[17] A. E. Griffiths. Process and apparatus for treating waste comprising mixed plastic waste US Patent 10760003/B2, 2020.
[18] E. J. Houdry. Catalytic materials and process for manufacture, US Patent 2078945, 1937.
[19] R. W. Hughes, D. Y. Lu, E. J. Anthony and A. Macchi, Design, process simulation and construction of an atmospheric dual fluidized bed combustion system for in situ co2 capture using high-temperature sorbents, Fuel Processing Technology, 86(14), 1523–1531, 2005.
[20] K. W. Hutchenson, C. L. Marca, G. S. Patience, J. P. Lavio- Lette and R. E. Bockrath, Parametric study of n-butane oxidation in a circulating fluidized bed reactor, Applied Catalysis A-General, 376(1–2), 91–103, 2010.
[21] K. Janmanchi, A. Coppernoll and D. Katsoulis, Two-step process for the synthesis of dimethyldichlorosilane using copper aluminate catalysts, Industrial & Engineering Chemistry Research, 59(8), 3321–3333, 2020.
[22] Y. Li, G. Zhai, H. Zhang, T. Li, Q. Sun and W. Ying, Experimental and predictive research on solids holdup distribution in a cfb riser, Powder Technology, 344, 830–841, 2019.
[23] Z. Ma, P. Perreault, D. C. Pelegrin, D. C. Boffito and G. S. Patience, Thermodynamically unconstrained forced concentration cycling of methane catalytic partial oxidation over ceo2 fecralloy catalysts, Chemical Engineering Journal, 380, 122470, 2020.
[24] M. P. Martin, P. Turlier, J. R. Bernard and G. Wild, Gas and solid behavior in cracking circulating fluidized beds, Powder Technology, 70(3), 249–258, 1992.
[25] G. Mei, R. Rinaldi, G. Penzo and M. Dorini, The Spherizone PP Process: A New CFB Reactor, in Circulating fluidized bed technology VIII, K. Cen, editor, 778–785, 2005.

[26] G. S. Patience and R. E. Bockrath, Butane oxidation process development in a circulating fluidized bed, Applied Catalysis A: General, 376(1–2), 4–12, 2010.

[27] G. S. Patience and D. C. Boffito, Distributed production: scale-up vs experi- ence, Journal of Advanced Manufacturing and Processing, 2(2), e10039, 2020.

[28] G. S. Patience and J. Chaouki, Solids Hydrodynamics in the Fully Developed Region of cfb Risers, in Fluidization VIII, J.F. Large and C. Laguerie, editors, New York: Engineering Foundation, 33–40, 1995.

[29] G. S. Patience and P. L. Mills, Modelling of Propylene Oxidation in a Circula- ting Fluidized-Bed Reactor, in new developments in selective oxidation II, volume 82 of studies in surface science and catalysis, V. Cort/'es Corber/'an and S. Vic Bell/'on, editors, Elsevier, 1–18, 1994.

[30] G. S. Patience, J. Chaouki, F. Berruti and R. Wong, Scaling considerations for circula- ting fluidized bed risers, Powder Technology, 72(1), 31–37, 1992.

[31] P. Perreault and G. S. Patience, Chemical looping syngas from CO_2 and H_2 O over manganese oxide minerals, The Canadian Journal of Chemical Engineering, 94(4), 703–712, 2016.

[32] M. Pretz. Dow fluidized catalytic dehydrogenation (FCDh): The fu- ture of on-purpose propylene production, accessed on 09 september 2020, https://refiningcommunity.com/presentation/dow-fluidized-catalytic-dehydrogenation-fcdh-the-future-of-on-purpose-propylene-production/, 2020.

[33] S. Rifflart, G. S. Patience and F.-X. Chiron. Method for producing synthesis gases, US Patent 8974699/B2, 2015.

[34] D. Sanfilippo, F. Buonomo, G. Fusco, M. Lupieri and I. Mirac- Ca, Fluidized bed reactors for paraffins dehdyrogenation, Chemical Engineering Science, 47(9–11), 2313–2318, 1992.

[35] A. Sasaki, N. Kikuya, T. Ookubo and M. Hayashida. Recovery method of pyrolysis product of resin, US Patent 8304573/B2, 2012.

[36] P. Sauriol, H. Cui and J. Chaouki, Gas jet penetration lengths from upward and downward nozzles in dense gas-solid fluidized beds, Powder Technology, 235, 42–54, 2013.

[37] H. Schoenfelder, J. Hinderer, J. Werther and F. J. Keil, Metha- nol to olefins – prediction of the performance of a circulating fluidized-bed reactor on the basis of kinetic experiments in a fixed-bed reactor, Chemical Engineering Science, 49(24, Part 2), 5377–5390, 1994.

[38] A. Shekari and G. S. Patience, Maleic anhydride yield during cyclic n- butane/oxygen operation, Catalysis Today, 157(1), 334–338, 2010.

[39] A. Shekari and G. S. Patience, Transient kinetics of n-butane partial oxidation at elevated pressure, The Canadian Journal of Chemical Engineering, 91(2), 291–301, 2013.

[40] A. Shekari, G. S. Patience and R. E. Bockrath, Effect of feed nozzle configuration on n-butane to maleic anhydride yield: from lab scale to commercial, Applied Catalysis A-General, 376(1–2), 83–90, 2010.

[41] H. Siegert. Method for producing hydrogen cyanide in a particulate heat exchan- ger circulated as a moving fluidized bed, US Patent 2011/0020207 A1, 2011.

[42] Sisas Group. The sisas group today; four production complexes and competitive techno- logy, accessed on 07 september 2020, https://old.audace.org/industry-dominant/basf-feluy/sisas/pantochim-history2.htm, 1997.

[43] D. Vitry, J.-L. Dubois and W. Ueda, Strategy in achieving propane selective oxidation over multi-functional mo-based oxide catalysts, Journal of Molecular Catalysis A: Chemical, 220, 67–76, 2004.

[44] J. J. Zacca, J. A. Debling and W. H. Ray, Reactor residence time distribution effects on the multistage polymerization of olefins – i. basic principles and illustrative examples, polypropylene, Chemical Engineering Science, 51(21), 4859–4886, 1996.

Jaber Shabanian, Gregory S. Patience, Jamal Chaouki

8 Case study III: Methanol to olefins

Abstract: The methanol to olefin (MTO) process has attracted significant attention in recent years to produce ethylene and propylene. MTO technology by the Dalian Institute of Chemical Physics, denoted by DMTO, is dominant among available technologies. The scale-up strategy of this process followed a conventional approach, i.e., a gradual process development from laboratory scale to pilot, demonstration, and commercial scales, that lasted over 25 years. The world's first MTO commercial plant started up based on this technology in 2010. This chapter summarizes the key challenges for the successful development and scale-up of the MTO process, with a principal focus on the DMTO process. It also presents how the iterative scale-up approach can help effectively scale-up the DMTO reactor, as the heart of the process, in an effective and considerably more timely manner. The new approach adopts critical background information and experimental data at laboratory scale and a decision-making procedure to select an appropriate gas–solid catalytic reactor based on weight hourly space velocity. Uncertainties of the new approach are identified and measures to address them by supplementary laboratory and pilot-scale results are discussed. Directions for future improvements in the profitability of available technology with the help of process simulation are discussed as well.

Keywords: methanol to olefins process, MTO, ethylene/propylene production, SAPO molecular sieve, scale-up, fluidized bed

Acknowledgment: The authors are grateful to Prof. Rahmat Sotudeh-Gharebagh for his constructive comments.

Jaber Shabanian, Process Engineering Advanced Research Lab, Department of Chemical Engineering, Polytechnique Montréal, C.P. 6079, Station Centre-Ville, Montréal, Québec, H3C 3A7, Canada, e-mail: jaber.shabanian@polymtl.ca
Gregory S. Patience, Canada Research Chair High Temperature, High Pressure Heterogeneous Catalysis, Department of Chemical Engineering, Polytechnique Montréal, 2500 Chemin de Polytechnique, Montréal, Québec, H3T 1J4, Canada, e-mail: gregory-s.patience@polymtl.ca
Jamal Chaouki, Department of Chemical Engineering, Polytechnique Montréal, 2500 Chemin de Polytechnique, Montréal, Québec, H3T 1J4, Canada, e-mail: jamal.chaouki@polymtl.ca

https://doi.org/10.1515/9783110713985-008

8.1 Introduction

Light olefins, e.g., ethylene and propylene, are among chemicals and intermediates in high demand that are conventionally produced from petrochemical feedstocks via naphtha thermal cracking and fluid catalytic cracking (FCC) processes [1]. Research and development had gained momentum by the first and second oil crises in 1973 and 1979 [2] and have been continuously growing to discover viable routes for producing light olefins from alternative and abundant resources, e.g., coal, biomass, waste, natural gas, and even carbon dioxide [1, 3]. Methanol to olefins (MTO) is known as a critical technology to produce light olefins and is believed to be a linkage between non-oil chemical industries, employing alternative resources, and the modern petrochemical industry [4]. It helps decrease the dependency on crude oil as methanol can be conveniently synthesized from synthesis gas (syngas), being derived from coal, natural gas, or any carbonaceous materials [5].

Methanol is a highly active C_1 compound, thus it is very sensitive to many acidic zeolites over which it can be converted to different hydrocarbons through complex reactions [2, 4]. The production of light olefins from methanol was first realized around 1977 during the development of Mobil's methanol to gasoline (MTG) process [6]. Scientists at Mobil showed that ZSM-5 molecular sieve catalyst is active in the MTO reaction and helps produce propylene and butylene as major products owing to its relatively large pore size, about 5.5 Å [1, 5]. In 1986, scientists at Union Carbide Corporation (UCC; now Honeywell Universal Oil Product (Honeywell UOP)) discovered that silicoaluminophosphate (SAPO) improves the light olefins selectivity [7, 8]. The smaller pore sizes of SAPO-34, about 4.3 Å, and optimized acidity explain its greater selectivity toward light olefins, especially ethylene and propylene. Since this discovery, several companies have intensely researched various aspects of the MTO process, including catalyst synthesis, reaction mechanism and kinetics, process development, and reactor scale-up. The Dalian Institute of Chemical Physics (DICP) has dedicated the most to the R&D of the MTO reaction. In August 2010, the world's first MTO commercial plant based on DICP's MTO (DMTO) process was successfully started up in Shenhua's Baotou coal to olefins plant in Inner Mongolia (North China) [1, 4]. The olefins production capacity was 0.60 million metric tons a^{-1}, while consuming 1.8 million tons a^{-1} of methanol. The methanol conversion was about 100% and the cumulative ethylene and propylene selectivity approached 80% [5]. The technical and economic success of the Shenhua Baotou plant was a breakthrough in the MTO process development that spurred a wave of MTO commercial plants [1, 5]. The main MTO technologies in addition to DMTO include the second generation of DMTO technology, known as DMTO-II, by DICP, MTO technology co-developed by UOP, Norsk Hydro and TOTAL, SMTO technology developed by SINOPEC Shanghai Research Institute of Petrochemical Technology and SHMTO technology developed by the Shenhua Group [5, 9]. Interested readers are invited to review references [1, 4, 10–16] for detailed information about the histories and processes of these technologies. A number of large commercial MTO plants are operating in Asia, in particular, in China, owing to the abundance of coal feedstock and government support [3]. A summary of MTO plants in China that were brought on stream by June 2016 as well as their capacities and technologies are presented in table 2 of Xu et al. [5]. It shows that DMTO is the dominant technology.

While the main goal of the MTO process is to produce ethylene and propylene, the methanol to propylene (MTP) process has also been carried out to produce propylene. Lurgi's fixed bed and Tsinghua University's fluidized bed MTP processes are the major MTP technologies. Since the propylene supply in the international market was increased and led to a more significant decrease in the propylene price than that of ethylene, the building of new MTP plants was halted [5]. Accordingly, we focus on the MTO process in this chapter.

Many scientific and technical problems should be addressed to develop and scale-up an applicable MTO technology. They include (i) developing an effective catalyst that selectively controls the formation of light olefins by a deep understanding of the complex reaction mechanism, catalyst deactivation and the interplay between catalyst property, synthesis method and reaction performance, (ii) scaling up an economic and effective catalyst synthesis process, which reliably and reproducibly produces catalyst on a large scale, with commercially available materials, (iii) determining the most suitable reactor type and corresponding operating conditions, as well as exploring the commercial availability, (iv) generating a complete process flow diagram of the process and establishing a solid reactor scale-up procedure, and (v) integrating the acquired knowledge from reactor scale-up with available and mature industrial experiences to prepare a basic design package for a commercial unit [4].

DMTO has the largest market share, particularly in China, and its burgeoning coal-to-olefin industry [15] and its technology development information is more readily available in literature. Hence, it will form the basis of subsequent sections of this chapter, while the selection of a proper DMTO reactor, as the heart of this process and its scale-up will be the core of the discussion. Readers are advised to see references [4, 17–21] for various proposed/adopted reactor configurations, including a multitubular fixed bed and dense, circulating, and riser fluidized bed reactors for the MTO process. A fundamental study of the MTO reaction and DMTO catalyst selection, which are essential for both conventional and iterative scale-up approaches, will be discussed first. The complete evolution of DMTO technology from laboratory experiments to the commercial plant took place over 25 years and through a conventional scale-up approach. In a subsequent section, we will show how effectively and at a lower cost the iterative scale-up approach helps select an appropriate commercial DMTO reactor in a shorter interval. A summary of key research and experiences in the successful development and scale-up of DMTO technology by DICP will be presented next. They help confirm the findings obtained by the iterative scale-up approach. Directions for future improvements, which may help in the industrial maturity of this process and improve its profitability, will also be recommended.

8.2 Fundamental study on MTO reaction and DMTO catalyst selection

Understanding the MTO reaction mechanism and shape selectivity toward desired products, i.e., ethylene and propylene, over zeolite catalysts forms the basis of reactor selection, process development, and optimization and the successful operation of processing units in the MTO process [1]. The MTO process involves a very complex reaction sequence and determining the governing reaction mechanism has been one of the most controversial in heterogeneous catalysis [3, 22]. According to previous investigations and the very recent findings on the formation of the first carbon–carbon (C–C) bond [23], the zeolite-catalyzed MTO reaction takes place in two steps: (i) an inefficient direct mechanism in the initial reaction period, also called the induction period and (ii) an indirect mechanism in the subsequent high activity period. The induction period during which an organic-free catalyst is transferred into a working catalyst is rather short [3, 16]. In addition, methanol conversion in the induction period is very low, thus coke formation is very slow [1]. Upon the formation of initial C–C bonds, the MTO reaction proceeds with the steady-state formation of hydrocarbons in the so-called autocatalytic dual-cycle concept. Despite new findings, further study is required to fully understand the actual nature/course of action of the direct mechanism [3, 16]. On the contrary, the indirect dual-cycle mechanism is more comprehensively understood. It involves an aromatic-based hydrocarbon pool (HCP) mechanism, proposed by Dahl and Kolboe [24–26] and the olefins methylation/cracking mechanism. HCP refers to cyclic organic species, as the active intermediates, that are confined in the zeolite cage/cavity or intersection of channels and act as co-catalyst for the assembly of olefins from methanol. Methylbenzenes and their protonated counterparts function as HCP in the MTO reaction [1, 4, 16]. These high molecular weight hydrocarbons grow through repeated methylation and subsequently spill off light olefins, leading to the regeneration of HCP species and closing the catalytic cycle [22]. While ethylene is predominantly produced from methylbenzenes with two or three methyl groups, higher hydrocarbons favor the generation of propylene and butylene [16]. The proposed MTO reaction network is presented in Figure 8.1 [27]. It includes the initial C–C bond formation, HCP species generation from the primary olefins, the highly efficient reaction step via the indirect mechanism, and the deactivation step owing to the production of heavier aromatic coke species [16].

The topology of zeolite particles, i.e., pore size, cavity, and pore network and their compositions, i.e., acid sites distribution and acidity strength, influence the catalyst performance, thus the product distribution [1]. SAPO-34 zeolite, DMTO catalyst, has shown the best performance among available catalysts to convert methanol to ethylene and propylene. It possesses a small eight-membered ring pore (0.38 nm) opening and a large CHA cage (0.94 nm in diameter) structure, which help show high

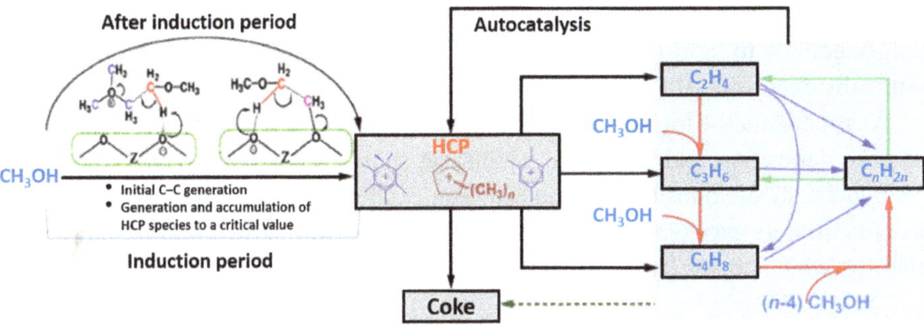

Figure 8.1: Proposed MTO reaction network. Source: Adapted from Xu et al. [27].

selectivity to light olefins in the MTO reaction [16, 28]. The narrow pore opening restricts the diffusion of large intermediate molecules out of the cages and makes short-chain olefins the dominant products (product-shape selectivity) [1, 3, 16]. In addition, the cage structure (shape of cavity) governs the configuration and reactivity of the active intermediates via a host-guest confinement effect. A schematic representation of the MTO conversion over the SAPO-34 catalyst with the HCP mechanism is presented in Figure 8.2.

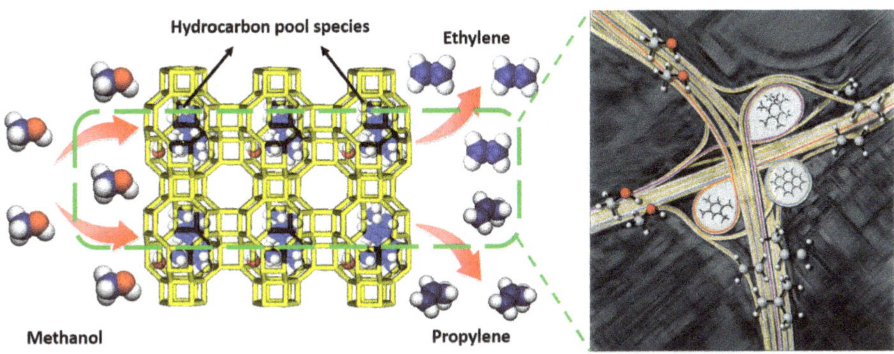

Figure 8.2: Schematic representation of the MTO reaction over the SAPO-34 catalyst with the hydrocarbon pool mechanism. Sources: Adapted from Hemelsoet et al. [22] and Sun et al. [28].

Over the course of the MTO reaction, active HCP intermediates form bicyclic and/or polycyclic aromatic hydrocarbons (coke) in the cage. They occupy most of the space in the cage due to their large volumes, thus preventing contact between the methanol feed and the active sites of catalyst and rapidly decreasing the catalyst activity [1]. The fast deactivation of SAPO-34 catalysts is identified as their main problem [28]. However, catalyst pore blockage owing to the coke formation prevents molecular diffusion in the pore channel, which, in turn, imposes diffusion limitations on large molecules and increases selectivity toward smaller molecules (light olefins).

Therefore, an optimal coke deposition of SAPO-34 catalyst is sought to achieve a high selectivity to light hydrocarbons, in particular, ethylene, while the catalyst is minimally deactivated.

A proper catalyst for the MTO reaction with a high olefin selectivity, low/moderate coke deposition and long catalyst lifetime should have a mild acidity and relatively low acid site density [16]. The SAPO-34 catalyst satisfies these conditions and benefits from its excellent shape selectivity and superior thermal and hydrothermal stability, which makes it an ideal MTO catalyst [28].

8.3 DMTO reactor selection and scale-up by iterative scale-up approach

In this section, we will show how the iterative scale-up approach helps select and scale-up the DMTO reactor, i.e., the selected case study here, on the basis of some critical background information and experimental data from a laboratory-scale DMTO fluidized bed reactor, as well as a reactor selection/design procedure. Some laboratory experimental data reported in literature for the DMTO process will be adopted here.

8.3.1 Critical background information and experimental data at laboratory scale for the selection and scale-up of the DMTO reactor

The MTO reaction is a heterogeneously catalyzed gas–solid reaction. Gas–solid fixed and fluidized bed reactors provide higher gas–solid contact than other options, thus are preferred for this reaction. The selected reactor should also help satisfy other conditions, including the presence of an optimal coke content of the catalyst that yields the highest methanol conversion and selectivity toward light olefins, convenient and simultaneous reaction heat removal from the reactor to keep the reactor temperature in the designed range (MTO reaction is highly exothermic; –196 kcal kg^{-1} of methanol feed when the reaction temperature is at 495 °C [4]), capability of online regeneration of the SAPO-34 catalyst in response to its rapid deactivation, convenient and concurrent coke combustion heat removal from the regenerator to keep the regenerator temperature in the designed range and prevent thermal damage of the catalyst, application of minimal catalyst inventory in the reactor, and straightforward operation and scale-up of the reactor [1, 4].

Fixed bed reactors have fewer pieces of peripheral equipment and are less complicated to construct and operate compared to fluidized beds. However, owing to the low heat transfer coefficient in fixed beds, heat transfer equipment in these reactors would be much larger than that in fluidized bed reactors. In addition, the total bed pressure drop in fixed beds could be much higher than that in fluidized beds as excess catalyst is required to compensate for the continuous catalyst deactivation. Although a near continuous catalyst regeneration can be achieved by a multi-fixed bed configuration, the increased cost of the process due to the increased number of reactors/regenerators and catalyst inventory, as well as additional risks during gas switching between reactor and regenerator makes it less favorable compared to a fluidized bed reactor/regenerator assembly. An excellent heat transfer performance and good fluidity of catalysts in the fluidization state help accomplish a convincing steady-state performance of catalyst with constant coke content. Hence, the fluidized bed reactor-regenerator configuration is superior to fixed bed reactors for the MTO process [3, 4]. This configuration is recommended and was adopted in the DMTO process [4]. Since fluidizable catalysts, which are generally fine particles ranging from 20 to 120 μm, are required for the fluidized bed reactor, they provide a significantly larger surface area [4] than relatively coarse catalyst particles that are employed in the fixed bed reactor to preserve the bed pressure drop in an acceptable range.

A laboratory-scale bubbling fluidized bed reactor with ~0.01 m I.D. and 0.01 kg of SAPO-34 catalyst inventory, capable of handling a methanol feed rate of 0.5 kg day^{-1} and without solids circulation was adopted to (i) evaluate the performance of the DMTO catalyst, (ii) scrutinize the optimal operation window, (iii) study the reaction kinetics of the DMTO process, and (iv) determine the optimal catalyst residence time. The most critical results will be discussed here. They will help identify the operating fluidization regime of the DMTO fluidized bed reactor. According to the collected experimental results about the influence of the coke content of the SAPO-34 catalyst on its performance, the catalyst can effectively function with a high selectivity to light olefins at high methanol conversion, >99%, at an optimal coke content of about 8%, 7.6–8.5% (Figure 8.3(a)). In order to achieve the optimal coke content in the reactor, depending on the reaction temperature and weight hourly space velocity (WHSV), a catalyst residence time of approximately 180 min, 160–195 min, is required (Figure 8.3(b)) [1, 4]. Accordingly, riser and downer fluidized bed reactors are inappropriate for the DMTO process since the catalyst residence time is only several seconds in these designs. This, in turn, results in a low methanol conversion and a great challenge in achieving the optimal coke content of the catalyst without recycling it. A dense fluidized bed reactor meets all the specifications and was selected by DICP [4].

The catalyst–gas contact time is a critical parameter in governing the overall performance of the MTO reactor and the design of the corresponding reactor on a larger scale. The obtained experimental results indicated that a nearly complete conversion of methanol can be achieved at even a very short catalyst–gas contact time of 0.04 s. The shorter contact time yields a higher selectivity to light olefins. Increasing the contact time enhances the side reactions and formation of byproducts, which promotes coke formation and decreases the selectivity to light olefins. Upon increasing WHSV in the range of 2–10 h^{-1}, methanol conversion slightly decreased from

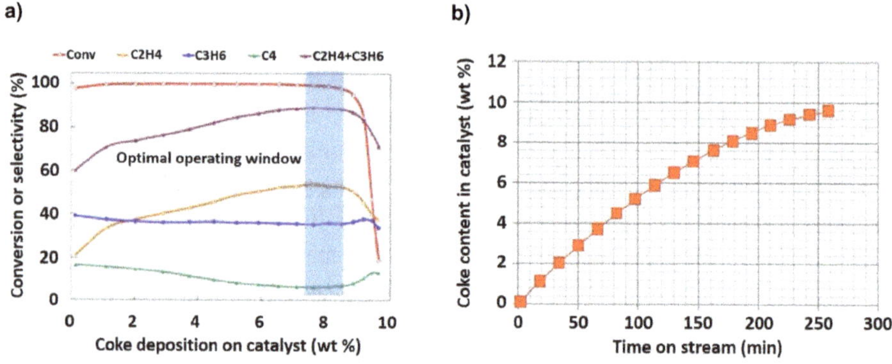

Figure 8.3: (a) Influence of coke content on the methanol conversion and selectivity to light olefins and (b) coke content in catalyst as a function of time on stream in a laboratory-scale DMTO bubbling fluidized bed reactor at reaction temperature 450 °C and WHSV = 1.5 h^{-1}.
Sources: Adapted from Tian et al. [4] and Ye et al. [1], respectively.

100% to 99.3%, while the selectivity to light olefins increased from 81.8% to 87.5%, most of which was at the expense of ethylene selectivity (Table 8.1). Although increasing WHSV decreases the size of the commercial reactor for a specific methanol feed rate, it makes the design of a fluidized bed reactor operating under this condition more challenging. Since a similar reactor performance can be obtained at a lower WHSV, which results into a simplified scale-up process of the MTO fluidized bed reactor, it is preferred. This was also selected for DICP's DMTO process development [1].

Table 8.1: Influence of WHSV on DMTO fluidized bed reactor performance. Source: Adapted from Ye et al. [1].

Contact time (s)		3.05	1.53	1.02	0.76	0.61
WHSV (h^{-1})		2	4	6	8	10
Methanol conversion (%)		100	99.9	99.7	99.6	99.3
Selectivity (wt%)	C_2H_4	31.2	36.0	37.4	37.3	37.8
	C_3H_6	50.6	48.8	49.8	50.1	49.7
	$C_2H_4 + C_3H_6$	81.8	84.8	87.2	87.4	87.5

Results of laboratory experiments indicated that an optimized catalyst–gas contact time of about 2–3 s is required to prevent undesired byproduct in the DMTO reaction. This short contact time implies that the superficial gas velocity U_g should be greater than 1.0 m s^{-1} for a commercial-scale fluidized bed having a dense bed height around 3 m. Otherwise, it will be very challenging to adequately and safely arrange internals, cyclone diplegs, and a catalyst draw-off bin in a shallower dense bed. The physical properties of DMTO catalyst particles are similar to FCC particles

[4] showing typical Geldart group A behavior at ambient conditions [29]. Similar to FCC particles, a fluidized bed of DMTO catalyst particles operate in a turbulent fluidization regime in a U_g range of 0.5–1.5 m s^{-1} [4]. In order to satisfy all of the conditions discussed above, the DMTO reactor should operate in the turbulent fluidization regime, corresponding to U_g range of 1.0–1.5 m s^{-1}. The same decision was made by DICP. A detailed comparison of the fluidized bed reactor–regenerator configuration in the DMTO process and that was employed in the FCC process is reported in Table 8.2 to obtain a better understanding about the DMTO process [4].

The translation of experimental results from the laboratory-scale fluidized bed to a larger scale circulating fluidized bed design is a critical step in scaling up the DMTO reactor. This can be achieved through the study of critical factors, including the coke content in the catalyst $C_{\text{cat}}(t)$, coke formation rate $\dot{c}(t)$, and catalyst-to-methanol (CTM) ratio; t is the time on stream. As alluded to earlier, the experimental results collected in the laboratory-scale DMTO fluidized bed reactor helped determine the optimal coke content of the catalyst. This information was required before designing/operating the larger scale DMTO reactor. In addition, Figure 8.3(a) shows that in a wide range of coke content, 0–8.7 wt%, the catalyst can maintain a high methanol conversion, >98.5%. This indicates that a small amount of active catalyst (active sites) in the MTO reaction can lead to a complete conversion of methanol. Employing the data reported in Figure 8.3(b), one can calculate $\dot{c}(t)$ and realize that it linearly decreases with $C_{\text{cat}}(t)$. Increasing the coke content in the catalyst leads to a higher coverage of active sites, a decrease in methanol conversion, and a decrease in the coke formation rate. The CTM ratio is another key parameter in controlling the solid catalyst circulation rate between the reactor and regenerator on a larger reactor scale. It was hard to directly measure the optimal CTM ratio in the laboratory scale, where there was no solids circulation. Assuming that the coke formation rate obtained at laboratory scale could be directly employed on a larger scale, the mass flow rate of catalyst $\dot{m}(t)$ to transport the desired amount of coke could be estimated by

$$\dot{m}(t) = \frac{\dot{c}(t)}{C_{\text{cat}}(t)} = \frac{1}{C_{\text{cat}}(t)}\frac{d}{dt}(W \cdot C_{\text{cat}}(t)) = \frac{W}{C_{\text{cat}}(t)}\frac{d}{dt}(C_{\text{cat}}(t)) \tag{8.1}$$

where W is the catalyst inventory in the laboratory-scale reactor. Hence, CTM ratio could be calculated as

$$\text{CTM} = \frac{\dot{m}(t)}{\dot{Q}_m(t)} = \frac{\dot{c}(t)}{\dot{Q}_m(t)\, C_{\text{cat}}(t)} \tag{8.2}$$

where $\dot{Q}_m(t)$ is the methanol feed rate. According to the reaction results collected at laboratory scale and the abovementioned assumption, the favorable CTM ratio was in the range of 0.1–0.2 [1].

Since the methanol conversion in the induction period is very low, it is crucial to properly identify this period and operate the reactor outside this range. The low

Table 8.2: Detailed comparison between DMTO and FCC processes. Source: Adapted from Tian et al. [4].

	DMTO	FCC
Feedstock	Methanol/dimethyl ether Pure and simple in composition No impurity	Large molecular, heavy oil Complicated in composition Sulfur, nitrogen, and heavy metals
Reaction feature		
Mechanism	$CH_3OH \rightarrow C_2H_4 + C_3H_6 \ldots$	CxHy → gasoline, diesel, LPG, dry gas, heavier distillate Cracking
	Chain growth First C–C formation through direct mechanism, HCP mechanism for chain growth	Carbenimum ion mechanism
Reaction rate	Very fast	Fast
Catalysis type	Auto-catalysis	No auto-catalysis
Coking rate	Relatively low (in hours)	Very fast (in seconds)
Deactivation	Fast (in hours)	Very fast (in seconds)
Catalyst	SAPO-34 Highly hydrothermal stable Avoid external impurities Geldrat group A behavior at ambient conditions	Re-USY Highly hydrothermal stable Impurities and metal resistance Geldrat group A behavior at ambient conditions
Products	Light olefins Simple molecules	Fuel Complicated in composition, in distillation
	Active molecular Lots of water	Similar to MTO No water formation
Process		
Flow pattern	Gas–solid flow	Gas–liquid–solid flow
Feeding system	Gas distributor	Feed nozzles
Reaction heat	Exothermal	Endothermic
Catalyst to feed ratio	Low (<1)	High (5–6)
Heat balance between reactor and regenerator	No	Yes
Gas–solid contact time	Shorter contact time is favorable	~2 s
Catalyst residence time	In hours	In seconds
Solid separation	Two-stage cyclones	Two-stage cyclones
Quench tower	Necessary	Necessary
Regeneration	Partial burn	Full burn
Reactor type	Turbulent fluidized bed	Riser
Regenerator	Bubbling fluidized bed	Turbulent fluidized bed
Stripping efficiency	>95%	75%

methanol conversion could lead to low coke formation. Hence, a complete conversion of methanol, which occurs after a certain amount of coke deposition on the catalyst, will be delayed if the reactor operates for an extended period of time in the induction period. The duration of this period depends on the reaction temperature. The induction period occurs at temperatures below 350 °C for the DMTO catalyst. Increasing the temperature shortens the induction period and increases the duration of steady-state methanol conversion over the catalyst. The SAPO-34 catalyst can show its optimal performance with a high methanol conversion, maximized selectivity to light olefins, and long lifetime above 425 °C. These findings were obtained in the laboratory-scale fluidized bed reactor and can be employed on larger scale DMTO fluidized bed reactors. This was also accomplished in the development of the DMTO process [1]. To properly operate a commercial DMTO reactor, the circulation of catalyst between the reactor and regenerator should not commence until a high reaction temperature and high methanol conversion are obtained [4].

8.3.2 Reactor selection for gas–solid catalytic reactions

According to Sections 8.2 and 8.3.1, critical parameters, e.g., reaction mechanism and kinetics, optimal coke content, WHSV, coke formation rate, and the CTM ratio that defines the catalyst circulation rate, were determined with the help of laboratory-scale experimental data. With this critical information collected at laboratory scale and a good reactor selection and scale-up strategy based on available knowledge and previous industrial experiences, the process scale-up could be accomplished with minimal risk, cost, effort, and time. In this section, we present a decision-making procedure for the selection of a proper gas–solid fixed/fluidized bed catalytic reactor based on WHSV. It can effectively help achieve the mentioned goals when integrated with the critical information being collected at laboratory scale.

WHSV is a critical reactor design parameter for catalytic reactions. It is defined as the mass flow rate of reactant $\dot{Q}_r$ (in kg h^{-1}) divided by the mass of required catalyst for a complete conversion of the reactant W_{RX} (in kg). $\dot{Q}_r$, W_{RX}, and WHSV can be defined as

$$\dot{Q}_r = 3,600 \rho_g U_g A \tag{8.3}$$

$$W_{\mathrm{RX}} = \rho_p (1 - \varepsilon) H A \tag{8.4}$$

$$\mathrm{WHSV} = \frac{\dot{Q}_r}{W_{\mathrm{RX}}} = 3,600 \frac{\rho_g U_g}{\rho_p (1 - \varepsilon) H} \tag{8.5}$$

where ρ_g and ρ_p are the gas and particle densities, A is the bed cross-sectional area, ε is the average bed voidage, and H is the bed height. In a fluidized bed with solids circulation, the average solids hold up $(1-\varepsilon)$ depends on U_g and the solids circulation rate [30]. The maximum average solids hold up, however, depends only on U_g. The correlation by King [31], originally proposed for calculating average bed voidage in turbulent fluidized beds, is commonly adopted. Patience and Bockrath [30] showed that the King's expression can adequately predict the average bed voidage in a wide range of superficial gas velocities from 0.2 to about 10 m s^{-1}. One can, hence, employ this correlation for nearly all fluidization regimes. The minimum average bed voidage according to King [31] is expressed as

$$\varepsilon = \frac{U_g + 1}{U_g + 2} \qquad \left(\varepsilon > \frac{U_g + 1}{U_g + 2}\right) \tag{8.6}$$

Substituting eq. (8.6) into eq. (8.5) results in

$$\text{WHSV} = 3,600\,\frac{\rho_g U_g (U_g + 2)}{\rho_p H} \qquad \left(\text{WHSV} > 3,600\,\frac{\rho_g U_g (U_g + 2)}{\rho_p H}\right) \tag{8.7}$$

Equation (8.7) can be further simplified as follows since ρ_g/ρ_p is typically around 1/2,000

$$\text{WHSV} \cong 1.8\,\frac{U_g (U_g + 2)}{H} \qquad \left(H > \approx 1.8\,\frac{U_g (U_g + 2)}{\text{WHSV}}\right) \tag{8.8}$$

For slow reactions, where the catalyst activity is low, reactions with intermediate rates and fast reactions with very active catalysts, WHSV is generally below 0.1 h^{-1}, between 0.1 and ~5 h^{-1}, and greater than about 5 h^{-1}, respectively. The relationship between WHSV, U_g, and H, as reported in eq. (8.8), for these three scenarios is plotted in Figure 8.4. The subplots show that for a given WHSV, multiple combinations of U_g and H are feasible. Hence, the best U_g and H should be wisely selected. For example, according to Figure 8.4(c), for a WHSV = 80 h^{-1}, typical for FCC with fast reactions (refer to Table 8.2), a minimum catalytic bed height of about 10 m is required when U_g=20 m s^{-1}. However, for the MTO process with a typical WHSV ≈ 5 h^{-1} (refer to Table 8.3), a minimum catalytic bed height of about 100 m is required if one wants to operate the corresponding catalyst particles, which have similar fluidization behavior to FCC particles, at the same U_g=20 m s^{-1}. This is clearly an improper selection. The subplots provided in Figure 8.4 can help determine the most appropriate U_g, H, and reactor type/operating regime for a specific WHSV. The final selections are influenced by many factors, including the reaction chemistry, catalyst deactivation, selectivity, particle attrition and agglomeration, bed hydrodynamics and pressure drop, gas dispersion, heat transfer, and electrostatic effects.

One of the critical selection factors is the bed pressure drop. A threshold value of less than 1 bar has been typically adopted for a safe operation and to avoid additional expenditure for a special compressor. By neglecting the pressure drop due to

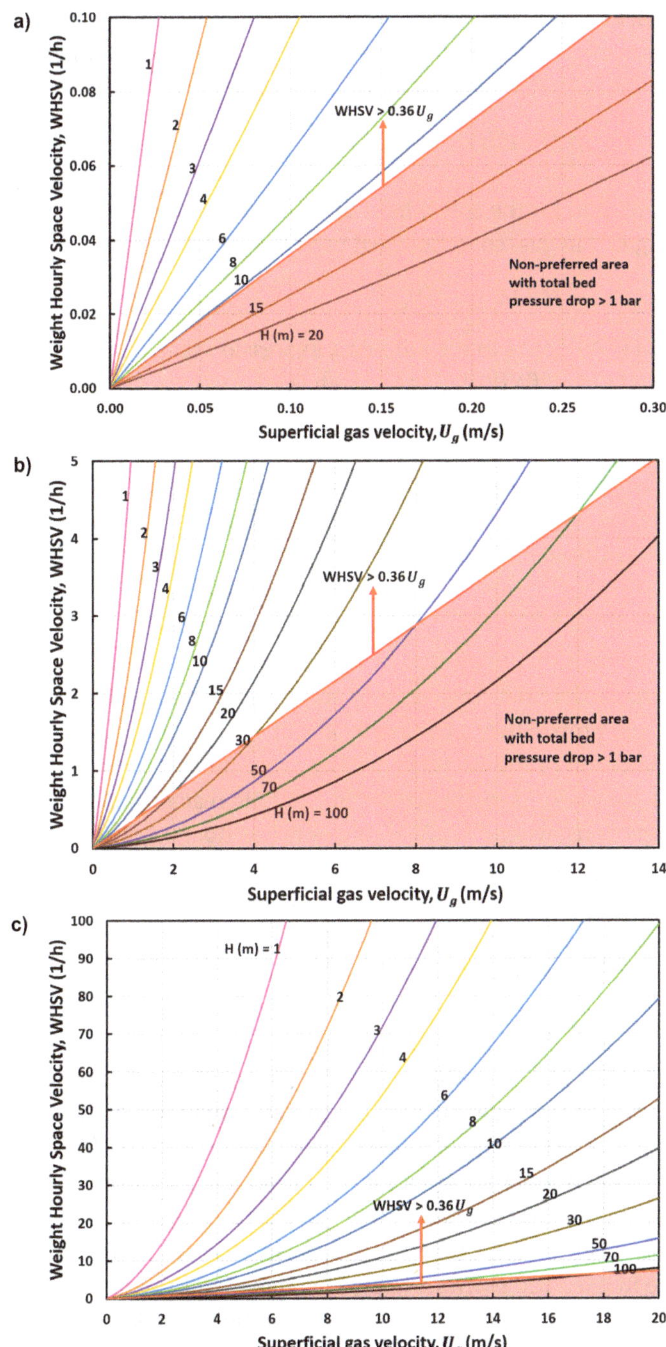

Figure 8.4: Variation of WHSV with U_g as a function of bed height for (a) slow, (b) intermediate rate, and (c) fast reactions.

acceleration and wall friction and assuming that the bed pressure drop is principally owing to the friction of gas on particles, the maximum allowable bed height H_{max} can be obtained as

$$\rho_p(1-\varepsilon)H_{max} = 100,000 \tag{8.9}$$

Applying the King's expression for the solids hold up and assuming a typical solid density of about 2,000 kg m^{-3} in gas–solid fluidized beds, H_{max} can be estimated as

$$H < H_{max} = 5(U_g + 2) \tag{8.10}$$

Combining eqs. (8.8) and (8.10) leads to the following expression that describes the minimum value of allowed WHSV (WHSV_{min}) at a given U_g

$$\text{WHSV} \geq \text{WHSV}_{min} = 0.36U_g \tag{8.11}$$

Equation (8.11) alternatively means that there is a maximum allowable superficial gas velocity $U_{g,\,max}$ for a given chemical reaction process having a specific WHSV in a fixed/fluidized bed catalytic reactor as follows

$$U_g \leq U_{g,\,max} = 2.8\text{WHSV} \tag{8.12}$$

The criterion described in eq. (8.11) is plotted by a straight line in Figure 8.4. All possible design solutions under this line, marked as the red area, have a non-preferred total bed pressure drop of over 1 bar in the reactor.

A schematic representation of the decision-making procedure for the selection of the most appropriate gas-solid fixed/fluidized bed catalytic reactor is presented in Figure 8.5. For slow reactions, where WHSV is generally less than 0.1 h^{-1}, the possible choice of U_g is governed by eq. (8.12) to <0.28 m s^{-1}, refer to Figure 8.4(a). Depending on the gas and catalyst particle properties, the reactor may operate in the fixed or bubbling fluidization states under the mentioned U_g range. For highly exothermic/endothermic reactions under this condition, a bubbling fluidized bed reactor is recommended to help satisfy an effective heat transfer and uniform temperature distribution throughout the bed. It also helps generate a reduced pressure drop. However, it suffers from undesired gas bypass and high gas dispersion. The narrow U_g range required for slow reactions additionally limits the process capability of the reactor. Therefore, for not highly exothermic/endothermic slow reactions, a fixed bed reactor is an ideal reactor type. A great advantage of this design is the presence of a high volumetric concentration of catalyst particles in the reactor vessel, which leads to an enhanced reaction yield toward desired products. In order to minimize the high-pressure drop issue in fixed bed reactors, relatively large catalyst particles, with a faster diffusion rate in the catalyst pellet than the reaction rate, should be loaded into the reactor.

Unlike the slow-reaction scenario, for reactions with intermediate rates having WHSV between 0.1 and 5 h^{-1}, there are fewer constraints for the possible choices of

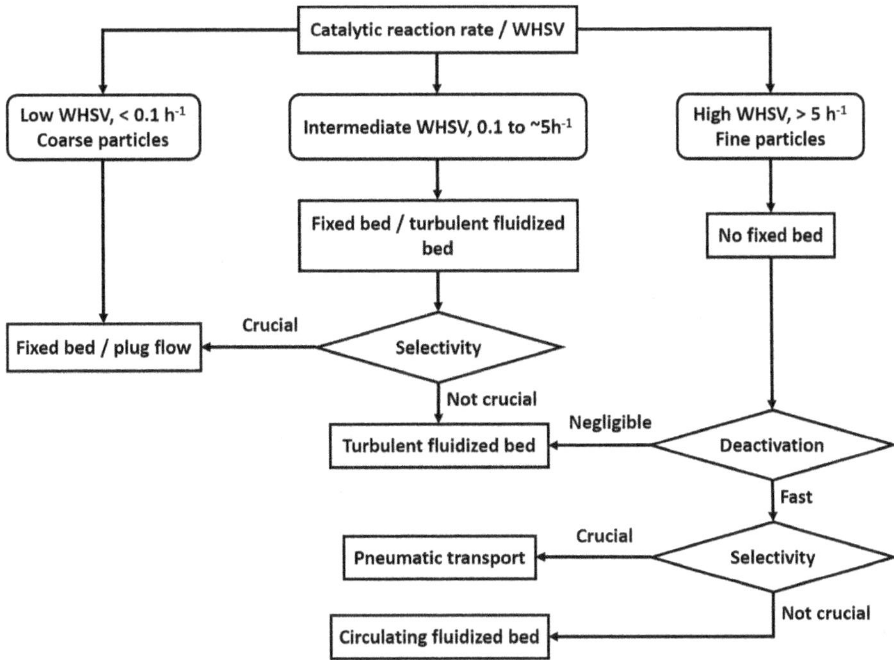

Figure 8.5: Classification of different gas–solid fixed/fluidized bed catalytic reactors.

U_g and H, thus the type of reactor. Fixed or turbulent fluidized bed reactors can be employed in this case. In addition, diffusion resistance within the particle may become critical. This makes the particle size an important parameter in fixed beds. Since particle size is in the order of 70 µm in catalytic gas–solid fluidized beds, the internal mass transfer resistance is negligible. Compared to fluidized beds, where the radial and especially axial dispersions are pronounced, the selectivity to desired products is generally higher in fixed bed reactors, where the gas is predominantly in plug flow. This deficiency can, however, be remedied in a properly designed turbulent fluidized bed reactor by the application of high fine (particles under 45 µm) content, e.g., 30 wt%, catalysts. The production penalty could be <2% with this manipulation.

When the reaction is not highly exothermic, one may consider multi-tubular fixed bed reactors as an option for intermediate rate reactions. This design, however, increases the reactor cost by an order of magnitude in comparison with either a single fixed bed or turbulent fluidized bed reactor. The maldistribution of reactant gas between parallel reactors is a common problem in a parallel design. An alternative option to multi-tubular reactors could be multiple packed beds with intercooling between stages. This type of reactor requires a catalyst that can adequately tolerate high temperature operation. This design may suffer from an increased pressure drop owing to the increased U_g and decreased reactor diameter to maintain the ideal axial dispersion. From an operational standpoint and equipment complexity, turbulent fluidized beds are the most adequate reactor choice for the highly exothermic reactions owing to their superior heat transfer characteristics and their ability to operate at nearly isothermal conditions. The isothermal operation, in many cases, may result in high selectivity to desired products.

For fast gas–solid catalytic reactions, where the catalyst is very active and WHSV is typically greater than 5 h^{-1}, the solids hold up in the reactor is less important. As depicted in Figure 8.4(c), the limitation from the total bed pressure drop is negligible in this case, thus the choice of reactor type is more versatile. The size of catalyst particles must be small, similar to those adopted in turbulent fluidized beds, for fast reactions to decrease internal diffusion mass transfer resistance. This requirement makes fixed bed reactors an impractical choice in this case owing to the very high total bed pressure drop that can be generated from a packed bed of fine particles. A turbulent or circulating fluidized bed is the preferred option. For the case of negligible catalyst deactivation, the turbulent fluidized bed is recommended as circulating catalyst has a limited value under this condition. The catalyst circulation is accompanied by a significant expenditure in terms of investment, i.e., more vessels, slide valves, standpipes, cyclones, and so on and operation, i.e., higher pressure drop, compressors, electricity, and attrition. When there is a significant catalyst deactivation, solid catalysts should be recirculated between a regenerator and a riser reactor to maintain the catalyst activity.

By adjusting ρ_p in eq. (8.8) to that of the SAPO-34 catalyst, i.e., 1,500–1,800 kg m^{-3} [4] and refer to Figure 8.4(b), one can realize that a commercial-scale gas–solid reactor with $H\approx3$ m could be an appropriate choice for a DMTO process with WHSV≈5 h^{-1} and U_g in the range of 1–2 m s^{-1}, refer to Table 8.3. This condition can satisfy other requirements discussed in Section 8.3.1. Since the MTO reaction is highly exothermic and the SAPO-34 catalyst has a fast deactivation rate, a fluidized bed reactor is recommended. When integrating these remarks with catalyst particle properties, the ideal fluidized bed reactor should operate in the turbulent fluidization regime. This is identical to DICP's finding for the best demonstration/commercial-scale reactor and its critical operating parameters for the DMTO process. The diameter of the commercial-scale fluidized bed reactor can be determined by knowing the feedstock rate, while preserving WHSV, U_g, and H at corresponding optimal/desired values. In this evaluation, we reached the same conclusion as what was achieved by DICP for its DMTO commercial-scale reactor through a safe and much simpler calculation than performing experimental activities at pilot and demonstration scales, refer to Section 8.4.

8.3.3 Identification of scale-up uncertainties and addressing them by supplementary laboratory and pilot-scale data

Upon following the reactor selection and design procedure described in Sections 8.3.1 and 8.3.2 for the DMTO reactor, some critical uncertainties, as listed below with corresponding mitigation plans, can be identified. They could be answered by supplementary experiments at laboratory and/or pilot scale to decrease the risks of the iterative scale-up approach.

- Despite the mentioned advantages for the fluidized bed reactor in Section 8.3.1, the diversity of fluidized bed hydrodynamics at different scales is a big challenge in the development of catalytic fluidized bed reactors [32]. A set of tests in a pilot-scale DMTO fluidized bed reactor-regenerator assembly, which operates under/around similar desired conditions, including the fluidization regime, as those of a commercial DMTO reactor can highlight the importance of bed hydrodynamics on the performance of a DMTO fluidized bed reactor. To minimize the wall effect on the bed hydrodynamics, it is recommended that the pilot-scale fluidized bed reactor has an internal diameter larger than 15.0 cm [33], larger than that adopted by DICP, refer to Figure 8.6. If bed hydrodynamics has a subordinate role, the scale-up of the DMTO fluidized bed reactor can be more conveniently accomplished with minimal uncertainty concerning the results of the adopted DMTO reactor selection and design procedure. A comprehensive integrated hydrodynamic and reaction modeling of the DMTO fluidized bed reactor with the appropriate hydrodynamic and reaction kinetic information can also help identify the importance of bed hydrodynamics on the performance of the DMTO fluidized bed reactor.
- Ensuring minimal attrition of catalyst particles is critical with respect to cost and proper operation of the fluidized bed reactor-regenerator configuration. The application of a properly designed jet cup [34] and the same pilot-scale unit mentioned above can help determine the extent of mechanical, thermal, and reaction–regeneration cycle induced attritions of catalyst particles. If the results of attrition tests are not satisfactory, proper improvements can be applied to the catalyst synthesis process or detailed designs of the reactor and/or regenerator.
- Controlling the residence time distribution of catalyst particles in the reactor is crucial since it is closely related to the coke distribution in the reactor, which influences the selectivity to light olefins in the DMTO process [4]. An optimal design of internals can be obtained with the aid of the same pilot-scale unit recommended above and an appropriate particle tracking technique, e.g., fluorescent, magnetic, or radioactive particle tracking technique. In addition, the heat transfer performance of the optimized internal design can be simultaneously studied and the design may be adjusted for the highest performance.
- Although particle agglomeration can help decrease particle entrainment in the DMTO fluidized bed reactor-regenerator configuration, it could be principally

deleterious for this process. Particle agglomeration (i) decreases the available cumulative surface area of catalysts, (ii) leads the bed hydrodynamics to deviate from its optimal conditions, which in extreme cases, yields bed defluidization and forced plant shutdowns, and (iii) progressively changes the residence time of particles in the bed when preserving operating parameters. An increase in the level of cohesive interparticle forces, mainly sintering in this case, that dominate repulsive forces in the bed originates particle agglomeration. One can determine the sintering propensity of catalyst particles by dilatometry tests [35] at laboratory scale. Alternatively, more comprehensive information about this phenomenon can be obtained in the same pilot-scale unit proposed above. One can vary the operating conditions, e.g., temperature, U_g, and bed height, and/or internal design to simultaneously delineate the effects of interparticle and hydrodynamic forces and internal design on the extent of agglomeration in the DMTO reactor-regenerator assembly.

Refer to Section 8.4, researchers at DICP tried to answer most of the abovementioned uncertainties during conventional scale-up of DMTO reactor. Based on the mitigation plans recommended above, one can minimally leave out the demonstration plant when scaling up the DMTO reactor through the iterative approach with minimal uncertainties.

8.4 DMTO fluidized bed reactor scale-up

DICP spent over 25 years and a lot of effort and money to successfully commercialize its DMTO process. While many scientific and technical aspects were thoroughly investigated before the complete commercialization, particular attention was dedicated to the reactor selection and scale-up owing to two main reasons: (i) the reactor is the key equipment in the production process and (ii) reactant conditioning and downstream fractionation train have been well developed. Before making the final reactor type selection, researchers at DICP proposed and tested different types of fluidized bed reactors, including two downer reactors and a dense fluidized bed reactor, for the DMTO process [4]. Other MTO reactor designs, as discussed in Section 8.1, were also recommended in literature. Detailed discussion as to why DICP selected a dense turbulent fluidized bed reactor for the DMTO process was provided in Section 8.3.1. The scale-up of the DMTO fluidized bed reactor was accomplished through experiments at four different scales, i.e., laboratory, pilot, demonstration, and commercial scales, as schematically shown in Figure 8.6 [4]. The scale factor between adjunct scales was 100 and 10, in terms of methanol feed rate and reactor diameter, respectively [1]. The most critical challenges to be properly addressed in a successful scale-up of DMTO reactor were (i) fluidization of catalyst particles by methanol gas at desired operating conditions, (ii) catalyst attrition, (iii) optimal catalyst residence time, (iv) optimal gas–solid catalyst contact time, (v) catalyst circulation, (vi) catalyst stripping, (vii) the effect of reactor size on the fluidization quality of catalyst particles, (viii) the effect of reactor size on the progress of the reaction, and (ix) the heat of reaction. A summary of operating conditions and main tasks at different scales of the DMTO fluidized bed reactor are summarized in Table 8.3.

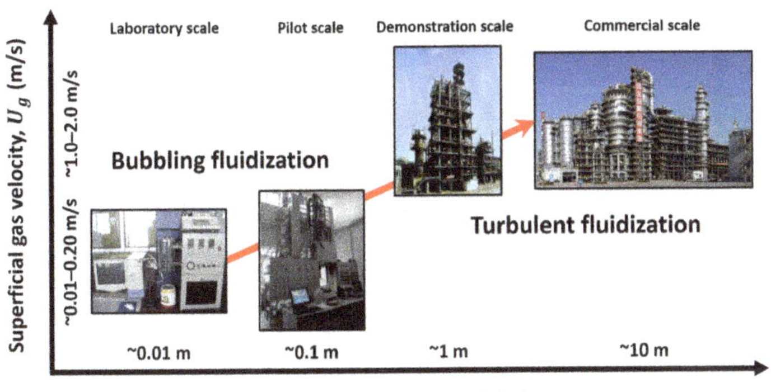

Figure 8.6: Schematic representation of DMTO fluidized bed reactor scale-up. Sources: Adapted from Tian et al. [4] and Cai [14].

In the laboratory, pilot, and demonstration scales, the methanol conversion was higher than 99% and the average selectivity to ethylene and propylene was about 80%. A slightly better reaction performance, i.e., methanol conversion of 100% and cumulative ethylene and propylene selectivity of slightly greater than 80%, was obtained at the commercial scale. This could be attributed to the good fluidization quality in the commercial-scale DMTO reactor [4]. As it can be inferred, the variation in bed hydrodynamics, which could happen at different scales, did not have a critical effect on the DMTO reactor performance. It, hence, made the scale-up of the DMTO fluidized bed reactor more convenient [1]. The commercial DMTO reactor was a shallow turbulent fluidized bed with 11 m diameter, to handle the target feedstock rate at the desired U_g and 3 m bed height, to satisfy the required WHSV, gas–solid contact time, and bed pressure drop. This is similar to what was achieved in Section 8.3.2 through the iterative scale-up approach.

For a successful scale-up of the DMTO process, DICP carried out a series of fluidization and reaction engineering research in addition to the abovementioned set of experiments at four different DMTO reactor scales. These additional studies aimed at obtaining a greater understanding about the fluidization of the DMTO catalyst, optimizing the DMTO reactor design, and enhancing the operation efficiency. Minimum fluidization and bubbling velocities of DMTO catalysts were studied first. In addition, by investigating the effect of internals on catalyst residence time distribution in a

Table 8.3: Operating conditions and main tasks at different scales of the DMTO fluidized bed reactor. Sources: Adapted from Tian et al. [4] and Ye et al. [1].

Reactor scale	Methanol feed rate (kg day^{-1})	Catalyst inventory (kg)	U_g(m s^{-1})	WHSV (h^{-1})	Fluidization regime (solid circulation?)	Main tasks
Laboratory	0.5	0.01	~0.01–0.1	2–3	Bubbling (no)	(i) Evaluate the performance of DMTO catalyst, (ii) scrutinize the optimal operation window, (iii) study the reaction kinetics of DMTO process, and (iv) determine optimal catalyst residence time in laboratory-scale fluidized bed reactor.
Pilot	50	1	~0.01–0.2	2–3	Bubbling (yes)	Study (i) catalyst stripping and (ii) catalyst circulation rate, (iii) evaluate the fluidization/ reaction performance of DMTO catalyst, and (iv) scrutinize the optimal operation window.
Demonstration	50,000	400	~1–2	~5	Turbulent (yes)	Study (i) catalyst circulation, (ii) catalyst stripping, (iii) catalyst attrition, (iv) heat of reaction and (v) fluidization behavior and reaction performance of DMTO turbulent fluidized bed reactor and (vi) provide data for commercial reactor design, as well as (vii) obtain exact product distribution data, including impurities, for the down-stream olefin recovery unit design.
Commercial	5,500,000	45,000	~1–2	~5	Turbulent (yes)	Solve some engineering challenges in the design of a large shallow fluidized bed reactor, including (i) catalyst mixing, (ii) gas–solid distribution, (iii) gas breakthrough, and (iv) propose and implement an effective start-up method.

cold-flow pilot-scale fluidized bed, the most effective internal design was developed for the DMTO reactor. It helped properly control the catalyst residence time in the reactor, thus the coke distribution in the reactor and, in turn, the selectivity to light olefins. Attrition tests of DMTO catalyst particles indicated that they are better than typical FCC particles in terms of attrition resistance. Researchers at DICP also highlighted that while fragmentation and abrasion are the governing attrition mechanisms of DMTO catalyst particles at room temperature, abrasion is the dominant attrition mechanism of these particles at high temperature of 500 °C [4]. Moreover, detailed kinetic studies resulted in a lumped kinetic model for the industrial DMTO catalyst [36]. Numerical modeling/simulation of DMTO fluidized bed reactor was further conducted to more precisely scrutinize the fluidization and reaction phenomena happening in the DMTO reactor [37–39].

8.5 Potential ways to do better by process simulation

Although the MTO process has been industrially implemented during the past decade at a number of large commercial plants operating in Asia, similar to other technologies, industrial maturity is not equivalent to a full understanding. Hence, researchers are intensely looking for new discoveries that can improve the process performance at a reduced cost and preserve the economic advantage of nonoil feedstock to olefins process [3, 9]. This becomes very crucial owing to the huge drop in crude oil price [9]. As outlined in Chapter 2, process simulation is a critical tool that can significantly help discover new findings and improve the profitability of a process. Despite numerous studies of the MTO process, there are a limited number of studies for the simulation of the whole process. The complete nonoil feedstock to olefins process is a complex process with multiple interconnections and restrictions in various units. Hence, modifications in operating parameters and/or process structure may have a chain effect. In addition, a detailed analysis of the relationship between the process stream and energy performance of the process is required for a complete understanding of the process and the influence of a proposed modification on the process performance. These explain the importance of a full process simulation for a thorough study of proposed changes to the process [9]. For the sake of brevity, we only discuss the critical results of a comprehensive study of a full MTO plant simulation here. Interested readers are invited to review other relevant references, e.g., Dimian and Bildea [21].

Decreasing energy consumption on the basis of a comprehensive energy analysis and integration of the process is an effective method in improving the profitability of a process. Pinch technology and a mathematical programming method are two categories of energy integration technology [9]. For energy integration, the pinch technology includes the integration within the unit, in distillation and between units, as well as the total site integration. Since large-scale MTO process is a multi-level utility and multi-pinch problem, a heat exchanger network (HEN) synthesis with the aid of a gradual/iterative optimization integration strategy based on an enthalpy–temperature (T-H) diagram [40] is required [9]. The flowchart of this strategy that was applied for a large-scale MTO process is schematically shown in Figure 8.7. Readers are referred to Liu et al. [9] for further details on this strategy.

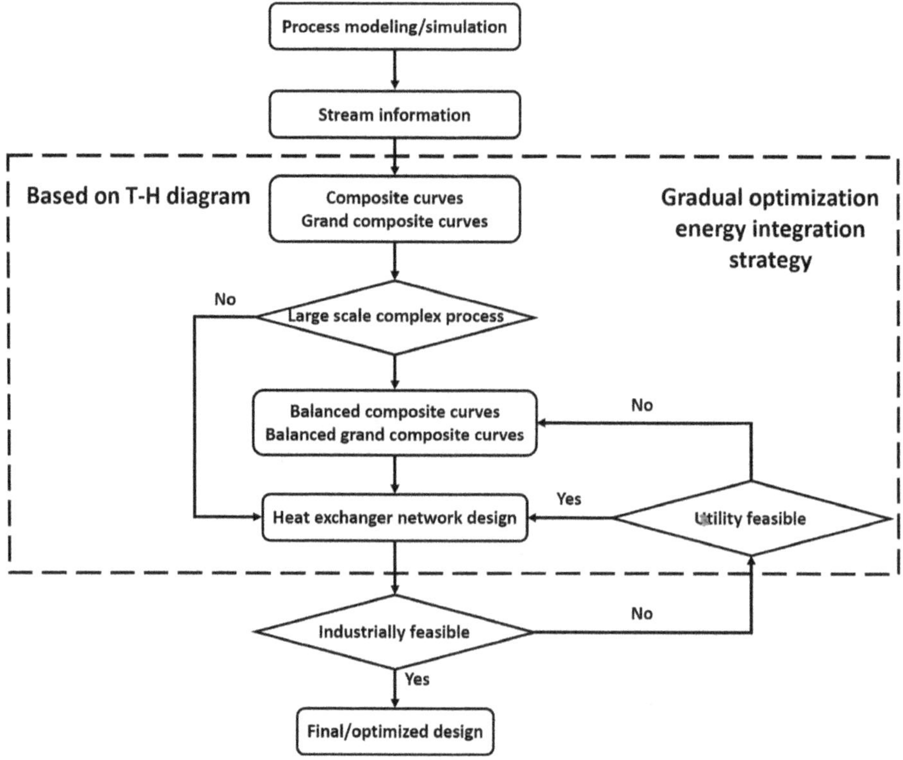

Figure 8.7: Schematic representation of gradual/iterative optimization energy integration strategy. Source: Adapted from Liu et al. [9].

Liu et al. [9] performed a complete MTO process simulation by Aspen Plus software. As shown in Figure 8.8, there are two main parts in the whole MTO process, i.e., MTO unit (established on DMTO technology here) and light olefin recovery (LOR) unit by LUMMUS. The energy integration of the MTO process accomplished by the integration within the MTO or LOR unit and the integration between MTO and LOR units. Results of energy integration of the MTO unit will be discussed here, while interested readers are referred to Liu et al. [9] for further results. According to the composite curve and grand composite curve of the MTO unit (results not shown here), the temperature distribution of the MTO unit is relatively wide, which requires multi-level utilities for heating and cooling. Based on the principle of energy cascade utilization and step-wise matching, where a hot stream should be cooled down by a cold stream at the corresponding level, Liu et al. developed a new grid diagram and synthesized the corresponding HEN for the MTO unit (Figure 8.9). Compared to the industrial design, the new design (i) could consume about 6.0 MW less energy, which accounts for 3.7% in energy savings, (ii) has one less heat exchanger, and (iii) requires a slightly larger heat exchanger area, by 84 m^2. Details of the main energy saving strategies for the MTO unit are discussed in Liu et al. [9]. By performing a similar energy integration for the LOR unit and proposing a new HEN design, Liu et al. showed that the new design (i) could consume about 6.0 MW less in energy, which accounts for 3.4% in energy savings, (ii) has one less heat exchanger and (iii) decreases the required heat exchanger area by 2,585 m^2, which accounts for a 17.5% area saving, in comparison with the typical industrial design.

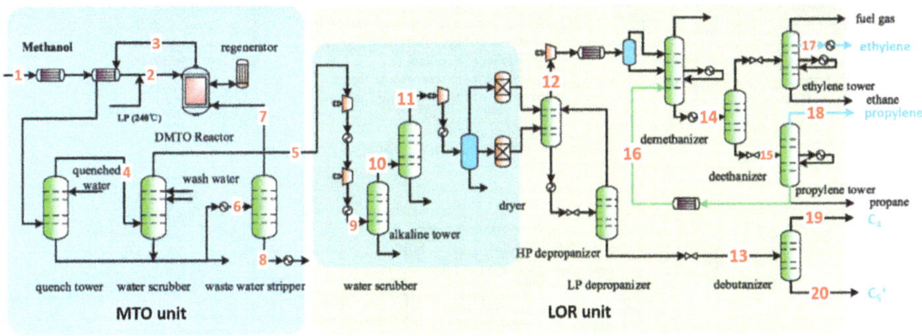

Figure 8.8: Detailed schematic representation of the methanol to olefins process (simulated by Aspen Plus Software). Source: Adapted from Liu et al. [9].

A techno-economic analysis, including the capital, operating, and total annual expenditures, of the new HEN designs for the whole MTO process in comparison with the industrial design was accomplished by Liu et al. (Figure 8.10). The total annual expenditure includes the cost of operating expenditure (OPEX) and capital expenditure (CAPEX) with the annualization factor. The CAPEX of the new MTO unit design increases by $0.05 million, while its OPEX decreases by $0.16 million, refer to the industrial design reference case (Figure 8.10). The CAPEX and OPEX of the new LOR unit design decrease by $0.78 and $0.26 million, respectively. The total annual expenditures of the new designs of the MTO and LOR units are $0.15 and $0.51 million lower than those for the industrial designs, respectively. For the whole MTO process/total site, the new design has $8.2, $14.6, and $17.3 million CAPEX, OPEX, and total annual expenditures, respectively, which are 8.3%, 2.8%, and 3.7% less than those for the industrial design.

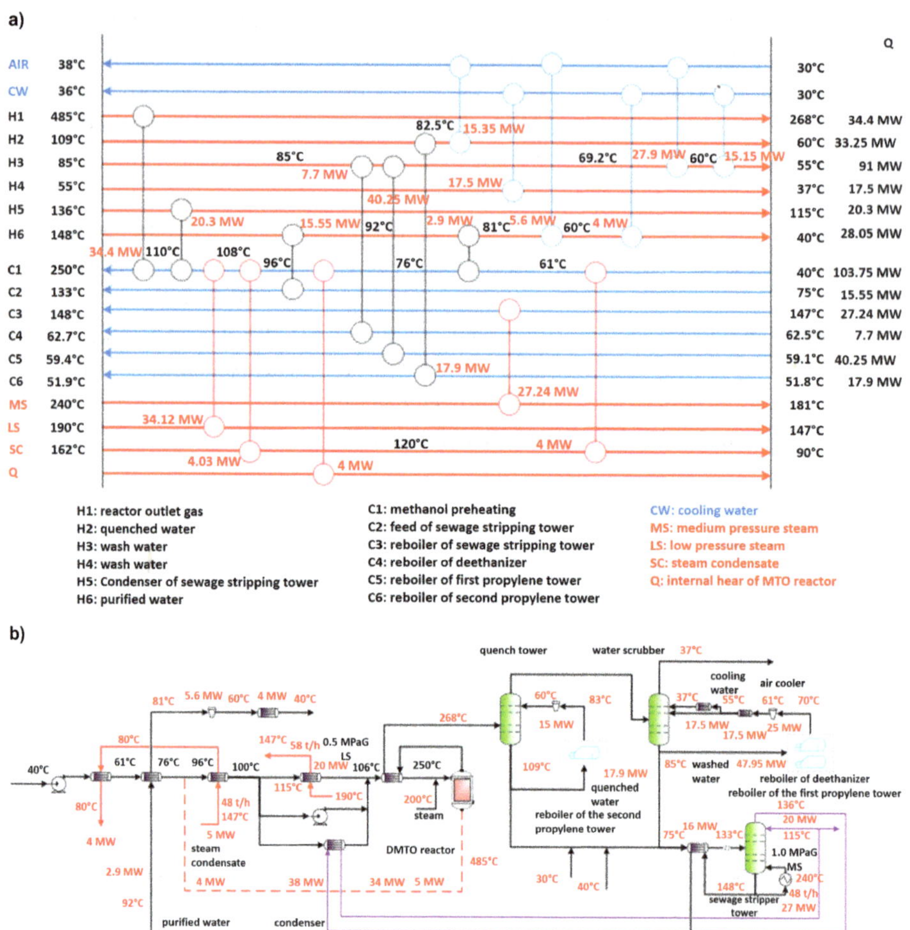

Figure 8.9: (a) Grid diagram and (b) schematic representation of proposed HEN for MTO unit. Source: Adapted from Liu et al. [9].

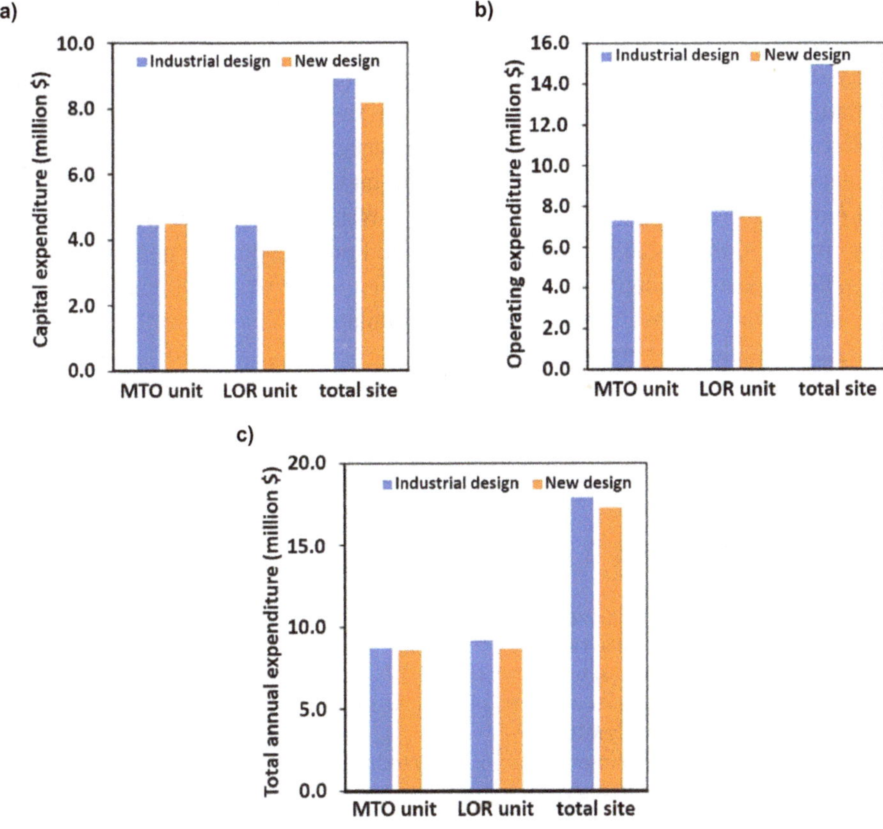

Figure 8.10: Comparative techno-economic analyses of the industrial and new designs of MTO and LOR units, as well as the whole MTO process/total site. (a) CAPEX, (b) OPEX, and (c) total annual expenditure. Source: Adapted from Liu et al. [9].

8.6 Conclusion

The production of light olefins from methanol has been industrially implemented over the past decade. Among major MTO technologies, the DMTO technology is dominant. Hence, available information about key research and experiences in the successful development and scale-up of this technology, in particular, its reactor, formed the core of this chapter. DICP successfully scaled up its DMTO fluidized bed reactor through experiments at four scales, i.e., laboratory, pilot, demonstration, and commercial scales. The fundamental research on the reaction and reactor selection and scale-up led to the world's first MTO commercial plant in 2010, operating DMTO technology, after over 25 years of research. The selection of the most effective reactor and its design are key challenges in the successful development of many chemical processes, particularly those with gas–solid catalytic reactions. With the aid of an iterative scale-up approach, we demonstrated how one can more effectively and less expensively select the most appropriate DMTO reactor and its critical operating conditions, i.e., WHSV, U_g, and H, in a shorter interval. To implement the new scale-up approach, we adopted critical background information and experimental data at laboratory scale and a decision-making procedure for the selection of a proper gas–solid catalytic reactor based on WHSV. To minimize the risks of the new approach, we identified its uncertainties and presented mitigation plans to address them by supplementary laboratory and pilot-scale results. Directions on how one can further improve the profitability of available technology with the help of process simulation was also discussed.

Notations

Abbreviations

CAPEX	Capital expenditure
C–C	Carbon–carbon
CTM	Catalyst to methanol ratio (–)
DICP	Dalian Institute of Chemical Physics
DMTO	DICP's MTO
DMTO-II	Second generation of DMTO technology
FCC	Fluid catalytic cracking
HCP	Hydrocarbon pool
HEN	Heat exchanger network
Honeywell UOP	Honeywell Universal Oil Product
LOR	Light olefin recovery
MTG	Methanol to gasoline
MTO	Methanol to olefins
MTP	Methanol to propylene
OPEX	Operating expenditure
SAPO	Silicoaluminophosphate
syngas	Synthesis gas
T–H	Enthalpy–temperature
UCC	Union Carbide Corporation
WHSV	Weight hourly space velocity (h^{-1})
$WHSV_{min}$	minimum value of allowed WHSV (h^{-1})

Symbols

A	Bed cross-sectional area (m^2)
$C_{cat}(t)$	Coke content in catalyst (–)
$\dot{c}(t)$	Coke formation rate ($kg\ s^{-1}$)

H	Bed height (m)
H_{max}	Maximum allowable bed height (m)
$\dot{m}(t)$	Mass flow rate of catalyst (kg s^{-1})
$\dot{Q}_m(t)$	Methanol feed rate (kg s^{-1})
$\dot{Q}_r$	Mass flow rate of reactant (kg h^{-1})
t	Time on stream (s)
U_g	Superficial gas velocity (m s^{-1})
$U_{g,max}$	Maximum allowable superficial gas velocity (m s^{-1})
W	Catalyst inventory in the laboratory-scale reactor (kg)
W_{RX}	Mass of required catalyst for a complete conversion of reactant (kg)

Greek letters

ε	Average bed voidage (-)
ρ_g	Gas density (kg m^{-3})
ρ_p	Particle density (kg m^{-3})

References

[1] M. Ye, H. Li, Y. Zhao, T. Zhang and Z. Liu, Chapter Five – MTO Processes Development: The Key of Mesoscale Studies, in Advances in chemical engineering, G.B. Marin and J. Li, Eds., Academic Press, 279–335, 2015.

[2] U. Olsbye, S. Svelle, M. Bjørgen, P. Beato, T. V. W. Janssens, F. Joensen, S. Bordiga and K. P. Lillerud, Conversion of methanol to hydrocarbons: how zeolite cavity and pore size controls product selectivity, Angewandte Chemie International Edition, 51(24), 5810–5831, 2012.

[3] I. Yarulina, A. D. Chowdhury, F. Merier, B. M. Weckhuysen and J. Gascon, Recent trends and fundamental insights in the methanol-to-hydrocarbons process, Nature Catalysis, 1(6), 398–411, 2018.

[4] P. Tian, Y. Wei, M. Ye and Z. Liu, Methanol to olefins (MTO): from fundamentals to commercialization, ACS Catalysis, 5(3), 1922–1938, 2015.

[5] X. Xu, Y. Liu, F. Zhang, W. Di and Y. Zhang, Clean coal technologies in China based on methanol platform, Catalysis Today, 298, 61–68, 2017.

[6] C. D. Chang and A. J. Silvestri, The conversion of methanol and other O-compounds to hydrocarbons over zeolite catalysts, Journal of Catalysis, 47(2), 249–259, 1977.

[7] E. M. Flanigen, B. M. Lok, R. L. Patton and S. T. Wilson, Aluminophosphate molecular sieves and the periodic table. In: Proceedings of the Seventh International Zeolite Conference, Tokyo. 1986.

[8] J. M. O. Lewis, Methanol to olefins process using aluminophosphate molecular sieve catalysts, Studies in Surface Science and Catalysis, 38, 199–207, 1988.

[9] S. Liu, L. Yang, B. Chen, S. Yang and Y. Qian, Comprehensive energy analysis and integration of coal-based MTO process, Energy, 214(119060), 2021.

[10] J. Liang, H. Li, S. Zhao, W. Guo, R. Wang and M. Ying, Characteristics and performance of SAPO-34 catalyst for methanol-to-olefin conversion, Applied Catalysis, 64, 31–40, 1990.

[11] B. V. Vora, T. L. Marker, P. T. Barger, H. R. Nilsen, S. Kvisle and T. Fuglerud, Economic Route for Natural Gas Conversion to Ethylene and Propylene, in Natural gas conversion VI/studies in surface science and catalysis, M. De Pontes, R. L. Espinoza, C. P. Nicolaides, J. H. Scholz and M. S. Scrurrell, Eds., Elsevier, 87–98, 1997.
[12] F. J. Keil, Methanol-to-hydrocarbons: process technology, Microporous and Mesoporous Materials, 29(1), 49–66, 1999.
[13] R. Jiang, J. Wang and P. Sun, Optimization of reactor operating factors a 600,000 TPY MTO plant, Petroleum Refinery Engineering, 44, 7–10, 2014.
[14] R. Cai, Recent R&D activities on clean coal technology in DICP. Dalian Institute of Chemical Physics, 45, 2017. Available from: https://usea.org/sites/default/files/event-/Thursday%20-%20EN%2010%20-%20Rui%20Cai%2CTechnology%20development%20of%20clean%20coal%20conversion%20of%20Dalian%20Institute%20of%20Chemical%20Physics%EF%BC%8CCAS%2CNOVEMBER%2030%2C%202017.pdf.
[15] M. R. Gogate, Methanol-to-olefins process technology: current status and future prospects, Petroleum Science and Technology, 37(5), 559–565, 2019.
[16] M. Yang, D. Fan, Y. Wei, P. Tian and Z. Liu, Recent progress in methanol-to-olefins (MTO) catalysts, Advanced Materials, 31(50), 1902181, 2019.
[17] N. Daviduk and J. H. Haddad, Fluid zeolite catalyzed conversion of alcohols and oxygenated derivatives to hydrocarbons by controlling exothermic reaction heat, Patent, US4238631A, 1980.
[18] A. V. Sapre, Conversion of methanol to olefins in a tubular reactor with light olefin co-feeding, Patent, US4590320A, 1986.
[19] L. W. Miller, Fast-fluidized bed reactor for MTO process, Patent, US6166282, 2000.
[20] S. Ruottu, K. Kääriäinen and J. Hiltunen, Method based on a fluidized-bed reactor for converting hydrocarbons, Patent, US6045688A, 2000.
[21] A. C. Dimian and C. S. Bildea, Energy efficient methanol-to-olefins process, Chemical Engineering Research and Design, 131, 41–54, 2018.
[22] K. Hemelsoet, J. Van Der Mynsbrugge, K. De Wispelaere, M. Waroquier and V. Van Speybroeck, Unraveling the reaction mechanisms governing methanol-to-olefins catalysis by theory and experiment, ChemPhysChem, 14(8), 1526–1545, 2013.
[23] Y. Liu, S. Müller, D. Berger, J. Jelic, K. Reuter, M. Tonigold, M. Sanchez-Sanchez and J. A. Lercher, Formation mechanism of the first carbon–carbon bond and the first olefin in the methanol conversion into hydrocarbons, Angewandte Chemie, 128, 5817–5820, 2016.
[24] I. M. Dahl and S. Kolboe, On the reaction mechanism for propene formation in the MTO reaction over SAPO-34, Catalysis Letters, 20(3), 329–336, 1993.
[25] I. M. Dahl and S. Kolboe, On the reaction mechanism for hydrocarbon formation from methanol over SAPO-34: I. Isotopic labeling studies of the co-reaction of ethene and methanol, Journal of Catalysis, 149(2), 458–464, 1994.
[26] I. M. Dahl and S. Kolboe, On the reaction mechanism for hydrocarbon formation from methanol over SAPO-34: 2. Isotopic labeling studies of the co-reaction of propene and methanol, Journal of Catalysis, 161(1), 304–309, 1996.
[27] S. Xu, Y. Zhi, J. Han, W. Zhang, X. Wu, T. Sun, Y. Wei and Z. Liu, Chapter Two – Advances in Catalysis for Methanol-to-Olefins Conversion, in Advances in catalysis, C. Song, Ed., Academic Press, 37–122, 2017.
[28] Q. Sun, Z. Xie and J. Yu, The state-of-the-art synthetic strategies for SAPO-34 zeolite catalysts in methanol-to-olefin conversion, National Science Review, 5(4), 542–558, 2018.
[29] D. Geldart, Types of gas fluidization, Powder Technology, 7(5), 285–292, 1973.
[30] G. S. Patience and R. E. Bockrath, Drift flux modelling of entrained gas–solids suspensions, Powder Technology, 190(3), 415–425, 2009.

[31] D. F. King, Estimation of dense bed voidage in fast and slow fluidized beds of FCC catalyst. In: J. R. Grace, L. W. Shemilt and M. A. Bergounou (Eds.), Proceedings of the Sixth Engineering Foundation Conference on Fluidization, Engineering Foundation, 1989, p. 1–8.
[32] T. M. Knowlton, S. B. R. Karri and A. Issangya, Scale-up of fluidized-bed hydrodynamics, Powder Technology, 150(2), 72–77, 2005.
[33] R. Mabrouk, R. Radmanesh, J. Chaouki and C. Guy, Scale effects on fluidized bed hydrodynamics, International Journal of Chemical Reactor Engineering, 3(1), 1542–6580, 2005.
[34] R. Cocco, Y. Arrington, R. Hays, J. Findlay, S. B. R. Karri and T. Knowlton, Jet cup attrition testing, Powder Technology, 200(3), 224–233, 2010.
[35] J. H. Siegell, High-temperature defluidization, Powder Technology, 38(1), 13–22, 1984.
[36] L. Ying, X. Yuan, M. Ye, Y. Cheng, X. Li and Z. Liu, A seven lumped kinetic model for industrial catalyst in DMTO process, Chemical Engineering Research and Design, 100, 179–191, 2015.
[37] Y. Zhao, H. Li, M. Ye and Z. Liu, 3D numerical simulation of a large scale MTO fluidized bed reactor, Industrial & Engineering Chemistry Research, 52(33), 11354–11364, 2013.
[38] J. Zhang, B. Lu, F. Chen, H. Li, M. Ye and W. Wang, Simulation of a large methanol-to-olefins fluidized bed reactor with consideration of coke distribution, Chemical Engineering Science, 189, 212–220, 2018.
[39] X. Liu, J. Xu, W. Ge, B. Lu and W. Wang, Long-time simulation of catalytic MTO reaction in a fluidized bed reactor with a coarse-grained discrete particle method – EMMS-DPM, Chemical Engineering Journal, 389(124135), 2020.
[40] I. C. Kemp, Pinch analysis and process integration: a user guide on process integration for the efficient use of energy, Second edn, Amsterdam, the Netherlands: Elsevier, 2007.

Mohammad Latifi, Mohammad Khodabandehloo, Jamal Chaouki

9 Case study IV: Hydropotash from potassium feldspar

Abstract: Using the iterative scale-up approach this chapter evaluates the commercialization potential of a roasting process wherein potassium feldspar ($KAlSi_3O_8$) of a syenite ore is altered through a reaction with calcium nitrate at a temperature of 550 °C. Lab-scale reaction kinetic investigations and greenhouse tests proved that high temperature processing can be a promising path to modify potassium feldspar, a mineralogically stable potassium aluminosilicate mineral, into a potash fertilizer called "HTP". This fertilizer releases potassium into the soil at a slow rate, which is suitable for tropical regions, like Brazil, where heavy rains are common. Despite the technical feasibility of the roasting process, two parameters challenge the economic viability of the process: the low K_2O content of syenite ore compared to competitive products (e.g., Muriate of Potash) and the relatively expensive price of calcium nitrate. Based on a techno-economic assessment, the total cost of HTP production for a 2 MTA throughput with a 220 USD ton^{-1} market price for calcium nitrate was estimated to be 282.2 USD ton^{-1}, which substantially exceeds a projected desired value of 43 USD ton^{-1}. The process scalability analysis with the iterative scale-up approach quickly illustrated that a direct scale-up from lab to pilot could have imposed significant research and development costs, e.g., in the order of several million dollars, for a process that was not commercially feasible. The iterative scale-up approach revealed however that an in situ regeneration of calcium nitrate could render the process economically viable depending on market demand.

Keywords: iterative scale-up, techno-economic assessment, potassium feldspar, fertilizer, potash, lab-scale experiments, roasting

Terminology

FTIR	Fourier-transform infrared spectroscopy
TGA	Thermogravimetric analysis

Mohammad Latifi, Department of Chemical Engineering, Polytechnique Montréal, 2500 Chemin de Polytechnique, Montréal, Québec, H3T 1J4, Canada, e-mail: mohammad.latifi@polymtl.ca
Mohammad Khodabandehloo, Department of Chemical Engineering, Polytechnique Montréal, 2500 Chemin de Polytechnique, Montréal, Québec, H3T 1J4, Canada, e-mail: mohammad.khodabandehloo@polymtl.ca
Jamal Chaouki, Department of Chemical Engineering, Polytechnique Montréal, 2500 Chemin de Polytechnique, Montréal, Québec, H3T 1J4, Canada, e-mail: jamal.chaouki@polymtl.ca

https://doi.org/10.1515/9783110713985-009

ISBL	Inside battery limit as all equipment and components that act upon the primary feed stream of a process
TCOP	Total cost of production
DNPV	Discounted net present value
DPBP	Discounted payback period
CDCF	Cumulative discounted cash flow
OSBL	Outside battery limit
CAPEX	Capital expenditures
OPEX	Operating expenditures
ST	Storage tank or vessel for liquid and solid raw materials and products
CV	Conveyors to transport solid material in powdery form
CR	Hammer mill crushers for size reduction of particles
SC	Screens to separate coarse and fine particles
HX	Air-cooled heat exchangers to cool down the product gases to 25 °C
P	Pump

Description of main simulation streams

S01	Syenite ore
S02	Calcium nitrate
S07	NO_2 and O_2
S08	HTP
S09	NO, NO_2, O_2
S11	$CaCO_3$
S12	H_2O
S14	HNO_3
S15	NO (>91%), NO2, H_2O and HNO_3 (~7%)
S16	H_2O
S17	NH_4NO_3
S19	CO_2 and H_2O
S20	HNO_3
S21 and S23	NH_4OH
S25	NH_4NO_3
S26	NO_2
S27	H_2O

Description of main simulation equipment

R01	Roasting reactor
AB01	Absorber
H01, H02	Hammer mills
R02	Reactor of calcium nitrate production
M01, M02, M03	Mixers
HX01	Heat exchanger for preheating raw materials

HX02, C01	Cooler
F01	Press filter
SEP01 and SEP02	Separators of NH_4NO_3 and H_2O

9.1 Importance of potash fertilizers

The population of the world is projected to be about ten billion by 2050, making it of great importance to secure food supplies [1]. Accordingly, the production of sufficient crops is a function of certain parameters, such as land expansion, cropping intensity, and yield growth. The latter constitutes more than 70% of this functionality. About 50% of the yield growth depends on the provision of fertilizers. Potassium (K) is one of the three key nutritional elements (i.e., along with phosphorous and nitrogen) for crop production. Potash, a water-soluble K-bearing material, is extensively applied to the K-deficient soils to improve the crop yield. The soluble K-bearing ores (SKO), such as sylvite (KCl) and Kaliophilite ($KAlSiO_4$), have been a primary source for potash production, i.e., 95% of the global production.

9.2 $KAlSi_3O_8$: an alternative potash resource

Due to the scarcity of SKO as well as a significant increase in global potash demand, there must be alternative resources of potassium-based fertilizers. Almost one half of the earth's crust is made of feldspars: $KAlSi_3O_8$ (K-feldspar, potassium feldspar or KFS), $NaAlSi_3O_8$ (albite), and $CaAl_2Si_2O_8$ (anorthite). Hence, the abundant K-containing minerals, such as KFS, are currently deemed a potential resource for potash production. KFS is a potassium mineral, wherein the Al–Si–O bonds form a highly stable tetrahedron structure leading to its slow dissolution rate in water [2]. The great abundance of KFS is identified in syenites and trachytes, which are mined at the earth's surface. In addition to potassium (~13% of K_2O), the ultrapotassic syenites can supply other valuable components to crops, for example, ~63% of SiO_2 [3].

Despite the potential of K-feldspar being an alternative potash resource, the associated fertilizers are in fact slow potassium releasers [4]. It is noteworthy that slow releasing K-fertilizer is one of those options that has received increasing interest in the past decade [5, 6] because it could be beneficial in tropical areas, like Brazil where heavy rains are common. However, the potassium availability for the plants is the main challenge of these fertilizers. The field test results of the K-feldspar showed that its K is almost completely unavailable for the plants [7]. Therefore, the development of novel approaches to increase the extraction rate of potassium has been investigated in recent years by many researchers [2, 3, 7–10]. To do so, an additive is employed to replace the potassium of KFS and, consequently, to disrupt its structure. Laboratory-scale approaches to modify the KFS structure to reach a desired level of potassium extractability (i.e., a potash fertilizer) could be grouped into two classes, each with pros and cons:

- Wet techniques (hydrothermal ion exchange and acid/alkali leaching): In spite of a high fraction of K extraction at a relatively low temperature, a high amount of waste in the form of gas and liquid adversely influences the local environment [3, 11–13].
- Dry techniques: They are assigned to the calcination of the KFS with sodium- and calcium-bearing additives at a high temperature to generate a soluble K-containing salt [14].

The ambition to overcome the obstacles associated with the hydrothermal technique led to proposing the high temperature potash production based on the solid–solid reaction between KFS and additives, to not only increase the released K from KFS, but also markedly reduce the liquid waste and energy consumption. Several research groups have implemented a series of KFS calcination tests using different Ca-containing additives, followed by water leaching to probe the fraction of K eliminated from KFS. The requirement of very high operation temperatures (i.e., from 800 to 1,200 °C) in previous KFS roasting studies makes it infeasible for a scaled-up process in terms of reaching a competitive production cost with respect to a lower K_2O content of KFS than in muriate of potash (MOP), which is a commercially traded product. Such high temperature processes require special attention to materials of construn, maintenance, provision of heat-up energy, and safety.

9.3 Roasting of KFS with calcium nitrate

Advanced Potash Technologies (APT) based in the United States produces potash products from its syenite ore. APT was interested in investigating a high temperature process to alter KFS to produce a new type of potash called "high temperature potash" or HTP at the minimum temperature. To reach a process with a much lower roasting temperature, new additives with a lower melting point were to be tested since a liquid film would form around KFS particles that convene the ion-exchange mechanism. Accordingly, we selected calcium nitrate ($Ca(NO_3)_2$) because it has a relatively low melting point of 561 °C. Furthermore, the final product might potentially contain both K and N, which are key nutritional elements for plants.

9.4 Iterative scale-up approach for roasting process scale-up

The iterative scale-up approach (Figure 9.1) is based on directly scaling up a process built upon lab-scale experimental data to a virtual commercial-scale process, that is, line L-C. The "virtual" term is employed because the commercial scale has not been physically built yet. Accordingly, we initially prepare a costing tool, which can be a series of spreadsheets and simulation software. Then, we generate flowsheets for the virtual commercial scale. Usually, one or more process flowsheets are considered to present the configuration of associated unit operations in different layouts for heat and mass calculations, process simulation and optimization, HAZOP studies, and social, environmental and economic assessments. The result of these evaluations is that some questions must be answered at the lab scale. Thus, new experiments will be designed for lab-scale tests to address the questions (line C-L), and eventually re-analyze the virtual commercial flowsheets (line L-C). Sometimes, performing lab-scale tests to answer the questions is not possible due to associated restrictions. In such cases, a comprehensive literature review could provide the necessary answers. Also, at times, new lab-scale experimental tests are required to generate a completely different process as an alternative to the previous one. Moreover, some questions require specific pilot-scale tests to find the answers (line C-P). These answers will be entered into the costing tool (line P-C) to improve and optimize the process flowsheet at the commercial scale.

We utilized the iterative scale-up approach to evaluate the scale-up feasibility of the roasting process to identify the technical and economic uncertainties and pertinent questions at a commercial scale, and to determine whether or not further research and development activities should be carried out in either lab or pilot scale to answer the questions. In this context, we started off by conducting a series of lab-

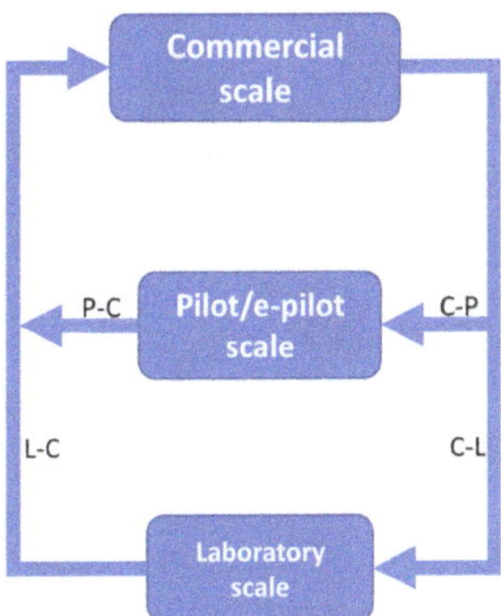

Figure 9.1: Scheme of iterative scale-up approach.

scale tests to investigate the kinetics of roasting reactions between KFS and calcium nitrate under various operating conditions. Subsequently, we determined a process flowsheet and developed a costing tool to evaluate the techno-economic features of the process at a virtual commercial scale of 2 million tons per annum.

9.4.1 CAPEX estimation methods

It is necessary to estimate the CAPEX of a process to provide a clear idea of the required investment. It can also serve as a guideline to focus on the most influential independent parameters to decrease the project costs. There are different methods for estimating the capital expenditure of a process. The main CAPEX estimation methods are [15]:

- Historical cost data: Extrapolation of CAPEX based on a reference project (about 50–60% uncertainty);
- Factorial method: Requires an initial estimation of main equipment, which is multiplied by Lang factors (about 30% uncertainty);
- Quotes from vendors/suppliers (5–10% uncertainty).

We employed the factorial method, which is a strategic choice when working with a project in the research step (TRL 1 to 4) because it is without too much uncertainty and it requires no commercial quotations.

The factorial method includes the following steps:

- Defining flowsheet, mass and energy balances for the desired capacity
- Estimation of the main equipment cost:

$$C_e = a + (b \times S)^n \tag{9.1}$$

where C_e is the direct cost of equipment; S is the equipment capacity in the unit specified; a, b, and n are factors tabulated for each specific piece of equipment. We extracted this data from [15]. It is worth noting though that the costs in this reference are for the year 2006; thus, we used the Chemical Engineering Plant Cost Index (CEPCI) of 2006 and 2019 to update the capital according to eq. (9.2), which are 499.6 and 619.2, respectively:

$$\text{Cost in year 2019} = \frac{\text{CEPCI in 2019}}{\text{CEPCI in 2006}} \times \text{Cost in year 2006} \tag{9.2}$$

After determining the direct cost of each main piece of equipment, the factors were applied.

- OSBL factors, which represent the modification to be made to the site infrastructure, are: Offsites (OS), design and engineering (DE), contingency (X), and location (L);

- For inside battery limits cost (ISBL), which represent the cost of the plant itself, the factors are: equipment erection (f_{er}), piping (f_p), instrumentation and control (f_i), electric component (f_{el}), civil engineering (f_e), structures and building (f_s), paint and lagging (f_l).

Such factors are tabulated in process design textbooks for each specific type of the process; that is, either fluids, solids–fluids, or solids [15]. These factors can be found sometimes in historical data of the industry to reduce the uncertainties. We employed eq. (9.3) to estimate the CAPEX of the process, where M is the total number of major equipment in the process, and L is a location factor selected for Brazil:

$$\mathrm{CAPEX} = (1+L) \times \left(\sum_{i=1}^{M} C_{e,i} \times \left[(1+f_p) f_m + (f_{er} + f_{el} + f_i + f_c + f_s + f_l) \right] \right) \times (1+\mathrm{OS}) \times (1+\mathrm{DE}+X) \tag{9.3}$$

9.4.2 OPEX estimation methods

The estimation of OPEX of a process is crucial as it provides essential information on production costs. Generally, OPEX represents a portion of the cost, which has more impact on the total project operation. It includes the cost of electricity, consumable materials, water, steam, and wastewater treatment.

To estimate OPEX, it is required to estimate the following terms:

- Electricity to supply the power of equipment, such as pumps, compressors, heaters
- Consumable materials, either the market price or the on-site production price
- Quantity of water used in the process and its supply cost
- Quantity of steam used in the process and its supply cost
- Quantity of wastewater generated by the process and its treatment cost

The required data can be found by research, contacting the client of the project, or using available industrial databases. If the data are available for a reference project, the OPEX of the reference process can be extrapolated using the rule of three, although it compromises results with more uncertainties.

9.5 Initial roasting tests at lab scale

9.5.1 Materials

We used an ultrapotassic syenite ore with 90% passing of 165 μm originally from Brazil. Chemical analysis of the syenite determined by X-ray fluorescence (XRF) is given in Table 9.1. The KFS contained (mass basis) 63.4% SiO_2, 17.1% Al_2O_3, and 13.4% K_2O. The roasting agent was a lab grade calcium nitrate tetrahydrate ($Ca(NO_3)_2 \cdot 4H_2O$) from Fisher Scientific (Canada).

Table 9.1: Chemical analysis of syenite obtained from XRF.

SiO_2 (%)	Al_2O_3 (%)	Fe_2O_3 (%)	CaO (%)	MgO (%)	TiO_2 (%)	P_2O_5 (%)	Na_2O (%)	K_2O (%)	MnO (%)	BaO (%)
63.4	17.1	2.34	1.09	0.45	0.07	0.105	1.54	13.4	0.03	0.43

9.5.2 Experimental plan

A series of roasting tests were carried out in a muffle furnace to probe the effect of operating conditions, such as temperature, time, and reagent/ore ratio on the K extraction, which are detailed in Table 9.2. The thermal decomposition characteristics of the mixture of syenite and $Ca(NO_3)_2 \cdot 4H_2O$ were also investigated using a thermogravimetric analyzer. Nitrogen gas at a flow rate of 40 mL min^{-1} was used as a carrier gas.

Table 9.2: Experimental conditions were examined during calcination experiments.

Parameters	Roasting temperature (°C)	Roasting time (min)	Ca/Si molar ratio	Particle size (μm)	Cooling condition
Roasting temperature (°C)	450–600	60	0.8	$D_{90} < 165$	At RM[1]
Roasting time (min)	550	5–60	0.8	$D_{90} < 165$	At RM
Ca:Si molar ratio	550	60	0.4, 0.6, 0.8	$D_{90} < 165$	At RM

[1]RM: room temperature.

9.5.3 Effect of roasting temperature on K extraction

Solid–solid reaction of syenite and calcium nitrate is substantially based on the diffusion of calcium ions through the crystalline structure of KFS particles. This is facilitated at a temperature close to the melting point of calcium nitrate where the diffusion and fluidity of the lattice atoms become significant. As illustrated in Figure 9.2, an increase in the calcination temperature up to 550 °C improved K extraction to 28.90% and a greyish clinker was obtained. The brownish color of clinker imparts the presence of calcium nitrate in the final product. Generally, the diffusion of calcium ions both into the crystalline structure of KFS and between newly formed phases plays a significant role in the KFS-$Ca(NO_3)_2$ reaction.

Increasing the roasting temperature to almost the melting point of calcium nitrate appreciably enhanced the reaction rate and, consequently, the K extraction percentage. A further increase in the roasting temperature to 600 °C slightly reduced the K extraction (25.10%), which was due to the decomposition of calcium nitrate taking place above 560 °C. For instance, calcium nitrate partially converted to calcium oxide (CaO) at 600 °C, and consequently, a lower fraction of calcium ions were accessible to take part in the chemical reaction with KFS. This referred to the high thermal stability of CaO between 450 and 600 °C. In other words, calcium ions could not easily move in the CaO's lattice and subsequently did not diffuse into the KFS surface at a temperature range of 500–600 °C, which is well-below the melting point of CaO.

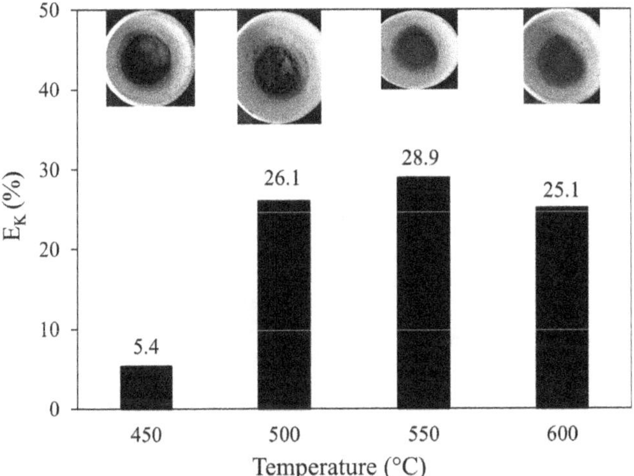

Figure 9.2: Effect of roasting temperature on the K extraction percentage (Ca:Si molar ratio of 0.8, roasting time: 60 min and $D_{90} < 160$ µm). The clinker cooled down at room temperature.

TGA and derivative thermogravimetric (DTG) analysis of the binary system of $Ca(NO_3)_2{\cdot}4H_2O$ and KFS were carried out at a Ca/Si molar ratio of 0.8 (Figure 9.3). The roasting reaction was characterized by three stages of weight loss in the temperature range of 25–700 °C. The first mass loss stage is referred to as the melting of $Ca(NO_3)_2{\cdot}4H_2O$, which melts in its own coordination water. The second stage is connected to the liberation of the water molecules from the crystalline structure of calcium nitrate, and subsequently, the formation of anhydrous calcium nitrate, which proceeds according to reactions (R1) and (R2). The next stage is described by the melting of anhydrous calcium nitrate and a chemical reaction taking place between calcium nitrate and KFS resulting in the formation of the K, Ca-aluminosilicate (K.Ca.Al.Si) and evolution of NO_x. NO_x represents the nitrogen oxide (NO) and nitrogen dioxide (NO_2) as presented in reaction (R3) where "a" denotes the mole of calcium nitrate used for KFS calcination tests:

$$Ca(NO_3)_2 + 4H_2O \rightarrow Ca(NO_3)_2 \cdot 3H_2O + H_2O \quad (R1)$$

$$Ca(NO_3)_2 \cdot 3H_2O \rightarrow Ca(NO_3)_2 + 3H_2O \quad (R2)$$

$$aCa(NO_3)_2 + KFS \rightarrow a(K, Ca, Al, Si, O) + (1-a)KFS + aNO_x + 1.5aO_2 \quad (R3)$$

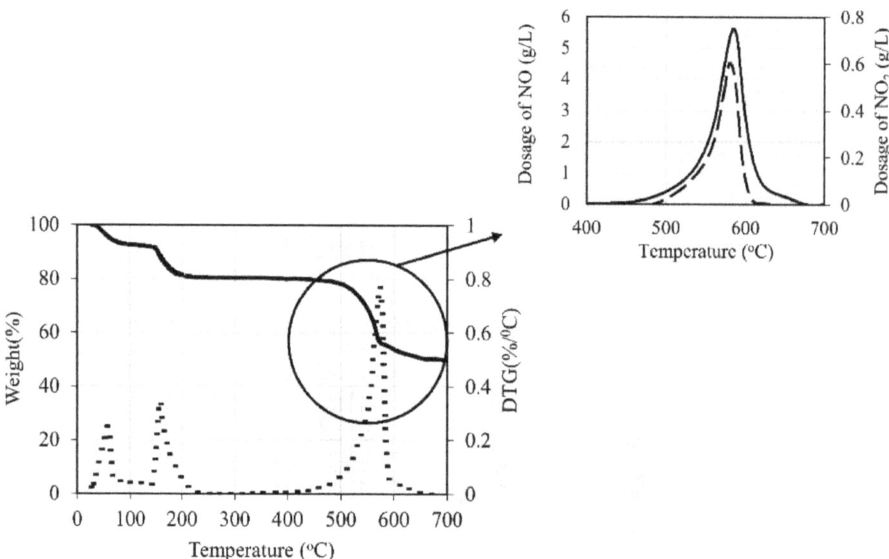

Figure 9.3: TGA of KFS + $Ca(NO_3)_2{\cdot}4H_2O$ system at the heating rate of 10 °C min^{-1} and dosage of NO and NO_2 gases evolved at the different temperatures.

FTIR affirmed the evolution of NO and NO_2 in the temperature range of 450–700 °C. As illustrated in Figure 9.3, the onset of the formation of NO was at a lower temperature (440 °C) compared to NO_2 gas at 480 °C. The dosage of product gases gradually increased and a maximum concentration of NO and NO_2 was attained nearly at 580 °C. However, an additional rise in the temperature to 620 and 690 °C drastically diminished the dosage of NO_2 and NO to zero, respectively.

9.5.4 Effect of roasting time on K extraction

Roasting time is an influential parameter, which plays a significant role in the solid–solid reaction of syenite and $Ca(NO_3)_2$ and eventually the release of K from HTP. It is because a sufficient length of time is required for the calcium ion to diffuse into the crystalline structure of KFS. Several experiments were carried out at

550 °C at different roasting times to investigate the extent of K extraction. The extraction kinetics of K in the presence of $Ca(NO_3)_2$ is shown in Figure 9.4. The equilibrium condition was achieved within 20 min. K extraction illustrated two distinct stages during solid–solid reaction: (i) steep slope or rapid K release (i.e., before the initial 15 min); and (ii) a low gradient or slow K extraction (i.e., between 15 and 20 min). The extraction rate of K was directly dependent on the extent of the surface area of KFS available for calcium ions and subsequently the interaction probability between Ca ions and the fresh surface of KFS. At the initial stage, a high fraction of KFS fresh surfaces were readily accessible for the calcium ions to diffuse and participate in the ion-exchange reaction with K. Thereafter, an increase in the contact time diminished the fraction of surface area of KFS particles available for Ca ions due to the formation of product layers around KFS particles. This resulted in the low gradient of K extraction.

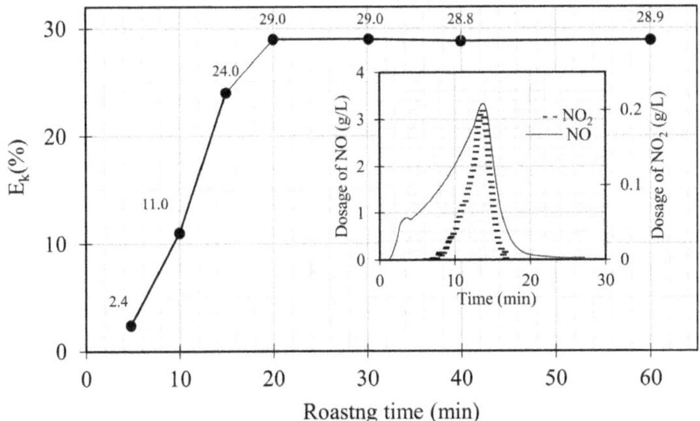

Figure 9.4: Effect of roasting time on K extraction (Ca:Si molar ratio of 0.8, roasting temperature of 550 °C and $D_{90} < 160$ μm). The clinker cooled down at room temperature. Smaller Figure, which is obtained from the FTIR analysis of gas products of the TG analyzer, illustrates the dosage of NO and NO_2 gases evolved at the different roasting times.

The onset of the formation of NO was at a shorter time compared to NO_2 gas after 7 min. The dosage of product gases gradually increased and a maximum concentration of NO and NO_2 was nearly attained after 14 min, whereas the concentration of NO and NO_2 was significantly reduced with a further increase in the roasting time to 20 min, implying the completion of the reaction between KFS and calcium nitrate. Figure 9.4 also confirms the equilibrium condition was reached within 20 min.

9.6 From lab scale to commercial scale: first attempt

APT has developed a hydropotash (HYP) from syenite ore through a hydrothermal process at 220 °C for 2 h [3]. It has shown that HYP provides a gradual release of potassium into the soil, which is beneficial for crop growth in tropical areas, such as in Brazil. HTP demonstrated during greenhouse tests a performance as a fertilizer that was quite competitive to HYP's performance. Therefore, it was desired to apply the iterative scale-up approach to evaluate the commercialization potential of our high temperature process. To evaluate the commercial potential of such a high temperature process at a 2 MTA scale, units of operation before and after the roasting reactor must be taken into account, such as the storage of materials, the comminution of solid reactants in terms of the size reduction of particles and liberation of KFS, transportation of materials, heat exchangers, and utilities.

We considered a flowsheet, including the necessary units of operation, to envisage a continuous process with an ISBL starting from the comminution of syenite ore and calcium nitrate to the storage of final products, which is presented in Figure 9.5. The flowsheet is about a process where calcium nitrate is supplied directly from a vendor. Both run-of-mine syenite ore and calcium nitrate are transported by a designated conveyor to a comminution processing unit to reach particles with a 90% passing size of 165 µm. A comminution unit is composed of a hammer mill followed by a shaking screen. Coarse particles at the outlet of the mill are separated from finer particles by the shaking screen before returning to the mill inlet. This loop continues until all particles reach the desired size distribution. Ground particles of syenite ore and calcium nitrate are transported toward the roasting reactor where they are fed into it with screw feeders. The roasting reactor is a rotary kiln with a speed of 5 revolutions per minute and a horizontal angle of 20° operating at 550 °C. The solids travel along the length of the kiln with a residence time of 90 min while they undergo radial cascade and cataract movements inside the reactor. The outgoing solids from the lower end of the roasting reactor are cooled down by natural convection over the surface of a grate-and-chain type conveyor, and they are subsequently collected in a storage silo. Product gases discharge from an outlet pipe; they are cooled down by an air-cooled heat exchanger to ambient temperature and stored inside a gas storage tank. It is most likely necessary to compress the gases through a compressor (i.e., up to 40 psig) to facilitate their storage.

Accordingly, we took the following assumptions to proceed:

- Raw materials, such as run-of-mine syenite ore and calcium nitrate, consist of particles with a 3–5 mm size, so they should undergo a suitable comminution processing to reach a 90% passing size of 165 µm.

- While the 90% passing of syenite ore is 165 µm, liberation of KFS is similar in all particles.
- Potassium content of syenite ore is only present in KFS.
- Silicate products after roasting are Ca_2SiO_4 (Ca-silicate), K_2SiO_3 (K-silicate), and Al_2SiO_5 (Al-silicate).
- Extractable potassium is proportional to the silicates in HTP. The remaining KFS is not reactive.
- Ca/Si ratio was estimated as the molar ratio of calcium in hydrated calcium nitrate ($Ca(NO_3)_2{\cdot}4H_2O$) and silicate of syenite ore (Table 9.1).
- According to the melting point and boiling/decomposition point of tetrahydrous calcium nitrate ($Ca(NO_3)_2{\cdot}4H_2O$), it is assumed that it was fully decomposed to the anhydrous form at a roasting temperature of 550 °C, and subsequently, calcium nitrate, $Ca(NO_3)_2$, participated in the reaction with the ore.
- Lab-scale investigations revealed roasting reactions took place between two solid powders of syenite ore and calcium nitrate (solid–solid contact with interfacial reactions under a molten state).
- Hydrated calcium nitrate will not melt during comminution operations. Its melting point is 43.9 °C.
- The following reactions take place presumably inside the smelting reactor according to the mass balancing of data from lab-scale tests:

1. Dehydration of calcium nitrate tetra hydrate:

$$Ca(NO_3)_2 \cdot 4H_2O \rightarrow Ca(NO_3)_2 + 4H_2O \tag{R4}$$

2. Alternation of KFS:

$$2Ca(NO_3)_2 + KAlSi_3O_8 \rightarrow KAlSiO_4 + Ca_2SiO_4 + SiO_2 + 4NO + 3O_2 \tag{R5}$$

3. Thermal decomposition of calcium nitrate:

$$12Ca(NO_3)_2 \rightarrow 12CaO + 22NO + 2NO_2 + 17O_2 \tag{R6}$$

- We assumed a rotating kiln could be a suitable reactor selection since reactants are in the solid phase. Reactants are introduced to the reactor in a co-current configuration. Product gases discharge from another side of the rotary kiln. The solid product is a mixture of converted reactants and new solid phases forming the HTP; it is collected in a storage tank without a downstream separation treatment.
- The required heat for roasting reactions is supplied by electrical heaters so that the temperature gradient is negligible along the reactor length.

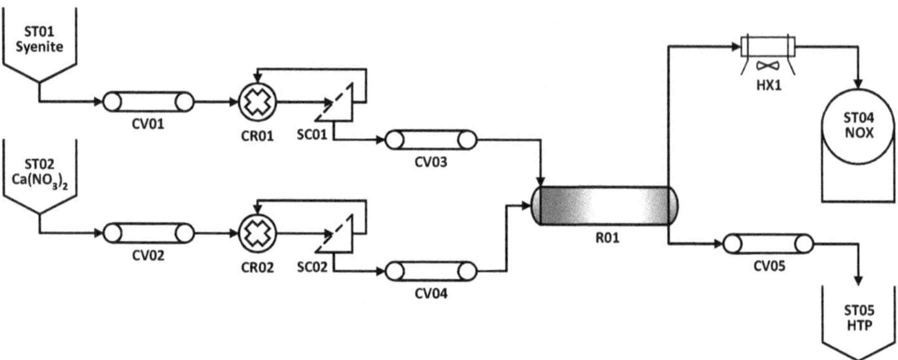

Figure 9.5: Process flowsheets for the roasting of syenite ore with calcium nitrate.

9.6.1 Process simulation and technical evaluation

We employed Aspen Plus™ to simulate the process by taking only the main process unit operations (Figure 9.6). In particular, we aimed to optimize energy consumption around the roasting reactor. We noticed it is better to take advantage of the heat content of the outlet gas from the roasting reactor to preheat its inlet raw materials (i.e., syenite ore and calcium nitrate) and save energy instead of directly purging the gas toward a storage tank. As illustrated in Figure 9.7, putting a heat exchanger to preheat the raw materials with the hot stream of generated gases during the roasting reactions could help significantly save energy costs. For instance, if the stream of gases were cooled down to about 35 °C (i.e., stream S07), by selecting a large enough heat exchanger HX01, an energy cost of more than 5 million dollars per annum would be saved. Under these conditions, the raw materials would be preheated to about 155 °C. We assumed raw materials would undergo a residence time of 90 min during preheating to estimate the capital cost of HX01. Consequently, we calculated the annualized capital cost of HTP would increase by only 5,000 dollars per annum. In other words, the idea of preheating the raw materials would be techno-economically a promising advantage. Also, stream S07 would be cold enough to be collected in a storage tank. It is worth noting that further experimental work is required to determine the preheating kinetics and the right selection of preheater, which could be a rotary drum or a fluidized bed providing an enhanced gas–solid heat transfer.

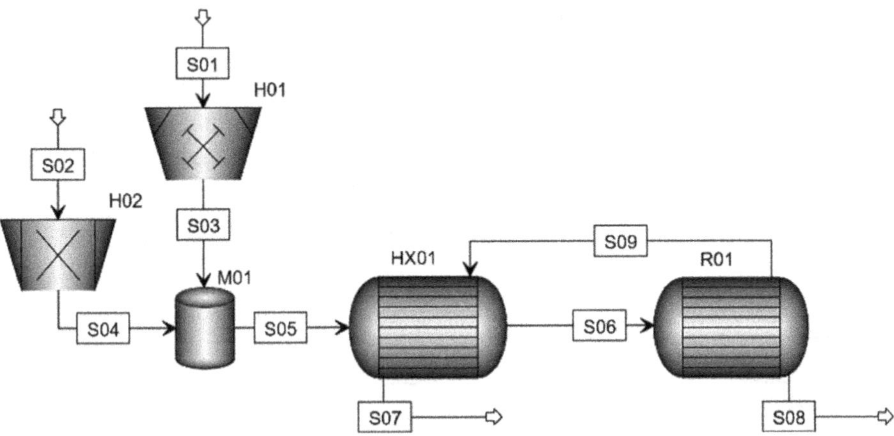

Figure 9.6: Simulation flowsheet of the roasting process.

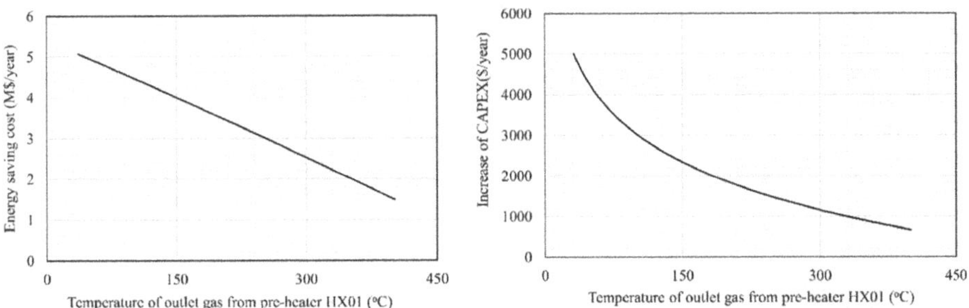

Figure 9.7: Impact of preheating raw materials by hot gases from roasting reactor.

9.6.2 Economic evaluation

An economic analysis at this stage of the process scale-up evaluation helps to verify whether the flowsheet in Figure 9.6 is economically viable. To proceed with the analysis, the potash market must be taken into account. It is worth noting that the price of MOP (muriate of potash or KCl) as of November 2020 was about 202 USD ton^{-1}. MOP is one of the most common potassium fertilizers in the world whose K_2O content is about 63%. While there are uncertainties regarding the advantages of other elements, such as SiO_2 in syenite ore, its K_2O content is 13.4% (Table 9.1); in other words, the market price of HTP should be below 43 USD ton^{-1} of HTP to be competitive with MOP with respect to its K_2O content.

Table 9.3 elucidates the CAPEX and OPEX calculations. We estimated a total CAPEX of 19,097,673 USD for HTP production. As per Figure 9.8, the investment associated with the roasting reactor and heat exchanger contribute to 31% and 23%, respectively, of CAPEX. To estimate the OPEX, the price of run-of-mine syenite ore (before comminution steps) was assumed to be a negligible price of 3 USD ton^{-1} while a market price of 220 USD ton^{-1} was assumed for calcium nitrate. However, as per Figure 9.8, around 96% of the total OPEX is associated with the purchasing price of calcium nitrate resulting in 540,847,405 USD per annum. Accordingly, the normalized production cost of the HTP process over a plant lifetime of 20 years was estimated at around 282.2 USD per ton of HTP. That is, the assumed HTP process cannot be economically viable with respect to the market value of MOP.

Table 9.3: A detailed list of estimated CAPEX and OPEX of the HTP process.

CAPEX	
Roasting reactor ($)	5,850,372
Hammer mills ($)	1,336,196
Heat exchanger ($)	4,387,779
OSBL ($)	578,717
Engineering costs ($)	3,472,304
Contingency ($)	3,472,304
Discount rate (%)	10
Plant life (year)	20
Capital Recovery Factory (CRF)	0.117
CAPEX ($)	19,097,673
Annualized-CAPEX ($ year^{-1})	2,243,205
OPEX	
Variable OPEX	
Raw material	
Syenite ore ($ year^{-1})	4,119,061
Calcium nitrate ($ year^{-1})	540,847,405
Utilities ($ year^{-1})	13,759,200
Other OPEX (fixed variable costs) ($ year^{-1})	3,405,618
Operating labor	
Supervision	
Direct Salary overhead	
Maintenance	
Insurance and property taxes	
Rent and land	
OPEX ($ year^{-1})	562,131,284
Total cost of production ($ year^{-1})	564,374,489
Total cost of production ($ ton^{-1} HTP)	282.2

9.7 Return to lab scale from commercial scale

We determined from the preceding section that despite the promising performance of HTP as a potassium-based fertilizer, the roasting process suffers from the low K_2O content of syenite ore and high cost of calcium nitrate. In fact, as per reactions (R5) and (R6), the outlet gaseous product contains toxic and corrosive gases of NO and

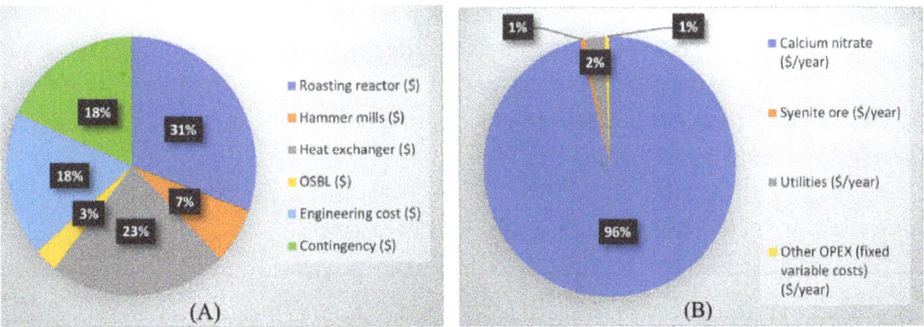

Figure 9.8: Estimated cost contribution of (A) CAPEX and (B) OPEX elements of the HTP process.

NO_2 mixed with oxygen. Storage of such a hazardous gas imposes additional costs and safety concerns to the process. On the other hand, calcium nitrate is a relatively costly and hazardous reagent. Its storage in a silo is risky as its exposure could cause health issues, such as headaches, dizziness, nausea, and vomiting. It is also a strong oxidizer that can promote the combustion of other species. Thereby, we decided to investigate the possibility of an in-site production of calcium nitrate. Since lab-scale calcium nitrate production has been reported in literature, we opted to apply a literature review to collect information to build a flowsheet of calcium nitrate production via published information.

Crystalline calcium nitrate is a highly hygroscopic crystal with a melting point around 45–50 °C. Calcium nitrate production through the continuous treatment of limestone with nitric acid within a laminar flow regime (with Reynolds number between 0.5 and 10) is a well-established technology. The laminar flow process is applied to maximize the reaction between calcium carbonate and calcium nitrate. In this process, firstly, the exothermic reaction (R7) occurs inside a packed bed reactor. Lime (CaO) can also be used to avoid producing CO_2, however, it is more expensive than limestone:

$$CaCO_3 + 2HNO_3 + 4H_2O \rightarrow Ca(NO_3)_2.4H_2O + H_2O + CO_2 \qquad \text{(R7)}$$

In another method, according to reaction (R8), nitric acid dissolves a phosphate material e.g., apatite mineral. Then, the produced calcium nitrate is crystalized and subsequently separated from the free phosphoric acid existing in the mother liquor.

$$Ca_3(PO_4)_2 + 6HNO_3 \rightarrow 2H_3PO_4 + 3Ca(NO_3)_2 \qquad \text{(R8)}$$

The calcium nitrate can also be produced by a reaction between calcium hydroxide and a mixture of nitrogen oxide gases. By exploiting a series of absorption reactions inside columns (at least two series of absorption reactors are required), calcium nitrate could be produced from lime slurry and nitrogen oxide gases. The general remarks of different pathways of calcium nitrate production are summarized in Table 9.4. Production via the reaction between limestone and nitric acid results in a product free of impurities, like ferrous components that are difficult to separate from water. Hence, we opted to consider this method for the in situ production of calcium nitrate in the HTP flowsheet.

Table 9.4: Overview of different calcium nitrate production methods.

Main raw materials	Remarks	Refs
Calcium carbonate and nitric acid	Through a simple mixing process and by carefully controlling the temperature and pH of the system, calcium carbonate is mixed with nitric acid. Ammonia is added for neutralization. The produced calcium nitrate is in close agreement with the required specifications.	[16]
Calcium cyanide, water, and solution of calcium nitrate	Calcium cyanamide is dissolved in water and mixed with the solution of calcium nitrate. Then, in the constant presence of excess air and by close contact of the produced mixture with nitrification bacteria, the calcium nitrate is produced.	[17]
Phosphate material and nitric acid	Phosphate materials, including phosphate rocks that contain $Ca_3(PO_4)_2$, could be dissolved with the nitric acid solution. From the mother liquor, calcium nitrate ($Ca(NO_3)_2{\cdot}4H_2O$) is crystalized followed by filtration.	[18, 19]
Phosphate material and nitric acid	By a combination of rudimentary stages, including subsequent heating, cooling, and filtration within a controlled pH and operating temperature, the production of calcium nitrate and calcium phosphate is rendered. The operating temperature of the process should not exceed 120 °C.	[20]
Dolomite and nitric acid	Dolomite is dissolved in nitric acid; calcium nitrate and magnesia are produced. A subsequent precipitating step is applied to separate the produced magnesia.	[21]
Lime slurry and nitrogen oxide gas	To produce calcium nitrate within a low amount of nitrate, a series of absorption reactions occurs. The lime slurry used in the process contains less than 20% of calcium hydroxide.	[22]

9.8 From lab scale to commercial scale: second cycle

The flowsheet in Figure 9.9 includes a conceptual regeneration (i.e., in situ production) of calcium nitrate using a cost-effective reagent, such as limestone ($CaCO_3$), which also has fewer hazardous risks than calcium nitrate. The calcium regeneration unit is composed of absorbing NO_x gases in a water tank (i.e., bubble cap absorption tower) to produce nitric acid. Water is fed to the absorption column with a set flowrate to ensure nitric acid with a 65% concentration is produced. Nitric acid is then fed to a subsequent reactor operating at ambient temperature where ground powders of limestone are continuously fed to produce calcium nitrate as per reaction (R7) [23]. The outlet reactor slurry enters a filter press to obtain a cake of about 10% moisture. The recovered water is recycled to the nitric acid production reactor. The cake of calcium nitrate is subsequently heated inside a furnace at around 105 °C to evaporate its non-bonded water content. The resulting calcium nitrate is recycled to the associated comminution unit to reach the desired particle size distribution.

9.8.1 Process simulation and technical evaluation

An advantage of the regeneration of calcium nitrate is the elimination of a storage tank for the mixture of NO_x and oxygen. A common practice for the production of nitric acid is the Ostwald process (Figure 9.10) [24] where ammonia is burned with oxygen through an exothermic reaction (R9) to produce NO, which is subsequently burned with oxygen to produce NO_2 (R10). NO_2 is then introduced to the bottom of an absorption column and water is introduced from the top to produce nitric acid (R11).

$$4NH_3 + 5O_2 \rightarrow 4NO + 6H_2O\left(\Delta H = -905.2\text{kJ mol}^{-1}\right) \tag{R9}$$

$$NO + O_2 \leftrightarrow 2NO_2 \ \left(\Delta H = -114\text{kJ mol}^{-1}\right) \tag{R10}$$

$$3NO_2 + H_2O \rightarrow 2HNO_3 + NO\left(\Delta H = -117\text{kJ mol}^{-1}\right) \tag{R11}$$

Measurement of product gases with a MKS multigas FTIR during our TGA tests revealed that the mass ratio of NO/NO_2 was about 14.5, under nitrogen atmosphere, whereas as per reactions R5 and R6, some oxygen is released during the reaction that should have oxidized NO to NO_2 (R10). Given the oxidation of NO to NO_2 is exothermic (i.e., $\Delta H = -114$ kJ mol^{-1}), NO tends to convert to NO_2 during cooling with sufficient time; however, it may require compression [24]. In fact, reaction R10 is usually at equilibrium. We carried out an equilibrium analysis of reaction R10 with the thermodynamic software FactSage© at pressures of 1–5 atm.

Figure 9.11 illustrates that at the onset of temperatures around 200 °C, the progress of reaction R10 should be in the right direction via decreasing the temperature, that is, production of NO_2, and the pressure does not have a significant impact on the equilibrium status. The dominant presence of NO in our tests could be because FTIR operates at 191 °C, and there was not sufficient time for the production of NO_2. This uncertainty requires further investigation at the lab to determine the kinetics of NO oxidation. Nevertheless, with respect to the thermodynamic analysis, we assumed the

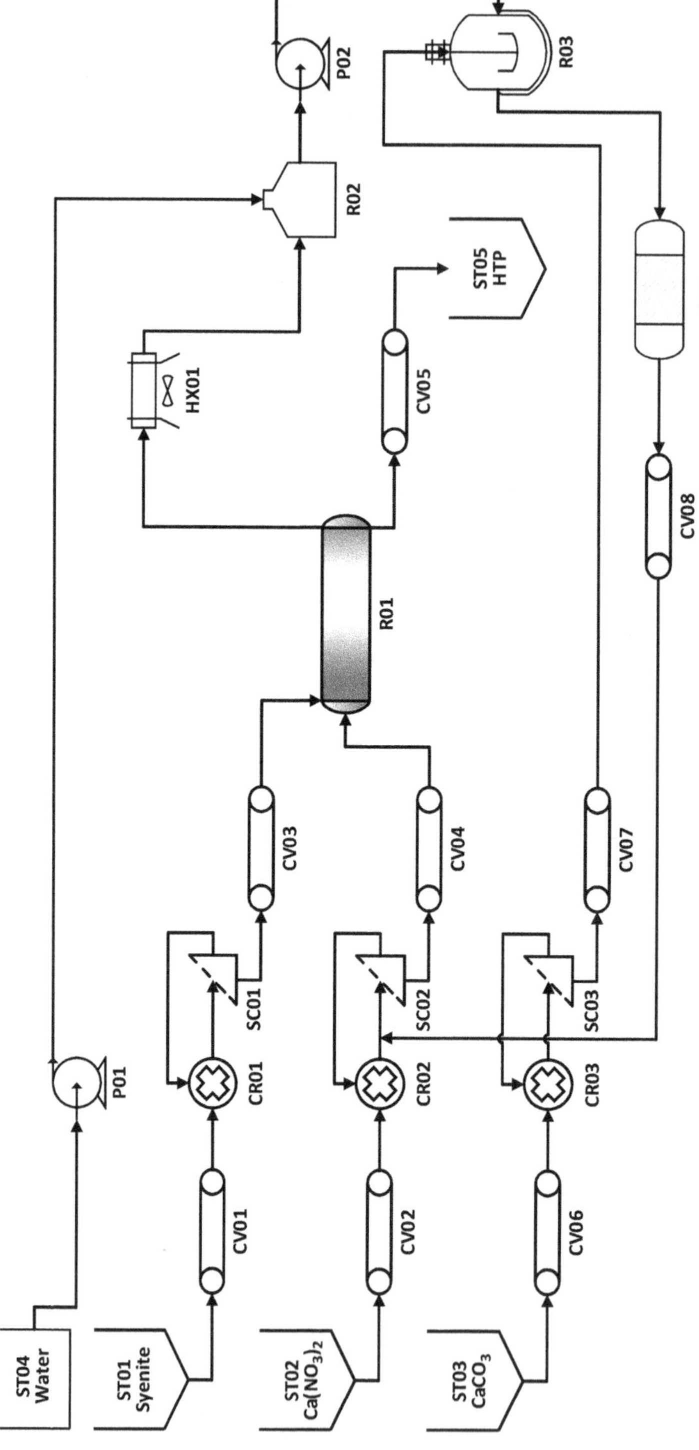

Figure 9.9: Process flowsheets for roasting syenite ore with in situ calcium nitrate regeneration.

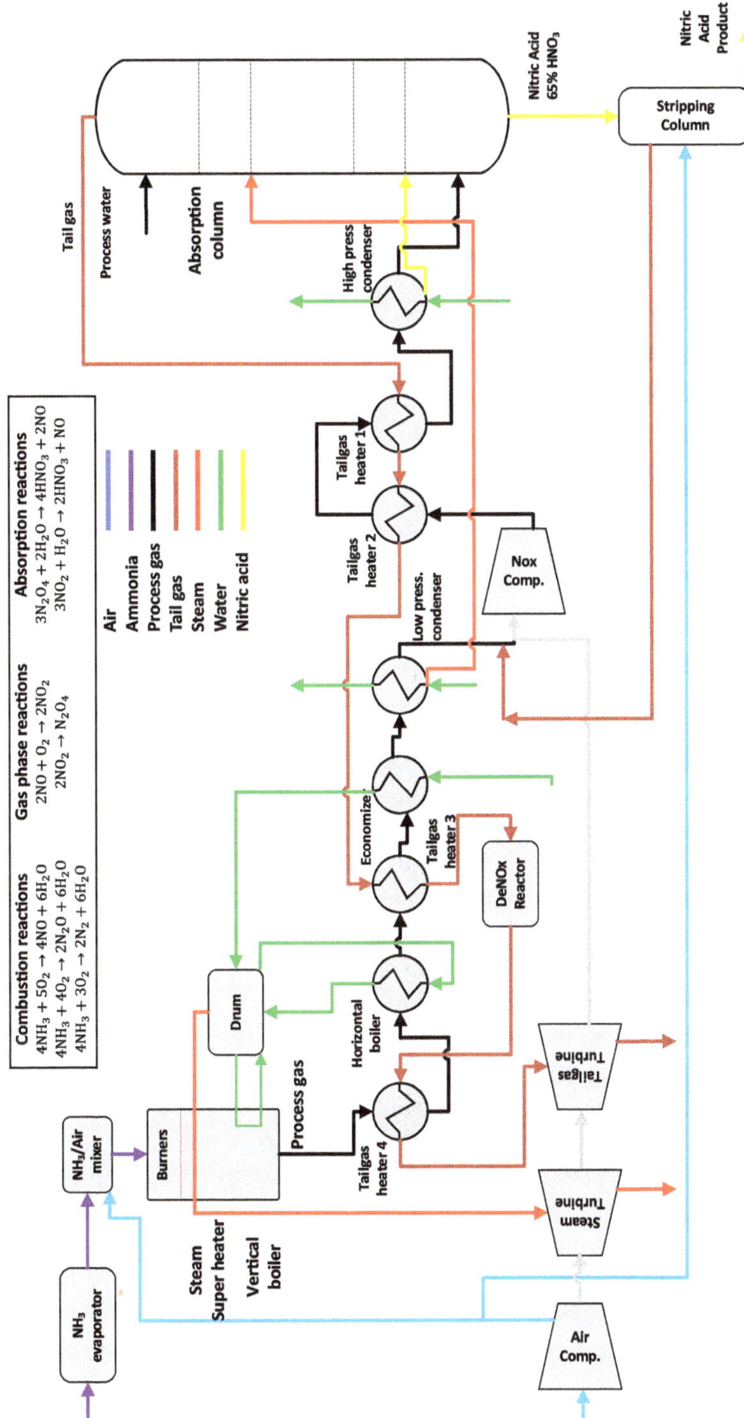

Figure 9.10: Ostwald process for the production of nitric acid from [24].

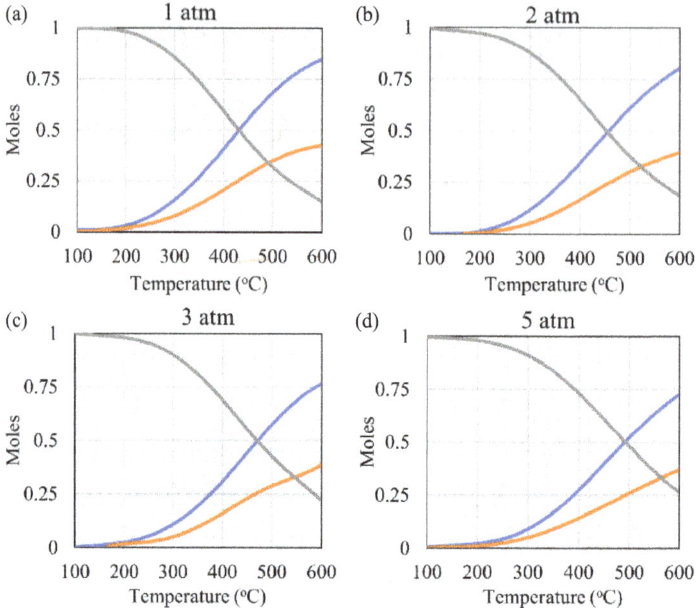

Figure 9.11: Thermodynamic equilibrium of NO + O_2 = NO_2 estimated by FactSageTM®.

outlet NO_x from the roasting reactor should be only NO_2 after passing through an air-cooled condenser. Thus, the cooled gas could be directly used for the production of nitric acid; that is, we opted to consider only the absorption column of the Ostwald process in the process flowsheet in Figure 9.9. There would be some NO in the outlet stream from the top of the absorption column according to reaction R11, which could be returned to the inlet of the absorption column.

A simulation with Aspen Plus™ revealed that about 3% of the total nitric acid would be captured by the outlet gases from the top of the absorption column. Thus, we considered a cooler to condense the trapped nitric acid as depicted in Figure 9.12(a). Further, we applied an economic evaluation of this option. However, as illustrated in Figure 9.13, despite an 80% recovery of nitric acid, which could be achieved at 2 °C, the annualized total production cost of HTP would increase by around 450,000 USD per annum, which is not economically promising. As an alternative, a neutralization tank was considered wherein ammonium hydroxide (NH_4OH) reacts with the nitric acid content of the outlet gas stream of the absorption column as depicted in Figure 9.12(b). As a result of neuralization, ammonium nitrate (NH_4NO_3) and water should be produced. The latter option of nitric acid recovery does not help to increase the nitric acid flowrate from the bottom of the absorption column (i.e., stream 14), but it is quite an economic option practiced in industry.

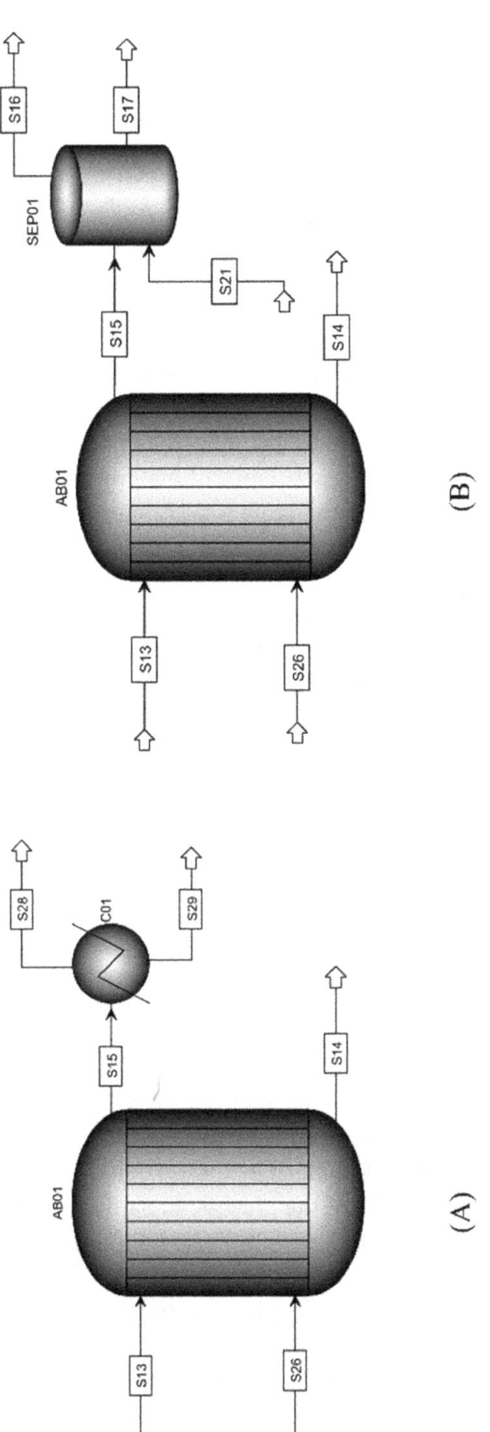

Figure 9.12: Simulation flowsheets of HNO_3 recovery from outlet gases of the absorption column via (a) a cooler; (b) neutralization with NH_4OH.

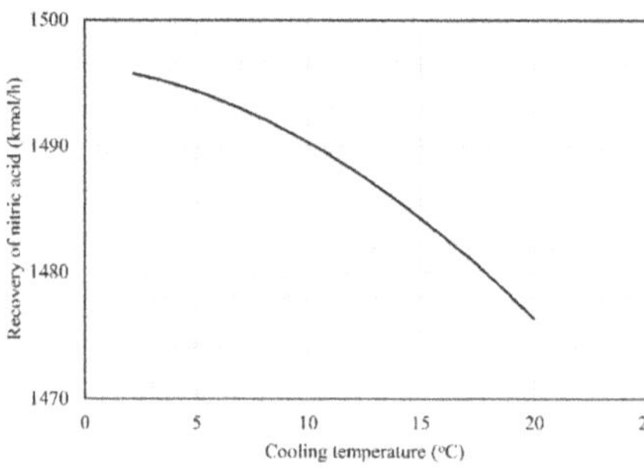

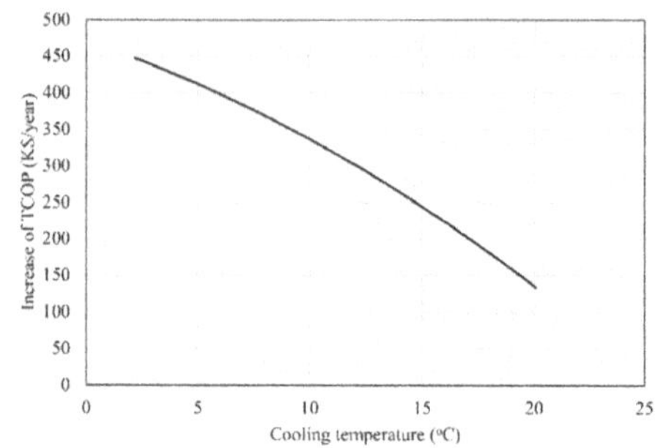

Figure 9.13: Impact of cooling on the recovery of HNO_3 and annualized total production cost of HTP.

We carried out a simulation of the roasting process, including the in situ regeneration of calcium nitrate to verify the heat and mass balance of the processes for the subsequent economic assessment. As depicted in Figure 9.14, calcium carbonate is continually reacted with nitric acid in a packed bed reactor (R02) within an exothermic reaction. As described in a preceding section, the reaction should occur in a laminar flow regime to maximize the reaction conversion. Followed by adding a flocculant (e.g., a cationic polymer) to the solution, the flocs are separated from the solution by a press filter. The collected cake (stream S02) is supposedly the produced calcium nitrate that should be transported to the hammer grinder H01 after the drying step. The acidic liquid stream (S22) from the press filter is neutralized with ammonium hydroxide in a separation tank (SEP02), where water is separated from ammonium nitrate and recycled to the inlet of absorption column.

9.8.2 Economic evaluation

Table 9.5 elucidates specifications of the main equipment introduced in the plant with in situ regeneration of calcium nitrate. Accordingly, Table 9.6 presents an economic analysis of the roasting process assuming a 40 USD ton^{-1} cost of calcium carbonate raw material. The contribution of each process parameter on CAPEX and OPEX is illustrated in Figure 9.15(a, b). Similar to economic calculations in Table 9.3, the OPEX is much more than the estimated annualized CAPEX of the process. In particular, the cost of calcium carbonate accounts for 74% of OPEX and 71% of production costs, even though it is much cheaper than calcium nitrate.

Our economic estimations revealed that the cost of calcium nitrate production would be 33.5 USD per ton. As a result, a HTP total production cost of 44.9 USD per ton was estimated. Following the discussion in Section 10.6.2 on the competitive price of MOP, the estimated production cost is still above 43 USD per ton of HTP indicating the process is not yet economically viable with respect to the fact that

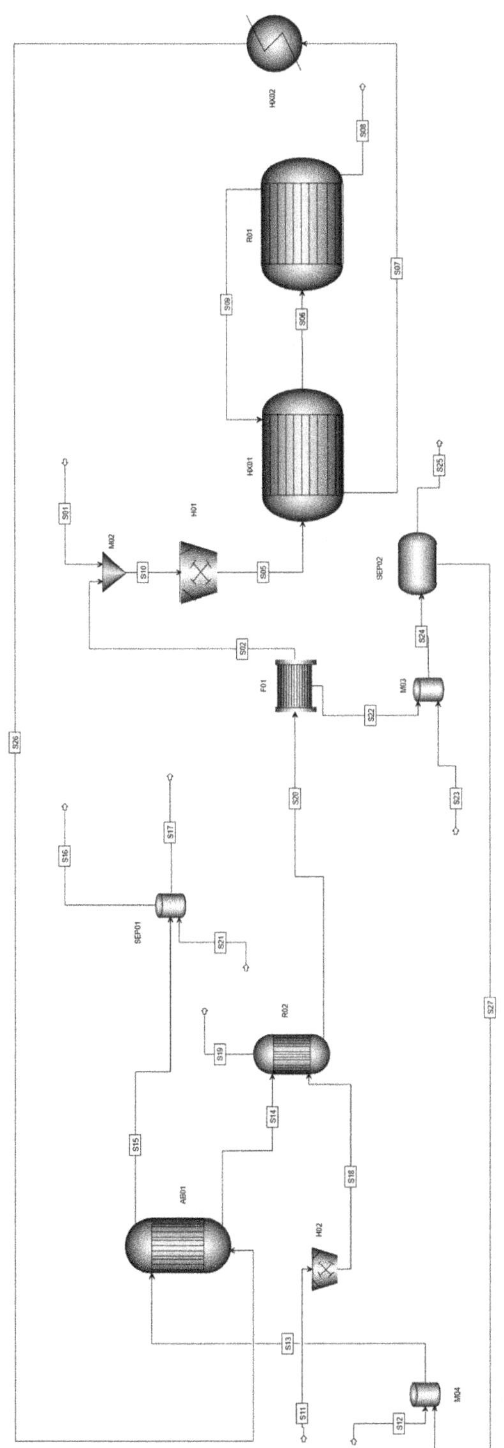

Figure 9.14: Simulation flowsheet of roasting process with in situ regeneration of calcium nitrate.

Table 9.5: Specifications of plant units with in situ regeneration of calcium nitrate.

Unit	Basic design remarks
Roasting reactor (R01)	Reactor type: rotary kiln reactor Retention time: 90 min Volume of the reactor: 100 m^3
Nitric acid production reactor (AB01)	Reactor type: bubble cap absorber tower No of trays: 22 Plate spacing: 60 cm
Calcium nitrate reactor (R02)	Reactor type: packed-bed reactor Reactor volume:30 m^3
Comminution (H01, H02)	Type: reversible hammer mil
Heat exchanger (HX01)	Exchanger type: rotary kiln dryer Retention time: 90 min Inlet temperature of the hot stream (S09): 540 °C Outlet temperature of cold stream (S06): 150 °C Heat exchange area:130 m^2
Press filter (F01)	Type: rotary drum filter Filtration area: 25 m^2

Table 9.6: A detailed list of estimated CAPEX and OPEX of the HTP process with in situ regeneration of calcium nitrate.

CAPEX	
Roasting reactor, absorber, calcium nitrate reactor ($)	14,023,177
Hammer mills and filter press ($)	1,938,949
Heat exchanger ($)	4,387,779
OSBL ($)	1,017,495
Engineering cost ($)	6,104,972
Contingency ($)	6,104,972
Discount rate (%)	10%
Plant life (year)	20
Capital Recovery Factory (CRF)	0.117
CAPEX ($)	33,577,343
Annualized-CAPEX ($ $year^{-1}$)	3,943,982
OPEX	
Variable OPEX	
Raw material	
Calcium carbonate ($ $year^{-1}$)	63,918,330
Syenite ore ($ $year^{-1}$)	4,119,061
Utilities ($ $year^{-1}$)	14,447,160

Table 9.6 (continued)

CAPEX	
Other OPEX (fixed variable costs) (\$ year^{-1})	3,405,618
Operating labor	
Supervision	
Direct Salary overhead	
Maintenance	
Insurance and property taxes	
Rent and land	
OPEX (\$ year^{-1})	85,890,168
Total cost of production (\$ year^{-1})	89,834,151
Total cost of production (\$ ton^{-1} HTP)	44.9

this assessment was based on the simplified flowsheet in Figure 9.14, and likely unrealistic assumed costs. However, the idea of in situ regeneration of calcium nitrate seems to be a very promising option depending on the market conditions and the price of operation parameters.

A sensitivity analysis was thus conducted on the independent parameters of CAPEX and OPEX to determine their impact on HTP production costs, assuming their price could vary within a margin of ±50%. Figure 9.15(c) illustrates that the cost of calcium carbonate is the critical uncertainty that could make the process of HTP production profitable. Therefore, to investigate its impact on the profitability of the plant with in situ regeneration of calcium nitrate, a cumulative discounted cash flow (CDCF) diagram was developed. As depicted in Figure 9.16, a lower calcium carbonate price reduces the production cost of HTP, which consequently leads to the enhancement of the discounted net present value (DNPV) of the plant. A calcium carbonate price as low as 20 USD ton^{-1} would result in a discounted payback period (DPBP) of less than 5 years, which means the DNPV after 20 years would reach \$101 M. On the other hand, the process would not be economical if the calcium carbonate price rose above 30 USD ton^{-1}.

9.9 Conclusion

In this chapter, the feasibility of commercial scale production of HYP from KFS was investigated using the iterative scale-up approach. Lab-scale data revealed that the produced HTP was a promising fertilizer to grow crops because it operates at a temperature much lower than that in similar high temperature processes using reagents, such as gypsum, calcite. On the other hand, applying the iterative scale-up approach from lab scale to commercial scale revealed that the scale-up process is not

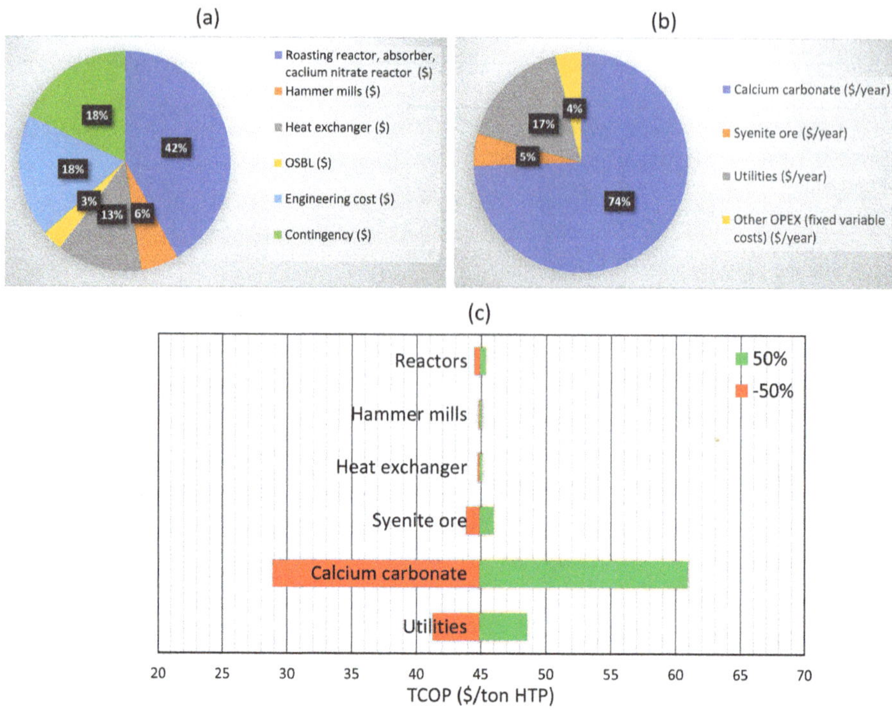

Figure 9.15: (a) Contribution of the items included in CAPEX, (b) contribution of the items included in OPEX, (c) sensitivity analysis of the plant with in situ regeneration of calcium carbonate and ±50% variations of independent parameters of CAPEX and OPEX.

economically viable at present due not only to the cost of the HTP process, but also limitations caused by the market. The economic evaluation of the process with in situ regeneration of calcium nitrate presented a potential pathway to a commercial scale process. However, based on results extracted by the iterative scale-up approach, it is still too early to investigate the process operation at a pilot scale, given the estimated cost was still above the criteria. For the in situ regeneration of the calcium nitrate unit, the process was only based on a flowsheet developed by extracting information from the literature and our simulation data. Thus, a new series of lab-scale tests is required to verify the feasibility of calcium nitrate regeneration after a systematic kinetic study. Based on the iterative scale-up approach introduced in this section, the scale-up of the HTP process could become feasible if the price of the fertilizer dramatically spiked due to demand or calcium nitrate was replaced with a cost-effective raw material, like calcium carbonate.

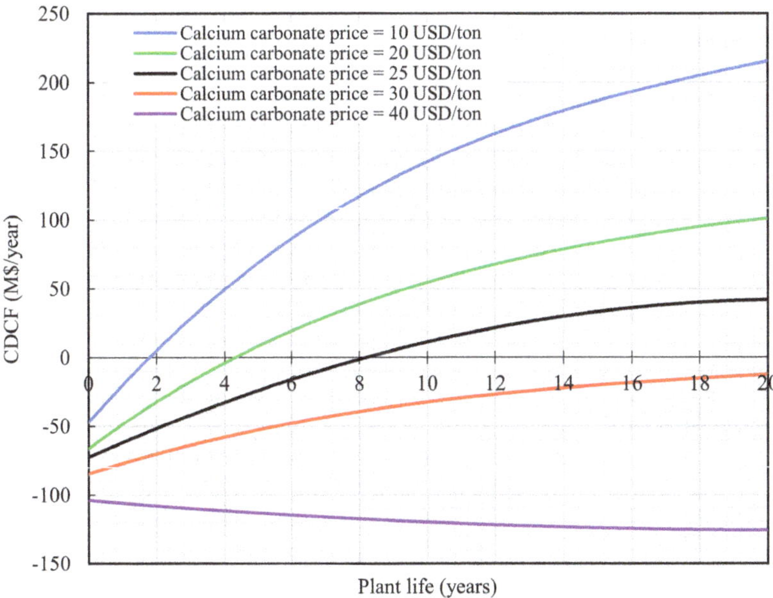

Figure 9.16: CDCF diagram of the plant with in situ regeneration of calcium nitrate (interest rate = 10%, plant life = 20 years, HTP selling price = 43 USD ton^{-1}).

References

[1] OECD/FAO, OECD-FAO agricultural outlook 2020–2029, Paris, 2020.

[2] T. Skorina and A. Allanore, Aqueous alteration of potassium-bearing aluminosilicate minerals: from mechanism to processing, Green Chemistry, 17(4), 2123–2136, 2015.

[3] D. Ciceri, M. De Oliveira, and A. Allanore, Potassium fertilizer via hydrothermal alteration of K-feldspar ore, Green Chemistry, 19(21), 5187–5202, 2017.

[4] M. T. Serdengeçti, et al., The Correlation of Roasting Conditions in Selective Potassium Extraction from K-Feldspar Ore, Minerals, 9(2), 109, 2019.

[5] A. K. Bakken, et al., Crushed rocks and mine tailings applied as K fertilizers on grassland, Nutrient Cycling in Agroecosystems, 56(1), 53–57, 2000.

[6] A. K. Bakken, H. Gautneb, and K. Myhr, The potential of crushed rocks and mine tailings as slow-releasing K fertilizers assessed by intensive cropping with Italian ryegrass in different soil types, Nutrient Cycling in Agroecosystems, 47(1), 41–48, 1996.

[7] A. K. Bakken, H. Gautneb, and K. Myhr, Plant available potassium in rocks and mine tailings with biotite, nepheline and K-feldspar as K-bearing minerals, Acta Agriculturae Scandinavica, Section B – Soil & Plant Science, 47(3), 129–134, 1997.

[8] C. Wang, et al., Mineralization of CO_2 Using Natural K-Feldspar and Industrial Solid Waste to Produce Soluble Potassium, Industrial & Engineering Chemistry Research, 53(19), 7971–7978, 2014.

[9] S. Su, et al., Preparation of potassium sulfate and zeolite NaA from K-feldspar by a novel hydrothermal process, International Journal of Mineral Processing, 155, 130–135, 2016.

[10] Y. Zhong, et al., Recovery of potassium from K-feldspar by thermal decomposition with flue gas desulfurization gypsum and CaCO3: analysis of mechanism and kinetics, Energy & Fuels, 31(1), 699–707, 2016.
[11] S. K. Liu, et al., Hydrothermal decomposition of potassium feldspar under alkaline conditions, RSC Advances, 5(113), 93301–93309, 2015.
[12] M. Hongwen, et al., 20 years advances in preparation of potassium salts from potassic rocks: a review, Acta Geologica Sinica-English Edition, 89(6), 2058–2071, 2015.
[13] J. Ma, et al., The leaching kinetics of K-feldspar in sulfuric acid with the aid of ultrasound, Ultrasonics Sonochemistry, 35, 304–312, 2017.
[14] K. Song, et al., Research progress on extracting potassium and preparing compound fertilizer from potassium feldspar, The Chinese Journal of Process Engineering, 18(2), 241–257, 2018.
[15] G. S. Towler and K. Ray, Chemical engineering design – principles, practice and economics of plant and process design, Elsevier, Editor, 307–328, Vol. 2008.
[16] M. Rodriguez and H. J. I. J. O. C. R. Zea, Evaluation of a synthesis process for the production of calcium nitrate liquid fertilizer., 7(4), 1960–1965, 2015.
[17] C. T. Thorssell and H. L. R. Lunden, Process for the production of calcium nitrate, Google Patents, 1918.
[18] J. Erling, Process of producing calcium nitrate and ammonium salts from phosphate rock and like phosphate material, Google Patents, 1932.
[19] I. V. F. Moldovan, M. Suciu, and E. Tomescu, Process for crystallizing calcium nitrate, Google Patents, 1977.
[20] M. H. Plusje, Production of calcium phosphates and calcium nitrate, Google Patents, 1952.
[21] K. Otto, Producing calcium nitrate from dolomite while simultaneously obtaining magnesia, Google Patents, 1933.
[22] S. L. Bean, et al., Manufacture of calcium nitrite solutions with low nitrate content, Google Patents, 1981.
[23] A. Nawrocki and R. Olszewski, Method for Calcium Nitrate Production 2006.
[24] C. A. Grande, et al., Process Intensification in Nitric Acid Plants by Catalytic Oxidation of Nitric Oxide, Industrial & Engineering Chemistry Research, 57(31), 10180–10186, 2018.

Mahdi Yazdanpanah, Jamal Chaouki

10 Case study V: Lactide production process development

Abstract: This chapter provides a practical example where iterative scale-up methodology is applied to develop a novel lactide production concept into a full industrial-scale process concept and evaluate it. Preliminary laboratory results were used to develop a reactor design. A process flow diagram and the corresponding process simulation were then developed step by step. Individual operation units were designed in the next step. The process was optimized using methods, such as heat integration, sensibility analysis, and intensification. Finally, operating and capital expenditures were estimated to compare the novel concept with conventional technologies and identify the main cost drivers for further optimization. The results were then used to design a continuous pilot plant for process demonstration at lab scale. The iterative nature of the current scale-up methodology permits continuous improvement and optimization of the technology under development.

Keywords: industrialization, scale-up, process development, polylactic acid, PLA, technoeconomic assessment, TEA

10.1 Introduction

This chapter illustrates an example of the iterative scale-up to industrialization and the development of a novel lactide production method [1] in the polylactic acid (**PLA**) process. Two main questions need to be answered while evaluating a new idea; first is its industrial feasibility and second is its production cost. The current case study shows how the iterative scale-up methodology can answer these questions while developing the novel technology.

PLA is a bio-based and biodegradable polymer derived from renewable resources, such as corn starch and sugarcane. PLA is one of the most used biopolymers with a fast growth rate worldwide [2]. It can be produced through two main routes: an indirect route via a lactide intermediate, and a direct polymerization by polycondensation. The indirect route is the most common in the industry. Indirect PLA production involves an important intermediate step of converting lactic acid (LA) to its cyclic dimer lactide (LD) as a monomer of PLA polymerization.

Mahdi Yazdanpanah, Research and Development, TotalEnergies, Harfleur, France, e-mail: mahdi.yazdanpanah@totalenergies.com
Jamal Chaouki, Department of Chemical Engineering, Polytechnique Montréal, 2500 Chemin de Polytechnique, Montréal, Québec, H3T 1J4, Canada, e-mail: jamal.chaouki@polymtl.ca

https://doi.org/10.1515/9783110713985-010

The conventional PLA production passes through three principal steps as illustrated in Figure 10.1. The first step involves the fermentation of glucose or similar sugars into LA. A series of purification steps is employed after fermentation to separate the biomasses and purify the LA. The produced LA has a concentration of about 12 wt%. A series of distillation steps is then used to increase the LA concentration up to 90 wt% (polymer grade) in case long transportation is required between the fermentation and PLA production sites. LD production from LA is the second principal step. This conversion consists of two steps, oligomerization and cyclization. In addition, a purification step is employed to separate L-Lactide from other isomers. The LD is then sent to the polymerization step to produce PLA.

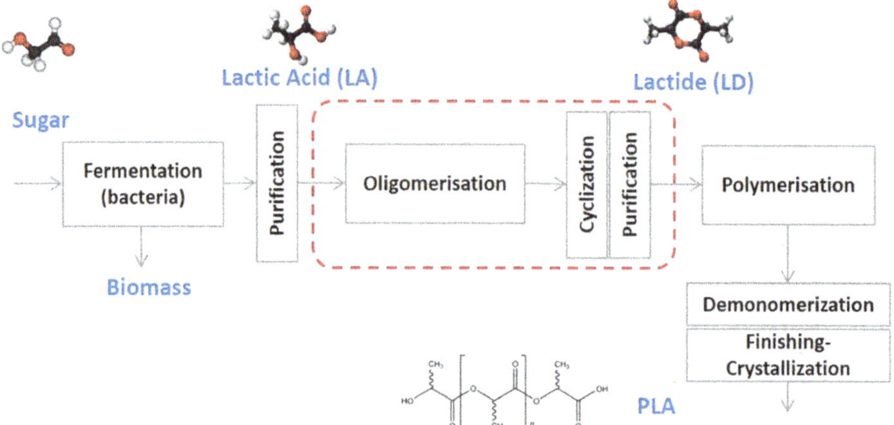

Figure 10.1: Main steps in conventional PLA production.

10.2 Concept generation (TRL1): novel lactide production method

An alternative LD production pathway was innovated [1] consisting of the use of a size-selective zeolite catalyst, which limits access to the catalyst's active site only to lactic acid and prevents larger oligomers from accessing it. Therefore, it enhances the LD production and reduces the probability of bigger oligomer production. In addition, a solvent is used to dissolve the produced LD, to minimize contact between LD and water to avoid its hydrolysis. In addition, it enhances the reaction by extracting LDs from the solution. The simplified reaction pathway is illustrated in Figure 10.2. LA is first converted into di-lactic acid (L2A). L2A has then two parallel conversion paths to form a LD or bigger LA oligomers, such as tri-lactic acid. These reactions are reversible, and a water molecule is produced as a co-product either in LD production or oligomerization. Therefore, water can hydrolyze LD or oligomers back into L2A and LA.

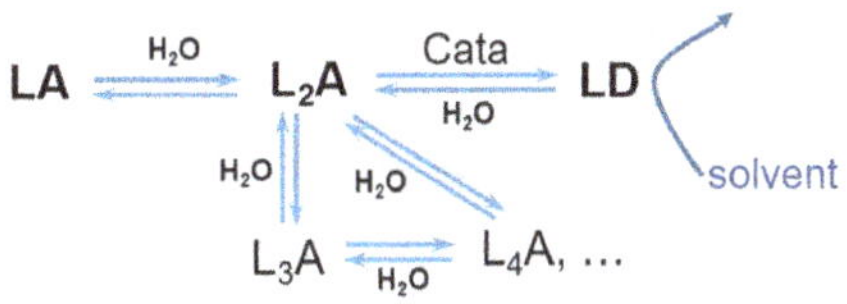

Figure 10.2: Simplified reaction scheme of lactide (LD) production using novel zeolite catalyst [1].

10.3 Concept validation

The first development step involves concept validation in small laboratory-scale setups mostly in batch mode. A reactor system was developed by Van Wouwe [3] to demonstrate the novel LD production concept with two operation modes: batch and semibatch modes. The batch reactor is initially filled with LA solution, catalyst, and solvent. It is then placed in a thermal bath and continuously mixed. The operating temperature of the system is the boiling point of the solvent. The vapors exit the reactor into a condenser at the gas exit of the system. The vapor phase is composed of water and solvent, which are separated in the bottom of the condenser by decantation. The evaporated solvent is continuously replaced by fresh solvent to maintain the level of liquid in the reactor. The experimental results derived from this study demonstrated the thermochemical feasibility of the developed concept [3] and made it possible to attain proof of the basic scientific concept. The experimental set-up described in this section was also used as a tool for further technology development.

10.4 Process development: iterative scale-up

Iterative scale-up methodology is used to develop a process scheme to evaluate industrial feasibility and commercial performance of the novel LD production method. Figure 10.3 presents the principal steps followed during this phase, the results of which will be used in the next phase to demonstrate the technology in a continuous lab-scale pilot plant. The concept is directly projected to industrial scale, which is used as a reference to define intermediate scale-up steps.

The first step is the definition of the design basis and benchmark scale. A preliminary reactor concept and design is defined. A preliminary process flow diagram (PFD) is then developed around the defined reactor by adding the main process units, step by step. This primary PFD helped evaluate each of the process constituents, including solvent, catalyst, and heat requirements. Process simulation is the next step to develop heat and material balances. The simulation results are used to design various process units, such as the reactor, heat exchangers, and separation units. The simplified design permitted to evaluate industrial feasibility of each step and its readiness. The main equipment and utility consumption lists are developed based on the design data. These lists are used as a basis for CAPEX and OPEX estimation and economic evaluation.

These development steps are employed in an iterative mode. For instance, after PFD development, there may be a need to return to the reactor design step and improve its design. Various critical process challenges are identified in each step and novel ideas will be identified to overcome each challenge. Therefore, the process development is dynamic with continuous improvement. In the current example, different questions were identified, which needed further experimental tests carried out in the experimental set-up described above.

Each of the steps used for process development are briefly described in the following sections with illustrative examples.

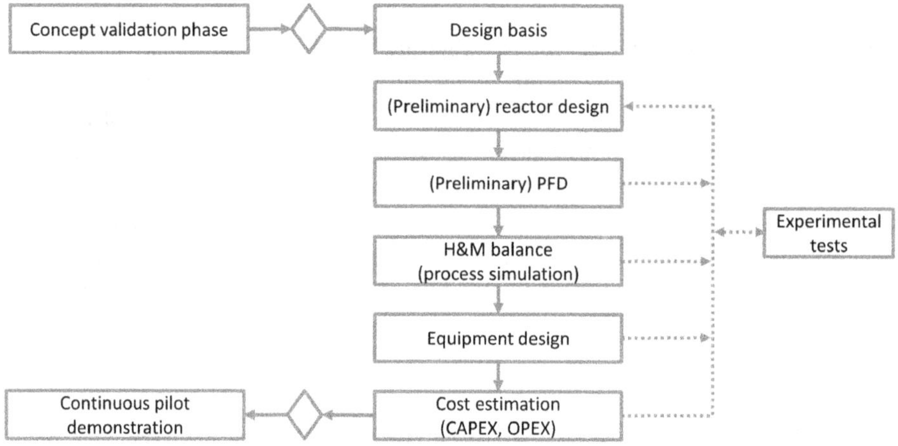

Figure 10.3: Process development flow chart employed in this study. Optimization and integration of the process is studied based on the process simulation and design.

10.4.1 Design basis

The design basis defines the basic information and data necessary for process development. It covers important information, such as unit capacity, product specifications, feed specifications, applicable unit of measurements (SI), geographical location, lifetime of the plant, battery limits, and unit availability. Table 10.1 lists the basis used for the current case study.

Table 10.1: Design basis example used for the lactide production plant.

Feed	90% LA
Product	60,000 ton $year^{-1}$ LD-L > 99.5%
Location	Western Europe
Lifetime	20 years
Availability	95%

10.4.2 Preliminary reactor characterization

Reaction kinetics is an important element to be developed for reactor design. A kinetic model was developed and validated based on the experimental results. The purpose was to achieve a simple but reliable kinetic model in the target operation range of the system to be used in the steady state process simulation carried out by ASPEN Plus.

The final reaction model used in this study is presented in Table 10.2 and compared with experimental results in Figure 10.4. The kinetic model needs to be improved as process development advances based on the objective of each step. For instance, activation energies need to be independently validated for heat integration and optimization study. A stirred-tank reactor model is currently considered for the simulation. More sophisticated hydrodynamic models may be required in case that hydrodynamic related diffusion is identified as the limiting step in the next development steps.

Table 10.2: Reaction scheme used for the reactor modeling.

Reaction	k	Reaction
1	*k*1	2 LA → L2A + H_2O
2	*k*2	L2A → LD + H_2O
3	*k*3	3 L2A → 2 L3A + H_2O (L3A representing all LxD)
4	*k*4	2 L3A + H2O → 3 L_2A

10.4.3 Preliminary Process flow diagram (PFD)

The preliminary PFD was developed based on the basic reaction information illustrated in Figure 10.5. The main operation units and streams were then added step by step to attain a preliminary PFD with essential unit operations as illustrated in Figure 10.6. The operating units were added to the reactor gradually to attain products according to specification. It is recommended to start with basic blocks, which can be used to develop the first simulation to ensure consistency of results. The

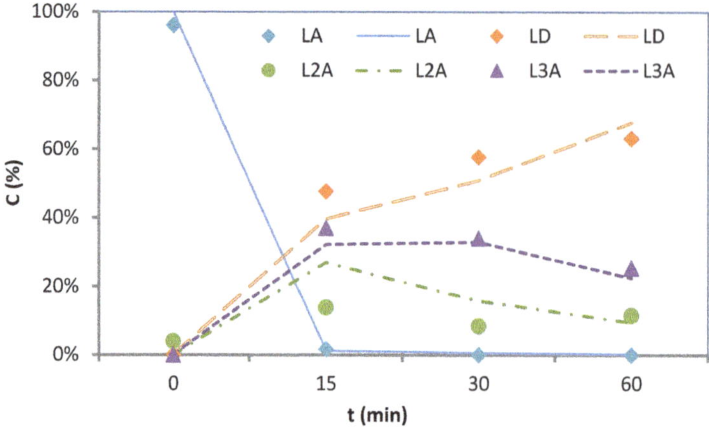

Figure 10.4: Comparison of the kinetic model (lines) and the experimental results (points) [3].

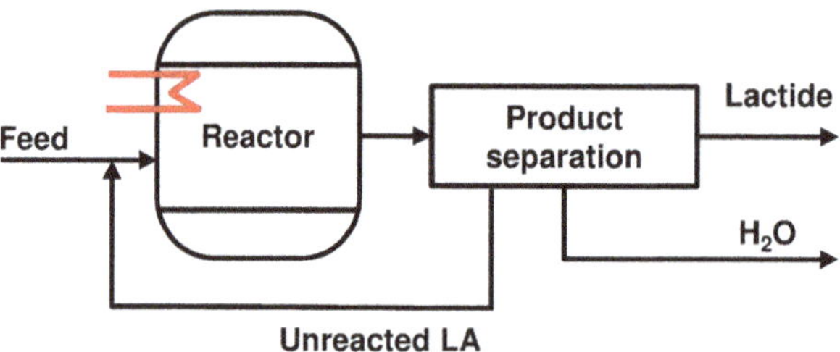

Figure 10.5: Basic PFD of the process including reaction, separation, and recycling.

scheme can be gradually improved by specifying conventional technologies for each block. In the subsequent iterations, the process can then be improved by integrating design and the use of potentially more advanced technologies to improve performance or intensify the system. Therefore, PFD development and process simulation are closely integrated and improve gradually. The preliminary scheme illustrated in Figure 10.6 was then used as the basis for simulation and the proposal of new experimental tests to develop the process aspects, such as separation, purification, reactors, and heating.

The employed unit operations and streams must be analyzed in order to define the required specifications of all components and units necessary for simulation, such as solvent and catalyst.

Solvent plays an important role in this process as described earlier. From a reaction point of view, the selected solvent should have a maximum solubility for LD and minimum solubility for water. In addition, the solvent should be nonmiscible with water to be able to separate it by decantation. It should preferentially have a high boiling point to enhance the reaction rate. Some general process aspects should be

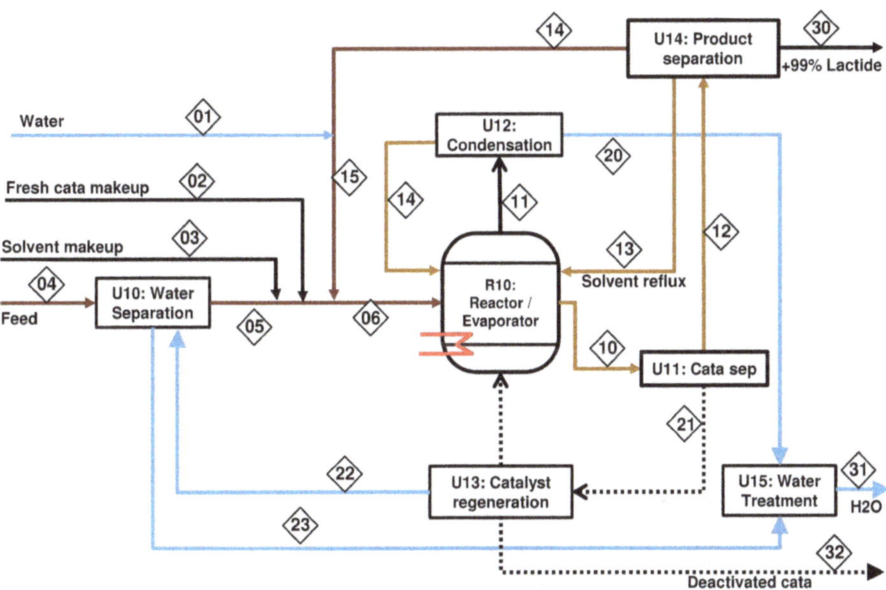

Figure 10.6: A preliminary PFD including main operation units.

considered, such as cost, availability, health, safety and environmental risks, and chemical stability. Various aromatic and nonaromatic solvents were experimentally tested. Decane and isobutylebenzene were selected as reference aromatic and nonaromatic solvents in this study. Various tests were then carried out to compare these two solvents and select the reference solvent for the process development [3].

Blank tests were carried out to evaluate solvent stability at reaction conditions without reactants. Therefore, the solvent was in contact with catalysts at its boiling point with continuous reflux in the reactor. Gas chromatography was used to analyze the solvent composition before and after 24 h of blank testing. Various problems may rise in the case of solvent decomposition. In the case of operations under high pressure, the formation of new components may cause the formation of a gas phase, which can result in various malfunctions in the system requiring a gas separation. A probable problem is the possible formation of toxic components, which may require specific disposal procedures. Some of the formed components may be poisonous for the polymerization catalyst in the PLA formation reactor downstream of the LD production unit. In addition, solvent make-up will increase in this case, which means higher operating costs. Flaring may be required to eliminate the newly formed species. This has an additional cost and environmental impacts. A detailed comparison was carried out between the selected solvents based on the abovementioned characteristics resulting in the selection of decane as the reference solvent.

Water and decane form a heterogeneous azeotropic mixture (not solution). This immiscible mixture forms a Heteroazeotrope where the gas phase is in equilibrium with two separate liquid phases. To be able to operate at higher temperatures, the system needs to be under pressure. This makes it possible to operate the system in the liquid phase and avoid complex azeotropic distillation. In this study a pressure of 10 bara was chosen as the operating pressure of the system.

The catalyst is another important component of the current process. Commercial H-beta zeolite is used in this study based on the experimental results [1, 3]. The SEM image of this catalyst is illustrated in Figure 10.7 for two different mean sizes. Various parameters were analyzed for the catalysts, particle size being an important parameter. Larger particles favor particle separation and handling. However, the particle size should be limited to avoid diffusion limitation and ensure that the system is in a reaction-controlled regime. The commercial zeolites have a mean size of 10 and 200 nm. Larger particles (1–2 μm) were produced by growing the particles using calcinations.

Figure 10.7: SEM images of calcinated zeolite catalysts [3].

Experimental tests were then carried out to evaluate the impact of catalyst size on the reaction rate as illustrated in Figure 10.8. There was no difference observed between the two sizes, showing that diffusion is not an issue in the tested range. Bigger catalysts are therefore selected as the reference particle size in this study. The optimization of catalyst size is an example of improvement from an industrial and process point of view without negative impact on the reaction rate.

Another important parameter is catalyst deactivation by either temporary or permanent mechanisms. Temporary deactivation is due to the formation of oligomers inside the catalyst pores, where they will be trapped and decrease catalyst activity by imposing diffusion limitations or isolating active sites. Permanent deactivation is due to the destruction of the active sites.

Temporary catalyst deactivation was analyzed by following LD yield during long-term tests as illustrated in Figure 10.9. The catalyst is highly active in the first hour of

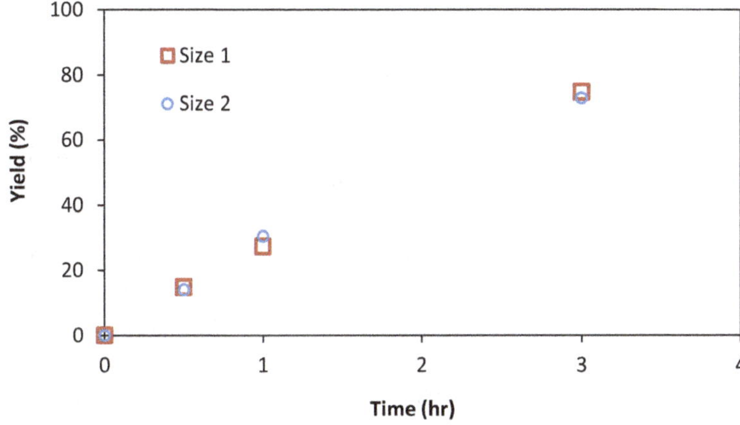

Figure 10.8: Impact of catalyst size on the LD production.

reaction. Its activity decreases by about 20% at 2 h of reaction compared to 1 h of reaction. Above this limit, the activity remains relatively stable. Therefore, there are two options, either limiting the particle residence time in the reactor below 1 h to maximize the catalyst reactivity or increasing the catalyst residence time in the reactor above 2 h while increasing the catalyst quantity in the reactor by 20% to compensate for the reactivity loss.

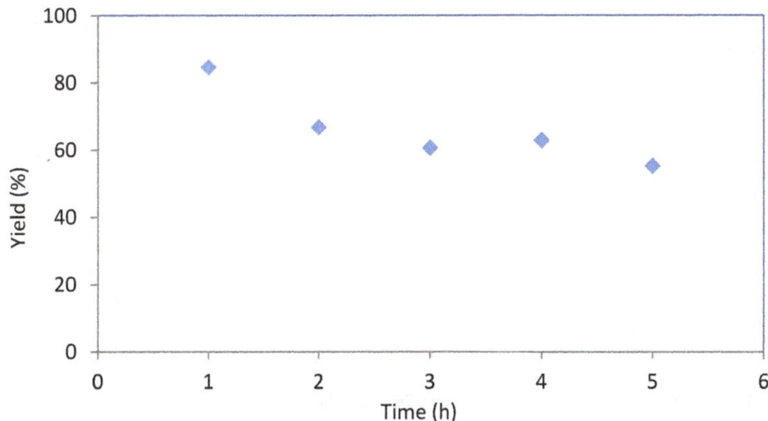

Figure 10.9: LD yield decrease as a function of time.

A catalyst regeneration unit was added to the system to overcome temporary deactivation. This unit has important impacts on the reactor and overall process design and, hence, on the costs (OPEX and CAPEX). Two possible methods were analyzed in this study. The first method is the calcination of catalysts to evaporate and burn out the hydrocarbons trapped inside catalysts. The second technique is catalyst regeneration by hydrolysis.

Detailed studies were carried out to compare the efficiency of catalyst regeneration using these methods. A similar reactivation performance was obtained from both techniques. Therefore, hydrolysis was selected as the best adapted technology. By using hydrolysis, the deposited hydrocarbons inside catalysts will be decomposed back to L2A and LA and can be sent back to the reactor to produce LD. Therefore, LA and solvent will be conserved inside the unit and reused after regeneration. This regeneration results in zero flaring, which reduces the negative economic and environmental impacts. In addition, catalysts can be easily regenerated by using the water produced during the reaction. There is no additional heat required since the reaction temperature is well above the required regeneration temperature.

Permanent catalyst deactivation is another important point to be considered in the process design. An average effective lifetime of 3 months was considered for the catalyst in the current study based on similar catalysts' lifetime in the industry. This assumption is acceptable for the current level of the study. However, it needs to be studied in more detail in the next development phases. Evaluation of the permanent degradation mechanism is more difficult than temporary deactivation depending on the nature of the catalyst. Long-term tests may be designed to measure the real deactivation mechanism. However, long tests are difficult to attain in laboratory particularly in early development stages without a continuous pilot. Specific blank tests can be designed with intensified degradation mechanisms (e.g., higher temperature, higher water content) to increase the degradation rate. The real catalyst lifetime will then be estimated based on the intensified degradation test results.

This section illustrated the approach to develop a simplified PFD. It should be noted that the PFD starts from a very simple concept and is modified by studying each item mentioned above. Selection of the main items depends on the nature of the process and may vary from one technology to another. However, the methodology of the development is similar between different cases.

10.4.4 Process simulation

Material and energy balance for the whole process is the next development step. The simulation can be done by different available simulators, such as Aspen Plus, Aspen Hysys, or Pro/II. The selection of the software depends on different factors, such as adaptability of the libraries to the application or available experience. Aspen Plus was selected for this study thanks to its wide database, including elements, such as LA, LD, L2A, and L3A. The NRTL-RK method is used for thermodynamic properties calculation. Solubility data of different elements in water and decane was not available and therefore carried out experimentally. Water solubility in decane was calculated according to the literature [4, 5]. The water solubility is an important parameter as the dissolved water can hydrolyze lactide dissolved in the decane. A more detailed thermodynamic study is required for the next process development step.

Concentrated industrial-grade LA was used as feed. The LA content of feed does not have any impact on the current design and can be optimized based on the available feed. The concentrations of different oligomers in the feed were calculated based on reference [6] and listed in Table 10.3.

Table 10.3: Properties of the feed stream.

Property	Unit	Stream 4 (Feed)
Flow rate	kg h^{-1}	10,120
T	°C	25
P	kPa	101
Concentrations		
LA	%	60.1
L2A	%	19.8
L3A	%	6.8
DLD	%	0.1
H_2O	%	13.3

The simulation was started from a simple design, including a reactor with a simple entry and heating system. The system was gradually completed by adding other sections, such as heat recovery, product separation, water separation, recycling, and regeneration sections. The addition of each new section requires a re-evaluation of the total process integrity and sometimes leads to modifications of previously developed systems. Figures 10.10 and 10.11 illustrate examples of intermediate designs where the process is gradually improved to attain the optimized system presented in Figure 10.15. This improvement is attained in an iterative approach as illustrated in Figure 10.3 in close relation with the equipment design, integration, and optimization as described in the following sections.

10.4.5 Process development and equipment design

This section illustrates examples of equipment designs, which make it possible to study them in more detail and further improve the overall process design.

10.4.5.1 Reactor design

The preliminary reactor design was based on three principal factors: required residence time, heat transfer, and mixing rate. An agitated reactor was selected based

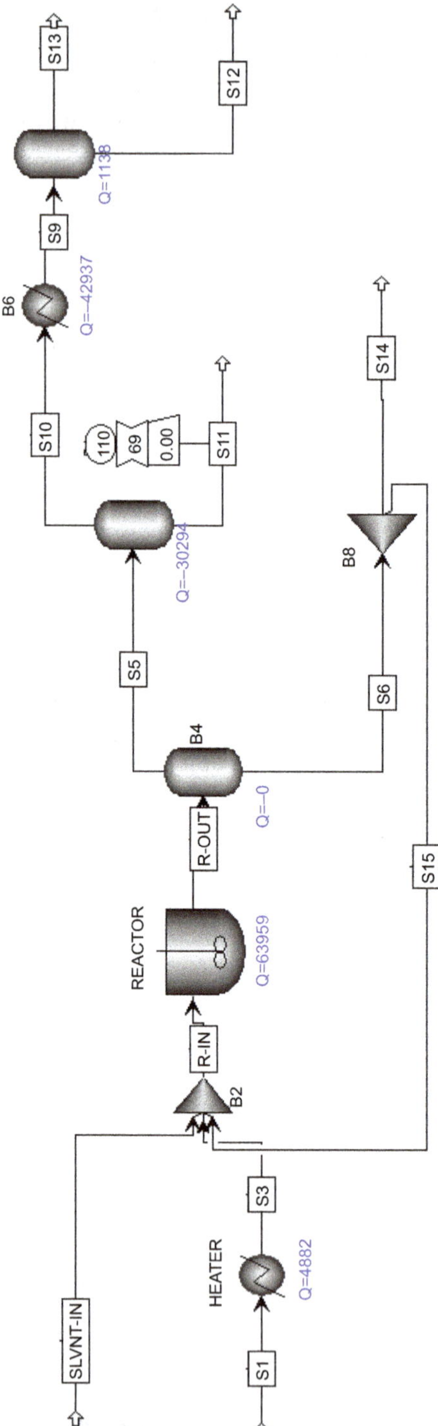

Figure 10.10: An example of intermediate process schemes where a single reactor is used, water and solvent separation is carried out in flash vessels, solvent and feed are mixed before entering into the reactor, and the reactor is heated directly.

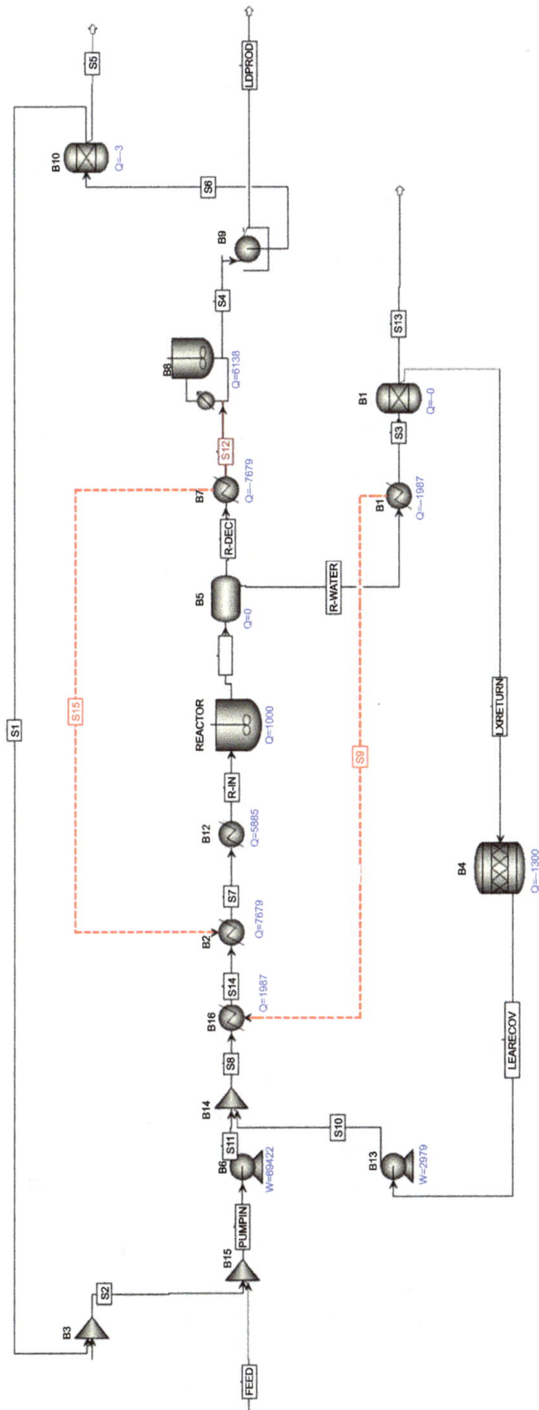

Figure 10.11: An intermediate PFD where the solvent and recycle line are mixed before the reactor, and feed enters directly into the reactor with an elementary heat integration.

on the experimental results. Different approaches may be used for this purpose, such as a mechanically mixed reactor or reactors mixed via liquid recirculation. Regarding the nature of the mixture in this study, mechanical agitation was selected with parameters obtained from the literature [7].

Recycling requires a dedicated separation system and additional heat exchange systems resulting in higher OPEX and potentially higher CAPEX. Therefore, the recycling rate needs to be minimized to obtain an optimized system. A sensitivity analysis was carried out by testing various configurations with a single reactor or two reactors in a series of variable sizes. Accordingly, more than 15 different cases were simulated to assess the tendency of results.

Figure 10.12 presents three simulation examples with variable reactor sizes. Results show that by increasing the reactor size, the LD yield increases and the required recycling flow rate decreases. In addition, fewer oligomers enter the reactor resulting in a higher yield. Some oligomers and LD enter the reactor (time 0) from the feed, recycle line or residues in the solvent after crystallization. To minimize this effect, a LA regeneration step by hydrolyzing is added to the system. This is an example of gradual process development where a problem is identified in each step and appropriate unit operations are added to improve the process design. It should be noted that a global optimization must be considered and not only a local optimization.

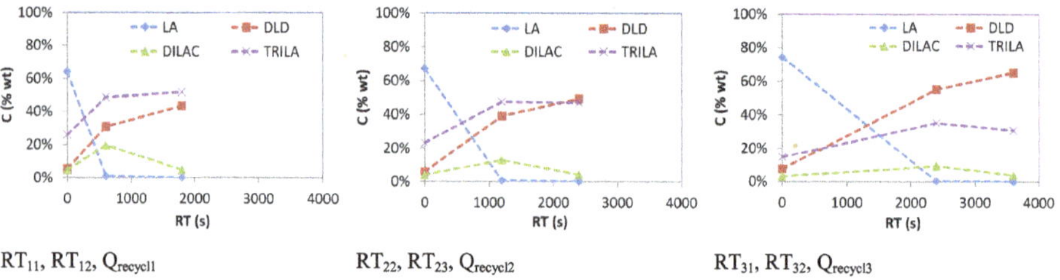

Figure 10.12: Examples of simulation with variable reactor sizes.

The use of two reactors provides other improvements in addition to the overall reaction yield [8]. For instance, each reactor can be a reasonable size, thus avoiding very large-scale reactors with construction limitations. It also makes it easier to independently control the temperature of each reactor. In addition, where there is a separate solvent entry, only the necessary fraction of solvent will be injected into the first reactor. This reduces the overall flow rate into the reactor and increases the residence time in the reactor for the same reactor volume. The minimum solvent flow rate depends on the required heat transfer into the reactor, and the LD solubility in the solvent. Figure 10.13 illustrates a typical reactor configuration and the inlet streams.

Water separation was the next unit operation to be designed. In situ water separation was initially considered in the design. The role of solvent is to avoid direct contact between water and the produced LD. On the other hand, the presence of water has a positive impact via the hydrolysis of oligomers. In the current design, the water concentration of the reactor inlet is optimized by injecting the feed before the water separation step. Therefore, water separation was separated out of the reactors using decantation technology. More experimental works in a continuous pilot plant are required to identify the impact of water content in the feed and inside the reactor on the overall LD yield and reaction conversion.

The reaction of LA conversion to LD is carried out at 174 °C (boiling point of the liquid phase). Therefore, the feed and solvent need to be heated up. Jacket reactors can be used to heat the reactor walls. However, the maximum heating rate is limited due to the limited surface of the reactor. Heat exchangers may be installed inside the reactors. However, this results in additional internal surfaces and dead zones inside the reactor, which reduces the mixing efficiency inside the reactor and increases the risk of fouling.

An alternative heating method was developed in this system where the decane stream was used as the heat carrier to the reactor (see Figure 10.13) instead of conventional reactor heating technologies. In addition, heat exchangers on the LA inlet steam could be eliminated. This results in a much simpler design without a direct reactor or LA heating, especially knowing that the LA stream contains catalysts in slurry form with enhanced erosion and fouling effects. This alternative design results in a reduction of CAPEX and OPEX, and increases reliability and process flexibility.

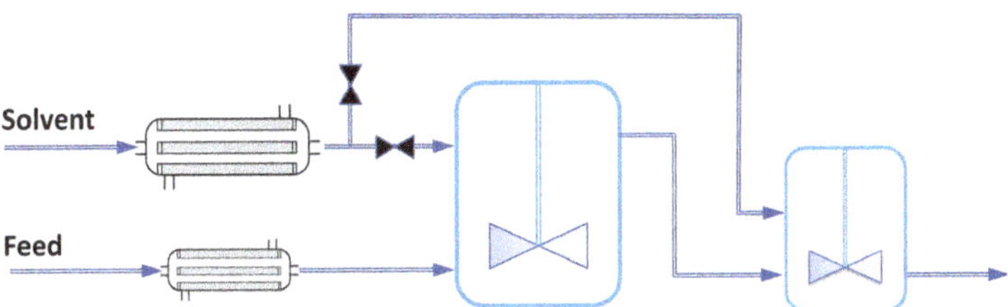

Figure 10.13: Reactor configuration and the heat exchanger system (the exchanger on the feed stream is optional).

10.4.5.2 Water–solvent separation

Decane and water form a heterogeneous azeotropic mixture due to the insolubility of the two phases. This causes distillation separation, which is a challenging approach for the current system. In addition, distillation is an energy intensive operation due to the evaporation of liquid phases. Therefore, the reactor operating condition is kept well below the azeotropic point of the mixture to maintain the whole system in liquid phase. Decantation separation is used to separate water and solvent thanks to the immiscibility of the two phases [9]. Various possibilities exist in this technique, such as separation between two reactors, inside each reactor or after the reaction system. The required residence time for design was measured based on experimental results. Further studies based on the average droplet size and settling velocity are needed to improve the unit design.

Another advantage of this configuration is inherent catalyst separation due to the hydrophilic nature of the current catalyst. The catalyst will be separated by water. Catalyst regeneration will therefore start in the decanter thanks to the direct contact between water and the catalysts. An additional filter or centrifugal separator may be installed on the decane stream, downstream of the decanter to separate possible traces of catalysts remaining in the solvent stream.

10.4.5.3 Catalyst regeneration

Other unit operations were defined and designed step by step, similar to the previous units. Process simulation was accordingly updated to achieve both local and global optimum designs.

Catalyst regeneration was designed using the hydrolysis reaction based on the experimental tests presented in the catalyst deactivation study. Catalyst regeneration will start from the decanter based on the fact that catalyst particles are immerged in hot water in this section. The regeneration will then continue in the conducting pipes of separated water – catalyst streams. Therefore, optimum regeneration conditions in terms of temperature, water concentration, and residence time were respected in the piping system so that no additional reaction vessel is needed for the regeneration.

10.4.5.4 Lactide separation and purification

Produced LD is dissolved in solvent (decane) and needs to be recovered. Distillation is a possible option for this purpose. It has the disadvantage of being energy intensive and may cause the thermal decomposition of LD in reboilers. Another possible method is the use of crystallization to selectively separate LD from the liquid stream. A comparative study between these options resulted in the selection of crystallization. The design of this step was carried out based on the literature [10] and experimental tests, such as the solubility of LD in decane as a function of temperature illustrated in Figure 10.14.

A sugar-type crystallizer design was initially proposed where solids will be in the form of solid crystals with a specific adjustable particle size. This design requires a solid separation step, which could be filtration or centrifugal separation. The use of filtration technology will be relatively challenging regarding the fact that the separated particles need to be continuously separated and recovered. This needs an

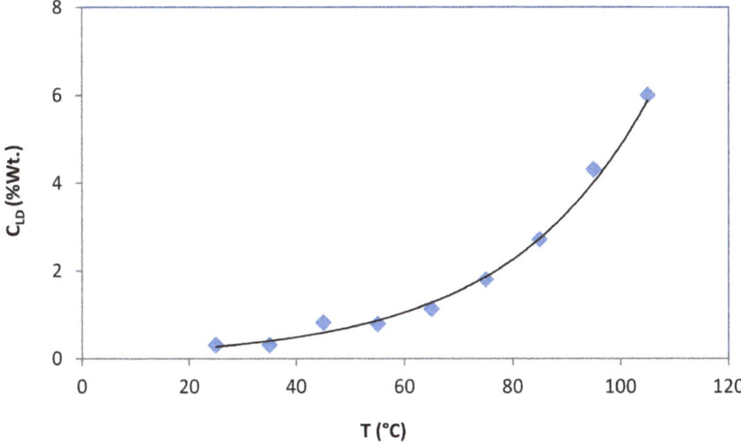

Figure 10.14: Apparent LD solubility in decane as a function of temperature.

automated filtration system or batch system. Centrifugal separation has the advantage of operating in continuous mode. The difficulty would be the agglomeration of LD particles and the formation of a cohesive cake. This can be resolved by controlling the outlet pressure and temperature of the centrifugation unit to fluidize the separated solid LD on the walls of the unit.

Another possibility is to use conventional static crystallizer systems where the solid LD will form over the walls of the crystals or over the cooling tubes. Solids will accumulate over these elements. The solvent will be drained out of the system and the LD will be recovered by melting it. This system will be in semi-batch and probably requires more than one unit in parallel. In this case filtration is not necessary.

Depending on the LD separation method, a trace of solvent may remain on the separated particles. Different technologies were identified and analyzed for this operation, such as distillation, solvent-solvent extraction, or vacuum flash. The last option was selected since it is simpler (and hence more cost effective), not energy intensive and results in a high product purity of 99.98% as listed in Table 10.4 based on process simulation results.

10.5 Heat integration

Heat integration is an important process design step, particularly in the case of energy intensive processes with heating and cooling operation units, such as the current case. Heat integration is a subcategory of process integration [11], permitting operating costs reduction, environmental performance improvement (e.g., by decreasing

Table 10.4: Resulting product stream from the lactide purification unit by vacuum flash.

Property	Unit	Stream 30 (product)
Flow rate	kg h^{-1}	7,280
T	°C	25
P	kPa	–
Concentration		
LA	%	0.00
L2A	%	0.00
L3A	%	0.01
DLD	%	99.98
Decane	%	0.01
H_2O	%	0.00

CO_2 emissions) and may result in process intensification. Different methodologies can be used for this approach, such as pinch analysis or pinch-exergy analysis.

Figure 10.15 illustrates an improved version of the process after about 50 iterations and tested designs. Heat integration permitted the recovery of a considerable amount of heat from products in different steps. The solvent inlet stream into the reactor is the principal heated section, which is above 200 °C. This stream is the heat carrier to provide the reaction heat. The solvent and separated water streams after the reactor have to be cooled down for the crystallization section to 50 °C and below 50 °C for the water separation section. The decane stream could not have been cooled below 110 °C inside a heat exchanger due to the risk of LD deposition on the walls of the exchanger. Therefore, crystallization was divided into two steps. The first crystallizer was designed to permit heat recovery. Heat exchangers were designed by internal design tools. However, conventional tools can be used for Heat Exchanger design, such as ASPEN Plus, Pro II, or specialized software, include HTRI suits.

10.6 Process intensification

Process intensification can be considered after the development of a complete process with all operation units. An example in the current system is the integration of reaction and water separation by the use of reactive distillation systems. Figure 10.16 shows some possible configurations. Depending on the nature of the solvent, other types of designs may be used. Some advanced distillation systems, such as a divided wall column, may be used to further integrate separation units to produce a concentrated LD outlet stream inside the reactor.

The impact of distillation will not be very high on the CAPEX in the case of using simple reactor configurations with an evaporation. However, due to a high

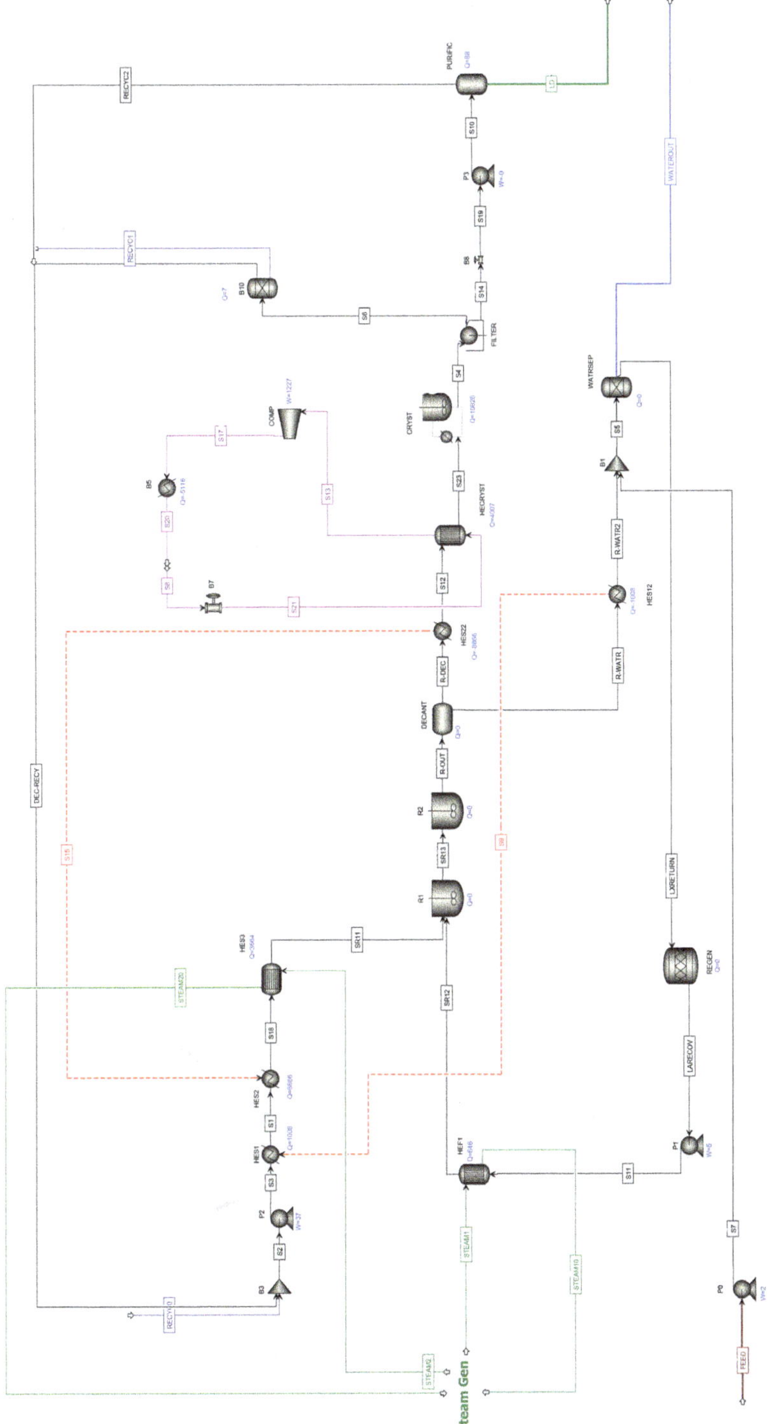

Figure 10.15: Improved process scheme and simulation results.

heating rate, it will increase OPEX and require a high heat integration effort to optimize the process and reduce OPEX. Considering water–decane solution with a heterogeneous azeotropic solution, the energy intensity of products will considerably increase, requiring a tighter heat integration.

In the current stage of development, a reference design is developed with identified alternatives to keep the flexibility and possibility of further improvements in the upcoming development steps.

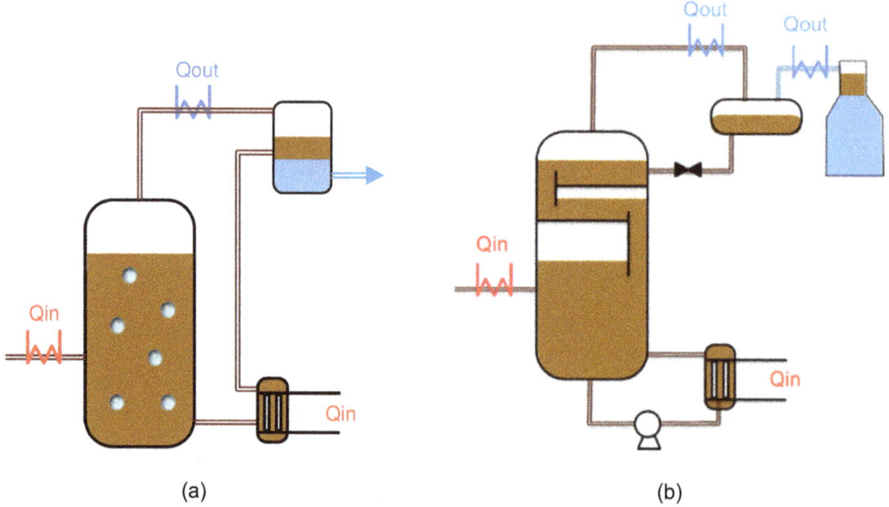

Figure 10.16: Examples of reactor configuration to integrate reaction and separation steps.

10.7 Economic evaluation

Economic evaluation is an important part of the process development and scale-up. It permits comparing the cost of the novel technology with benchmark industrial technology and similar alternatives. In addition, the breakdown of costs helps identify major cost constituent elements of the technology to orient more efforts toward those elements to reduce overall costs.

10.7.1 OPEX evaluation

Variable operating expenditures (OPEX) were calculated based on the energy and chemical consumptions based on the process simulation. Feed, chemicals, and utility consumptions are listed in Table 10.5.

Table 10.5: Utility and chemical consumption list (values are indicative).

Feed	ton year^{-1}	75,000
Solvent	ton year^{-1}	20
Catalyst	ton year^{-1}	10
Electricity	MW h yr^{-1}	15,000
Steam	ton year^{-1}	65,000

The resulting OPEX in terms of typical variable and fixed costs are listed in Table 10.6. Steam and electricity consumptions are the major variable costs while solvent and catalyst do not have a major impact. This shows the importance of heat integration in this system. As an example, the addition of a distillation step without heat integration results in 5.8 t h^{-1} of steam consumption and an additional 1 300 k€ annual OPEX.

Fixed operating costs [12] are calculated based on the principal elements listed in Table 10.6. The calculation of fixed costs is largely influenced by the plant location and specific project considerations.

Table 10.6: Indicative values of estimated operating costs.

	Cost (k€)	Cost (k€ ton^{-1})
Feed	85,100	1,420
Variable costs		
Solvent	4.4	0.07
Catalyst	195	3
Electricity	1,17	19
Steam	1601	27
Sum – variable costs	2,917	49
Fixed costs		
Maintenance	678	11.3
Operating labor	189	3.2
Lab, supervision, overheads	208	3.5
Capital charge	1,606	26.8
Insurance and taxes	1,388	23
Royalties	0	0
Sum – fixed costs	4,068	68

Sensitivity analysis is an important study to identify the influence of the main parameters to be further optimized in order to reduce product costs. Figure 10.17 illustrates the resulting sensitivity of OPEX to the main variable costs. Catalyst and solvent have a minor impact on the overall costs. This means that the consumption of these elements can be increased without a tangible impact on the overall costs. Electricity and steam consumptions have a considerable impact on costs. A 10% increase in electricity consumption increases OPEX by about 3.5%. A 10% increase in steam consumption increases OPEX by 6%. This shows the importance of heat integration and a possible pressure reduction on operating costs.

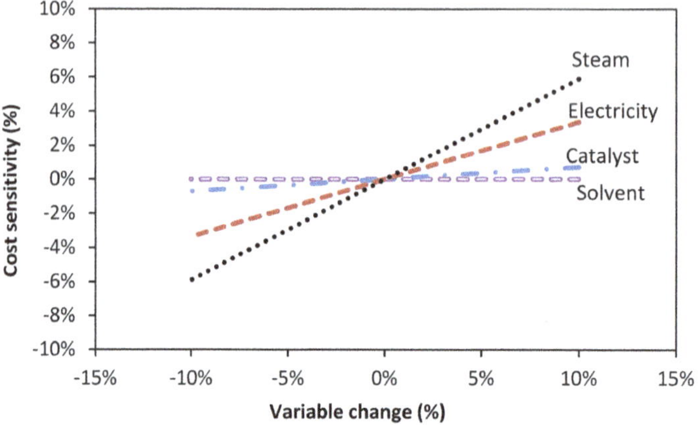

Figure 10.17: Sensitivity analysis of OPEX to principal variable costs.

10.7.2 CAPEX evaluation

The employed mythologies of CAPEX evaluation will evolve and become more comprehensive as process development advances. Two methods were used in this study: the Hill method as an example of a parametric method and the fixed factor method of Lang.

The first method was chosen from a class of economic evaluation methods, which are based on the process scheme (sequence of operational units) and do not need design data nor a detailed material balance (only capacity is required). The important point in these kinds of methods is to have all of the process steps, including auxiliary units. The Hill method [13] is one of the fastest available estimation methods. Other alternative methods with more sophistication are Zevnik and Buchanan [14] or the Wilson method [15]. The Hill method is applicable for systems with moderate pressure and only liquid flows. This method is applied directly into the PFD by giving a coefficient to each unit:

- one for carbon steel columns, reactors, evaporators, cyclones
- two for stainless steel columns, furnaces, centrifuges, compressors
- one additional unit for storage of feed or product results and two for secondary solid or gas storage

The base cost is considered for a production capacity of 4500 ton year^{-1} and a Chemical Engineering Plant Cost Index (CEPCI) [16] of 185 (1954). For larger units an extrapolation factor of 0.6 is used. The installed equipment cost is then calculated by multiplying the number of units calculated above and the obtained price. Table 10.7 shows results of the Hill method for the current study. The total installed equipment cost is 18 M€ by updating the costs based on the M&S index for chemical plants.

The estimation of the main equipment costs is the most important element for the second evaluation method. Costs can be obtained from Engineering databases (e.g., Match database [17]), available software solutions, such as Aspen Capital Cost Estimator [18], conventional literature resources [19], or a market inquiry. Table 10.7 illustrates the list of main equipment and estimated costs. This list is based on the sized equipment list (see reference [20] for a typical list), which is an outcome of the design step. It should be noted that the Hill method is much less reliable than the direct equipment cost estimation. The estimated costs need to be further improved in upcoming development phases.

Table 10.7: Main equipment list and indicative estimated costs for the Lang method and equipment unit for the Hill method.

Equipment		Cost (k$)	Equip. unit (Hill method)
101	Pumps P0	25	0
102	Pumps P1	80	1
103	Pumps P2	180	1
104	Pumps P3	70	0
201	Heat exchanger 1	300	1
202	Heat exchanger 2	400	1
203	Heat exchanger 3	200`	1
204	Heat exchanger 4	50	0
205	Heat exchanger crystallization	150	1
301	Compressor	500	2
302	Air cooler	70	1
401	Reactor R1	1,800	1
402	Reactor R2	1,200	1
501	Decanter	250	1
502	Crystallizer 1	500	1
503	Crystallizer 2	500	1
504	LD purification vacuum vessel	180	0
601	LD filter	1,100	1
602	Water membrane	2,100	1
701	Feed storage tank 1	200	1
702	Feed storage tank 2	200	1
703	Product storage tank	10	0
Sum		10,065	18

In order to obtain the investment cost from the total equipment cost, various parameters should be considered, including piping, instrumentation, buildings, auxiliary services, engineering, and unforeseen costs (contingencies). Various methods exist to calculate CAPEX from equipment costs. Lang proposed a simple method based on a single coefficient for the whole system varying from 3.1 for solid processing to 4.74 for liquid processing units. Table 10.8 gives the typical cost breakdown of the CAPEX constituents. A factor of 4.4 is therefore considered in this study resulting in a CAPEX of 44 M€.

Table 10.8: Typical structure of fixed capital costs in a plant [19].

	Range as ratio of fixed capital (%)	Typical as ratio of main equipment (%)
Direct cost		
Main equipment	20–40	100
Installation	7.3–26	38
Instrumentation and control	2.5–7	13
Piping	3.5–15	29
Electrical installation	2.5–9	18
Buildings	6–20	35
Site preparation	1.5–5	10
Services	8.1–35	56
Land	1–2	5
Indirect costs		
Engineering and supervision	4–21	40
Construction costs	4.8–22	45
Loyalties	1.5–5	9
Contingencies	6–18	41
Capital	100	439

The Lang method was modified by various works [21–23] proposing variable coefficients for each piece of equipment in the system. However, the CAPEX coefficients for these methods remain in a range of 2.5 to 4.8. In the current stage of development, the Lang method is a reasonable approach, which will be improved in the following phases.

10.8 Technology validation at lab scale (TRL4)

The next step is the validation of technology in a continuous lab-scale pilot. A series of hypotheses were considered in the current stage of study requiring further investigation. In addition, a series of questions were identified, which should be answered in the next phase of the study. These identified items cover different sections, such as catalyst deactivation, activation energies, thermodynamic solubility information, water content optimization, and interphase material transfer. It is important to come up with a clear list of hypotheses and questions as a guide for the next development step.

The unit operations used in the process can be divided into two categories: units using existing conventional technologies and novel operation units, which do not exist on an industrial scale. The examples of conventional units are distillations, decantation, heat exchangers for which considerable knowledge exists and little development is required. The reactor is categorized in the second category even if the employed technology is conventional, but further investigations are needed. Therefore, the continuous pilot plant will be focused on the reaction section.

The pilot plant must have a large flexibility to test different parameters, such as variable solvents, different catalysts types and concentrations, residence time, pressure, and temperature. The unit needs to have a representative scale in order to be able to attain a stable flow (small flows can cause instabilities), evaluate hydrodynamics, energy consumption parameters and to ease online analysis. A first proposal is a flow rate of 5 l/h depending on the equipment used, such as pumps, valves, and analyzers. The catalyst is proposed to be on a continuous single pass design. Catalyst regeneration can be done in offline mode while it can be added into the system in later development steps. Figure 10.18 illustrates a simplified scheme of the proposed pilot. Thanks to the continuous operation and high flexibility, various parameters can be adjusted separately in the system. The use of multiple reactors is not necessary at this stage. The second reactor can be studied by simulating the inlet into second reactor by adjusting concentrations in the feed.

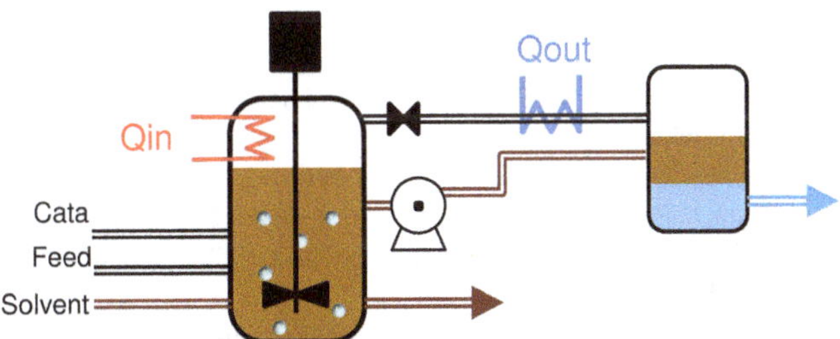

Figure 10.18: Pilot plant configuration with a single pass reactor working at variable pressure and temperature conditions.

10.9 Key learning points and conclusions

This case study illustrated the application of the iterative scale-up methodology to develop an industrial process from a novel idea and to evaluate and benchmark it. The process development started from a simple and basic flow diagram. Then, the technology was improved step by step to attain a level covering all the necessary steps from feed treatment to product purification, including heat integration. Different options were proposed and studied for each step, in order to find the optimum local selection while considering overall performance optimization. Experimental tests and modeling were used for the design of different sections. Heat and process integration and intensification are recommended to identify potential improvements. Finally, economic evaluation permitted comparing the novel technology with existing and alternative processes and to identify the main cost constituents of the technology for further improvement in the next development steps. One of the advantages of this iterative method is the speed with which the industrial design is carried out. In this example, four patents have been filed for the industrial-scale process design [8, 9, 24, 25].

A series of steps were followed in an iterative procedure, including the definition of Design Basis, the development of a preliminary PFD, the detailed study of the necessary process elements, process simulation, equipment design, and CAPEX and OPEX estimation. These steps are interdependent and require an integrated continuous improvement in the course of the development. Finally, a steady state lab-scale pilot design is proposed for the next development stage.

References

[1] M. Dusselier, P. Van Wouwe, A. Dewaele, P. A. Jacobs and B. F. Sels, Shape-selective zeolite catalysis for bioplastics production, Science, 349, 78–80, 2015.
[2] K. J. Jem and B. Tan, The development and challenges of poly (lactic acid) and poly (glycolic acid), Advanced Industrial and Engineering Polymer Research, 3(2), 2020.
[3] P. V. Wouwe, Stereocontrol and monomer synthesis for PLA production, Leuven: KU Leuven, 2015.
[4] IUPAC-NIST, Solubility Data Series. 81. Hydrocarbons with Water and Seawater – Revised and Updated. Part 9. C10 Hydrocarbons with Water.
[5] I. Smallwood, Handbook of organic solvent properties, Elsevier, 1996.
[6] D. T. Vu, A. K. Kolah, N. S. Asthana, L. Peereboom, C. T. Lira and D. J. Miller, Oligomer distribution in concentrated lactic acid solutions, Fluid Phase Equilibria, ISSN 0378-38, 236 (1–2), 125–135, 20 September 2005.
[7] W. McCabe, J. Smith and P. Harriott, Unit operations of chemical engineering, 7th edn, McGraw Hill Chemical Engineering Series, 2004.
[8] M. Yazdanpanah, J. Chaouki, P. Van Wouwe and B. Sels, Single step lactide production process with separate entry for solvent. PCT Patent WO2017220519A1, 2017.
[9] M. Yazdanpanah, J. Chaouki, P. Van Wouwe and B. Sels, Single step lactide production process with recovering water by decantation. PCT Patent WO2017220524A1, 2017.
[10] J. Couper, R. Penney, J. R. Fair and S. M. Walas, Chemical process equipment, Third edn, Butterworth-Heinemann, 2012.
[11] M. M. El-Halwagi, Introduction to Process Integration, in Process systems engineering, Vol. 7, M. M. El-Halwagi, Ed., Academic Press, 2006.
[12] J. Chaouki, Lecture on process economics, in École Polytechnique de Montréal, 2015.
[13] R. Hill, What petrochemical plants cost, Petroleum Refiner, 35, 106–110, 1956.

[14] F. Zevnik and R. Buchanan, Generalised correlation of process investment, Chemical Engineering Progress, 59, 70–77, 1963.

[15] G. Wilson, Capital investment for chemical plant, British Chemical Engineering & Process Technology, 16, 931–934, 1971.

[16] Chemical Engineering Plant Cost Index (CEPCI), [Online]. Available: https://www.chemengonline.com/pci-home.

[17] Matches, 2015. [Online]. Available: http://www.matche.com/.

[18] AspenTech, Aspen Capital Cost Estimator (ACCE), [Online]. Available: https://www.aspentech.com/en/products/engineering/aspen-capital-cost-estimator.

[19] A. Chauvel, G. Fournier and C. Raimbault, Manuel d'evaluation économique des procédés, Paris: editions technip, 2001.

[20] H. Baron, The oil & gas engineering guide, 3rd edn, Paris: Editions Technip, 2018.

[21] W. Hand, From flow sheet to cost estimate, Petroleum Refiner, 37, 331–334, 1958.

[22] R. Turton, R. Bailie, W. Whiting and J. Shae, Analysis, synthesis and design of chemical processes, 3rd edn, U. S. R. P. Hall, 2009.

[23] R. Sinnott and G. Towler, Chemical engineering design: principles, practice and economics of plant and process, 2nd edn, Oxford: Elsevier, 2012.

[24] M. Yazdanpanah, J. Chaouki, P. Van Wouwe and B. Sels, Single step lactide production process with heat recovery. PCT Patent WO2017220522A1, 2017.

[25] M. Yazdanpanah, J. Chaouki, P. Van Wouwe and B. Sels, Single step lactide production process with hydrolysis of oligomers and catalyst by recovered water. PCT Patent WO2017220521A1, 2017.

Naoufel El Bahraoui, Saad Chidami, Rached Rihani, Gabriel Acien, Jamal Chaouki

11 Case study VI: CO_2 sequestration in microalgae photobioreactors

Abstract: In this chapter, we explore the main guidelines and methods helping researchers, photo-bioreactors designers and technology providers, and algae-based projects developers to achieve adequate scale-up for their solutions through the implementation of the iterative scale-up methodology, providing a better assessment of the scaling risks of failure.

First, we tackle the strong physico-chemical complex phenomena resulting from multi-way coupling between transport processes (mass transfer and hydrodynamics essentially) and light assimilation inside the reactor and provide the appropriate tools to generate sufficient and relevant information, weigh risks and deal with them with anticipation.

We then focus on generic, manageable and appropriate computational strategies aiming to facilitate the design of a scalable process from ideation, to simulation, sizing and optimization. Engineering approach to the physical phenomena taking place during the biomass production phase will be provided as well as the wider framework of projects acceptability and life cycle assessment. Finally, industrial scale-up case studies will highlight learning points from both successes and failures in the algae-based industry.

Keywords: Photo-bioreactors, algae, photosynthesis, computational modeling, life cycle analysis, iterative scale-up

Naoufel El Bahraoui, MINES ParisTech, PSL Research University, CES – Centre d'efficacité énergétique des systèmes, 60 Bd St Michel, 75006 Paris, France, e-mail: naoufel.el_bahraoui@mines-paristech.fr
Saad Chidami, Department of Chemical Engineering, Polytechnique Montréal, 2500 Chemin de Polytechnique, Montréal, Québec, H3T 1J4, Canada, e-mail: saad.chidami@polymtl.ca
Rached Rihani, Department of Chemical Engineering, Polytechnique Montréal, 2500 Chemin de Polytechnique, Montréal, Québec, H3T 1J4, Canada, e-mail: rached-2.rihani@polymtl.ca
Gabriel Acien, Department of Chemical Engineering, University of Almeria, Carr. Sacramento, s/n, 04120 La Cañada, Almeria E04120, Spain, e-mail: facien@ual.es
Jamal Chaouki, Department of Chemical Engineering, Polytechnique Montréal, 2500 Chemin de Polytechnique, Montréal, Québec, H3T 1J4, Canada, e-mail: jamal.chaouki@polymtl.ca

https://doi.org/10.1515/9783110713985-011

11.1 Introduction

Public concern about the sustainability of the implementation of technological solutions, specifically targeting natural resources preservation, pollution mitigation and used resources regeneration, has for three decades now been a challenging framework for process scale-up. Meanwhile, it has also been a tremendous opportunity for emerging technologies, combining process viability, efficiency, and productivity.

Algae-based processes and products are one of the promising niches trending to provide an acceptable answer from an industrial ecology perspective, meaning the confluence of fundamental profitability and the sustainability imperative.

Photosynthetic microorganisms, such as eukariotic macroalgae and microalgae or prokaryotic cyanobacteria, are microscopic unicellular organisms that exist mainly in aquatic ecosystems where they thrive as individuals or chains. These microorganisms are the primary producers of biomass for aquatic systems and contribute to ecosystems by mainly transforming inorganic CO_2 into organic O_2.

They have a high potential as a bioresource and their production at industrial scale provides large upsides, such as a low quantitative and qualitative feedstock impact (cost and environmental impact), a relatively high productivity (growth as fast as 200% per day), and their consistency with solar and renewable energy [1]. They can therefore be developed into a sustainable bioprocess, yet the challenge remains in creating a large-scale process, with a proper understanding and also mastering of the complex reaction conditions to monitor and control, while preserving, the beneficial contribution of it and preventing the limitations from turning into major downsides, economically and environmentally.

11.1.1 Health and food applications

Microalgae are a source of proteins, lipids, and carbohydrates and have been reportedly used as sources of food in Asia and Africa for their high levels of protein. They can reach up to 70% composition in proteins and are a known source of fatty acids and essential vitamins [2]. The emergence of new microalgal technologies provides the possibility of developing products of high economic value as well as potentially improving identified and marketed products worldwide.

Since the 1960s, *Chlorella* has been in production and is currently grown industrially in multiple Asian countries by fermentation. It is reported that the annual production of *Chlorella* reaches around 5,000 tons of dry biomass, with 50% of this production occurring in China [3]. *Chlorella* is now marketed as a food, in the form of a *Chlorella* and protein mix and as a personal-care product in the form of an oil. Another important microalgae, *Haematococcus*, produces astaxanthin when under oxidative stress [4]. Astaxanthin is used as a supplement to the human diet or a food coloring agent [5]. The global production of astaxanthin is estimated at 1,000 tons of astaxanthin-rich biomass annually [3]. The microalgal species *Spirulina* is the producer of one of the most commonly sold microalgal foods and its biomass has been mainly used for the extraction of phycocyanin as a food supplement [2].

11.1.2 Aquaculture and agriculture applications

Incorporating microalgae into fish diets improves their health and growth, it also reduces the need for medication in fish by improving the general immunity of the fish population and improving the digestive system [6]. The applications for microalgae in the agricultural field are numerous. The health properties of microalgae make it an excellent supplement when added in low percentages to feed young farm animals and other domestic animals. However, limitations exist in the cost and the scale of production of these feeds. It is reported that the current total soy oil and meal production exceeds 200 million tons per year and fish oil and meal production is around 7 million tons per year. The cost of these two methods is estimated to be under the equivalent of 3 USD per kilogram [3]. In contrast, the microalgal production is estimated at approximately 25,000 tons per year.

The microalgae biomass also provides an alternative to chemical fertilization. Biofertilizers are products of a relatively lower price and the natural compounds present within them improve soil fertility as well as stimulate plant growth [7]. Researchers studying the parameters of lettuce plant growth while using *Chlorella vulgaris* as a biofertilizer reported enhanced germination in comparison to the unfertilized sample. The use of biofertilizer is also reported to reduce the amount of fertilizers required for a certain amount of crops, enhancing agricultural sustainability.

Certain molecules present within the microalgal biomass can act as biopesticides, having bactericidal, antifungal, and insecticidal activities [2]. *Chlorella*, *Nannochloropsis*, and *Scenedesmus* are part of the green microalgae that can generate the molecules required to be characterized as a biopesticide. These microorganisms can generate phytohormones, cytokinins, and antibiotics [8].

11.1.3 Wastewater bioremediation and CO_2 fixation

Microalgae have shown promising results in wastewater treatment as well as CO_2 fixation but have no known commercial process in the market [3]. Studies show that microalgae can be used to treat many types of wastewater, whether it is urban or agroindustrial. The treatment involves the oxygenation of water by the microalgae as well as the fixation of CO_2 gas to avoid its release into the atmosphere [9]. The nutrients present in these wastewaters are therefore converted into biomass, which can be isolated and used for many purposes, such as biofertilizers. The dominant microalgal species for wastewater treatment include *Chlorella* and *Scenedesmus*, as these two also are dominant in natural populations [10].

A process for carbon capture and storage (CCS) technologies has been developed for flue gases and involves CO_2 being biocaptured by certain species of microalgae and cyanobacteria. The CO_2 collected from certain facilities is stored in the presence of microalgae at very high temperatures (40 °C) and with added chemical compounds, such as NO and SO_2, but technological advances are required for this technology to become more efficient and produce larger scale results [3].

11.1.4 Production of biofuels and bioplastics

The triglycerols required to produce biodiesels are commonly obtained from oils and animal fats. Microalgae have recently been considered as a source of triglycerol to produce biodiesel. It is reported that microalgae can produce up to 80% of their dry weight in oil and the methods for converting microalgal biomass into biofuels include direct combustion, thermochemical and biochemical conversion, and transesterification. Producing biodiesel using microalgal biomass possesses many benefits: the biodiesel produced does not contain sulfur nor aromatic compounds [2, 11].

Bioplastics, since they have the same chemical properties, are an alternative to petroleum-based plastics. They are currently made from agricultural products, such as corn, soy, and wheat [12]. However, the use of these agricultural products raises the issue of the sustainability of water and land competing with the resources necessary for human consumption [13]. Therefore, when technological advances will be made, microalgae can emerge as a potential source to manufacture bioplastics since algae can be cultivated on land not suitable for agriculture.

11.2 Major phenomena-related issues in photobioreactors

Recently, photobioreactors (PBRs) have been considered an emerging technology for the removal and/or neutralization of undesirable components from fluid flows. It couples photosynthetic properties with bioaccumulation characteristics in microorganisms, thereby overcoming many of the downsides and limitations known in conventional air and water treatment solutions. They are also used for different biomass production strategies, such as agricultural, nutritive, and health purposes. Most of the agricultural use cases of PBRs tend toward low constrained contexts and therefore use simple and relatively well assessed technologies (open raceways, tubular, etc.). On the other hand, implementing PBRs in an industrial setting, limited by space, input typology and operational continuity imperatives, still deals with important scale-up issues. Indeed, to this date, no major viable PBR based plant scale solution has been deployed [14].

At large scale, the assessment of the culture medium characteristics themselves is hardly achievable with accuracy. The culture medium is host to multiphase and multiphysics interactions, encompassing gaseous bubbles, droplets, aqueous medium, dust and suspended matter, and finally algae cells. More importantly, light irradiation is involved in the biological reaction: photon energy, distribution and behavior on the one hand and culture medium radiatively participating and its reactive nature on the other hand are key parameters impacting – through cell exposure time to photons excitation – the total efficiency of the technology (Figure 11.1).

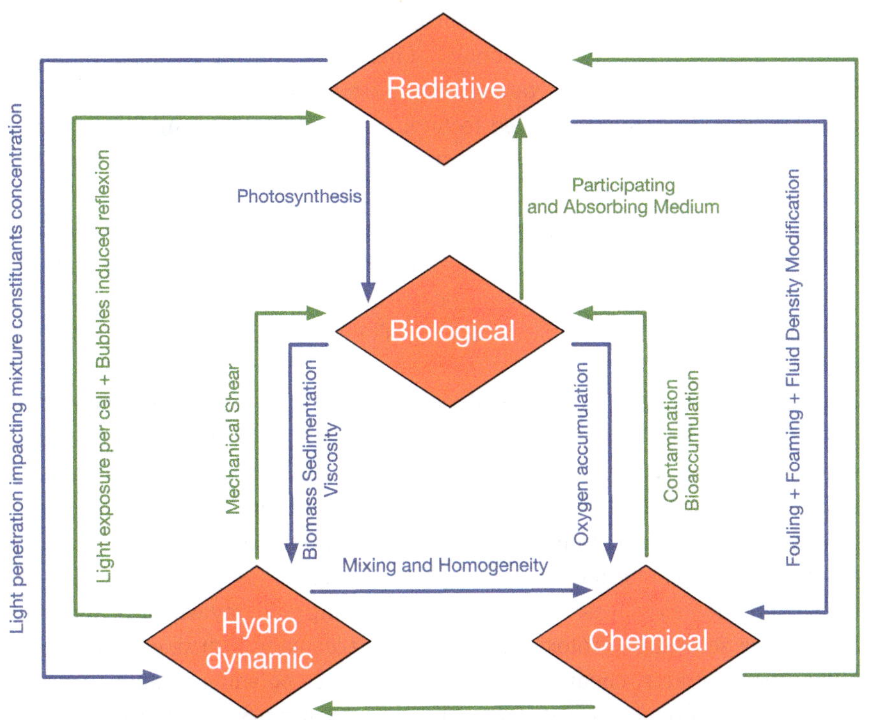

Figure 11.1: Major physics coupling inside a PBR.

11.2.1 Light availability and light exposure modulation

Scale-up must hence guarantee large contact between the reactants of the different reactions (mainly gas–liquid absorption, liquid–liquid carbon reactions, bioaccumulation in algal tissues, oxygen dissolution, and degassing) and at the same time provide the appropriate combination of design and operating conditions to achieve sufficient light irradiation exposure by algal cells [15]. In accomplishing this task, engineers will be challenged by two major obstacles:

1) An efficient PBR must provide a homogeneously distributed large amount of photons, which is, by definition, in contrast to a highly productive – hence dense and absorbing – reactor. Solar irradiation provides a thin illuminated layer in the range of 0.02–0.05 m, limiting the reactor's effective depth and increasing the need for land occupation. In the context of industrial effluent treatment (flue gas treatment, carbon dioxide mitigation, etc.), the reactor's footprint will merely impede the implementation of scaled technologies. Artificial lighting solutions are currently being developed in order to unblock this bottleneck. In this case, the cell-photon effective surface area is larger and can be optimized with proper operating conditions. Taking into account, however, the energy and carbon cost of lighting, process sustainability and life cycle become potential sources of scale-up failure. Also, light distribution systems add to the design and operating complexity and the phase interactions (mainly bubble wall behavior and light swarm-induced reflection).
2) While the first obstacle can be relatively overcome from a process perspective, the second one is inherent to the biological reaction of cell growth: the reaction rate is considerably slow in comparison to most conventional chemical conversions. This is partially due to the low concentration of carbon dioxide in the culture medium, however, radiative properties of photosynthetic microorganisms are the main reason for the slow biokinetics. The use of fast growing species, including engineered strains, is reportedly a promising pathway where current limitations are related to ecotoxicity and public acceptance considerations. With highly robust strains, chemical engineering explores potential solutions for pollutant densification in the gas phase (mainly carbon dioxide, but also nitrogen, phosphorus and sulfur compounds) or via mass transfer enhancement (mainly gas–liquid absorption and pH control).

It is worth noting here that a conventional scale-up method would unequivocally mislead the engineer to overestimate productivity rates. As we will see, the iterative scale-up method will prevent a simple multiplication of lab-scale performance and allow a reasonable yield prediction.

With these considerations in mind, let's attempt to break down the main complexity sources into simple and distinct parameters:

Photosynthetic activity in algal cells uses radiative energy *hν* as its energy source for carbon conversion. It is mainly governed by light availability and distribution, which is, unlike pH, temperature or nutrients, barely homogeneous. This process can be oversimplified and put in the following equation:

$$CO_2 + H_2O \xrightarrow{h\nu + Chl_a} (CH_2O)_n + O_2 \tag{11.1}$$

The accurate prediction of light distribution is a prerequisite for designing, operating, assessing, and optimizing PBRs. Whereas this prediction is easily achieved in lab scale PBRs, reproducibility of light intensity being properly monitored (allowing high productivity rates) is hardly achievable at large scale.

In a range delimited by a minimal radiation needed for photosynthesis activation, the photolimitation limit and a maximal radiation beyond which cell tissues are damaged, the photoinhibition (when damage is reversible) or photo-oxidation (when non reversible) limit, the photosynthetically active radiation (PAR) reaches the cells. When this radiation reaches the cells, it provokes pigment molecules excitation and triggers chain reactions (light dependent, Calvin cycle, and adaptive response reactions). Four photons will excite one electron, then three electrons will be needed to fix 1 mole of CO_2. Maximal photosynthesis rates are often observed in a plateau, the saturation area, that would vary in function of strains, design, and operating conditions [16].

Once again, the simple multiplication of lab-scale performance is likely to be rejected in large scale cultures [17]. A higher cell number results in a higher absorbance rate in the culture medium and in cell shadowing effects, which enhances incident light decay and limits the scattering ratio. Also, the cell shadowing effect locally impacts the effective amount of photons interacting with each cell. Reactor surface:volume (S/V) is then introduced as a critical parameter used here to quantify the light exposure per unit cell. An excessive S/V ratio would result in a deteriorating light antenna in cells, leading to limitations on the photosynthesis efficiency and lowering the productivity. A deficient S/V ratio on the other hand would cause light limitation, leading to a larger respiration occurrence volume ratio, autophagy (potentially predating) and ultimately lower productivity or arrest of the culture. Hence, at large scale, high light intensity gradients can occur, making the relation between radiation and the photosynthetic response more dynamic and highly dependent on strain adaptivity to the operating conditions.

Now, for a known suitable light intensity, light exposure modulation reportedly impacts cell adaptivity and, hence, culture yield. Alternating the cell exposure to light and a lack of light (under the photolimitation value) has been richly studied and different strategies aiming to cycle light and dark periods or to fraction light period and ultimately achieve instantaneous flash alternation are recommended for different time frames and period ranges [17–19]. Again, when in lab scale, the light period is easily predicted in proportion to the total illuminated volume, however, it becomes highly dependent on the mixing pattern at large scale.

Finally, let's stress the large variation of light exposure probability for a cell as a function of its initial position in the reactor and its pattern inside it. Depending on the modeling objective and the requested accuracy, the engineer would choose to implement local models, with inconsistent error permissiveness levels, to be later integrated in the whole reacting volume in accordance with the hydrodynamic behavior. From an iterative scale-up perspective, it would be rather more relevant to consider an average amount of light (or flux of photons) received (or hit) by the cells while being in a random motion inside of the reactor [20]. This average will be progressively and iteratively improved during the scale-up design process.

11.2.2 Temperature control

As we have seen above, cells use light as the energy source for photosynthesis-based metabolism, although, a large part of the incident radiation is not used by photosynthesis. When it comes to effective radiation, literature reports a recovery yield of around 5% from total received energy [21]. Hence, most of the light reaching PBR contributes to the Joule effect inside the culture medium.

The culture temperature rises as a result of the combined effect of different fluxes that can be properly shortlisted to the main light source radiation (direct, diffuse, and reflected), air radiation (direct and reflected), ground radiation, convection, evaporation, and gaseous phase contribution (cooling or, in the case of flue gas, heating) [22]. Specifically, when it comes to closed artificially illuminated PBRs, centralized or distributed light systems are source of hot spots that are hardly cooled due to a limited degassing surface area. Except for certain thermophilic species showing higher growth rates above 40 °C, most of the strains are mesophilic and grow in the range of 20–35 °C, with an optimal segment specific to every species [23]. Culture growth is therefore often exposed to overheating-related issues, where major seasonal changes and daily fluctuations add to the complexity.

At microscopic scale, temperature change implies a change in the photosynthetic apparatus response, which is responsible for the equilibrium between energy supply through light reactions and consumption through Calvin cycle. In the presence of high temperatures, excitation pressure limits electron transport inside the thylakoid, preventing CO_2 fixation from receiving sufficient energy and ending in slower kinetics. Thus, overexposure to light occurs and leads to photoinhibition. Similar phenomena are often observed in extremely low temperatures and also lead to photoinhibition or photo-oxidation. In large reactors, external conditions have a higher impact and the homogeneity condition is not verified. Indeed, important temperature gradients, up to 10 °C, can be observed in different regions of the reactor in a scale of one day. This fast modification in such a short time can hardly be handled by cell generational adaptation. Moreover, at this scale, cell temperature adaptivity is reported to be largely dependent on culture medium composition, notably nutrient content and salinity [24, 25].

Also, when it comes to defining biomass valorization strategies in the case of carbon capture and utilization setup, for instance, temperature becomes a vital factor in the accumulation process in the biomass. Many investigations reported the influence of temperature on the phenotypic and genotypic behavior of the cells through stressful growth conditions, ultimately to anti-oxidants containing molecules to be generated in higher quantities. Combined with other environmental factors, temperature management is then a strong tool to target high value compound production [26–28]. More than being growth limiting or accumulation driving, a poor internal temperature management, mostly in artificially illuminated PBRs, may have a direct implication on reactor robustness and material shelf-life (for acrylic glass or reinforced glass, for instance, widely used in PBRs) and, consequently, a critical impact on the technology life-cycle, especially in the context of depollution or emissions mitigation.

Let's finally not forget instrumentation difficulties when it comes to controlling temperature in volumes of 10 m^3 magnitude and the limited accuracy of any temperature measurement at this scale. When addressing the scale-up problem, the engineer should be aware of the crucial need of the iterative assessment of temperature in PBRs: at every scaling step, heat exchange solutions must be considered and validated to manage hot-spots and hot periods day to day and season to season. The iterative approach will be integrative, inserting the system in a grid perspective that allows heat recovery and heat recycling within the industrial facilities. It will be preventive, as it will include cooling and aeration in every design step. Most importantly it will be holistic, as it will consider, right from the beginning, the radiative flux as a whole, providing the energy needed for the biochemical conversion to happen, but also a source of heat proportional to the global input and the loopholes enabled by both system design and operating conditions.

11.2.3 Carbon

Notwithstanding the fact that photobioreactor productivity is mainly determined by light exposure and temperature conditions, the effects of CO_2 concentration on the lipid composition of the membrane, in addition to effects on photosynthesis, have also been highlighted in some studies [29, 30]. CO_2 concentration within the process can become a limiting factor within a microalgal mass production bioreactor [31].

It is reported that only 5% of the carbon requirements for algal cultures can be directly transferred from the atmosphere, showing a requirement for a secondary insertion of carbon [32]. As some bacterial requirements are known for certain species, it is important to assess the optimal concentrations of carbon (as well as other nutrients) for any unknown microalgal strain. This is done in order to maximize the growth of said microalgae. Carbon can be supplied throughout multiple methods, such as flue gas or air injection in the form of carbon dioxide.

The amount of CO_2 necessary for growth depends on the strains and the type of PBR. A reactor of the aerodynamic type (the gas phase is solely the driving force of the flow) has, for example, lower CO_2 requirements for given species than reactors of the bubble column or air-lift type, which implies a reduced cost of air compression and CO_2 requirements [33].

The need for reevaluation of the carbon requirement can be monitored by monitoring the pH, which indicates the amount of carbon available to the growing samples [34]. Studies have shown that carbon efficiency does not necessarily implicate high concentrations, it is rather important to maintain constant and adequate pH as well as reduce carbon limitation [35]. On a laboratory scale, CO_2 injection is performed in function of a pH set-point in order to optimize CO_2 consumption and avoid excess carbonic acid concentrations in the growth medium. On a larger scale, a non-homogeneous injection of carbon could cause irregularities within the growth medium [31]. Concentration differences cause local equilibrium displacement leading to sedimentation, unwanted biomass flocculation, and growth rate limitation [31]. The injection of CO_2 in gas form can be rather difficult and involve losses of carbon between 80% and 90% through bubble formation. This would potentially increase the required injection of gas, hence an increase in cost. Oxygen removal is also a tool in order to avoid photooxidation of algal cells and reduced growth [32].

Finally, although a higher CO_2 concentration can lead to increased biomass productivity, the effect of low pH on microalgae physiology, and the increased possibility of contamination should not be ignored [36].

11.2.4 Mixing and aeration

Mixing and aeration are crucial embedded parameters when it comes to reducing the spatial heterogeneity inside the culture. In addition to ensuring the carbon dioxide input, aeration provokes bubbly flow enabling particles (cells, bubbles, nutrients . . .) to follow bubble induced turbulence patterns, thus creating the medium mixing. It allows achieving a high mass transfer ratio, maintaining modulate cell light exposure, ensuring medium homogeneity, and preventing anaerobic or sedimental "dead" zones inside of it. As we have seen previously, light exposure modulation makes it possible to achieve higher photosynthesis yields for a known light intensity. Furthermore, algal metabolism-related photosynthesis is the root to CO_2 uptake and O_2 accumulation. As a result, the culture pH is constantly subject to alteration and, when not removed properly, the produced oxygen tends to accumulate in a dissolved form. As we will see later, high concentrations of dissolved oxygen cause photoinhibition, even at lower light intensities.

Mass transfer is mainly impacted by gas holdup, gas–liquid interface distribution, and the aeration regime. While the aeration regime governs the gas flow into the reactor, gas holdup informs on total gaseous phase content of the culture mixture, including carbon dioxide and dissolved oxygen in equilibrium with the gas phase. As a function of the driving force, the oxygen desorption rate represents the system capacity to liberate the accumulated oxygen. The distribution of the gas–liquid transfer interfaces (bubbles sizes, shapes, and swarm behavior) affects the gradients locally observed in the absorption and desorption rates and leads to the existence of chemical and biochemical kinetics singularities, such as local anaerobic behavior or sedimentation phenomenon, that a good mixing aim to reduce.

Anaerobic zones are host to the accumulation of certain culture-deteriorating compounds, such as H_2S, Mn^{2+}, and Fe^{2+} sedimentation, in addition to biomass decomposition loss, causing nutrient release to the medium (mainly phosphorus) and then negatively affecting production strategies [37]. Therefore, from the quantitative productivity perspective, maintaining high mass transfer capability through good mixing will be the first design requirement the engineer should consider while scaling a PBR system.

Now for continuous large scale production, mixing can be used as a strong density control or contamination control strategy. For mono-culture applications, mixing is specifically solicited to achieve a shear level beyond the unwanted species acceptance threshold, specifically cyanobacteria [38, 39]. Bubble bursting and coalescence, resulting from the interaction with turbulent eddies due to a high aeration regime, cause stress and cell damage on targeted species known for high sensitivity. The mixing implication on temperature and medium oxygen content increase can also be useful for the same purpose. Then, from a qualitative perspective as well, mixing will be a central tool for the engineer while scaling up the system.

As is well known, the mixing demands tremendous amount of energy and except for artificially illuminated PBRs, remains the most energy-consuming and one of the most important process cost or product price constructing elements. Depending on the energy source, a scale-up carried out too quickly can result in noteworthy CO_2 emissions and, in the context of flue gas treatment or carbon mitigation, may lead to solution failure at large scale. The iterative scale-up approach contributes to a progressive optimization for design and operation, an appropriate choice for the mixing solution (pumping, compressed air, etc.) and more importantly a careful study of the species' mechanical response toward the hydrodynamic forces produced by the aeration-mixing complex under different conditions, scales, and production strategies [40].

11.2.5 Growth medium components and nutrients

Within the bioreactor, the choice of growth medium becomes a key factor in controlling the major components of microalgal growth. The scaling of any reactor-based process involves the establishment of the required nutrients and a relatively cost effective method of feeding said nutrients. One challenge that the up-scaling of any process faces is the expensive cost of growth medium.

Studies have shown that artificial mediums of growth are too expensive and that wastewater may provide the vital algal nutrients at a lower price [31]. Wastewater provides the required nutrients and can be treated for phycoremediation in the microalgal growth process. Chlorinating the wastewater is another step that has been documented to reduce the risks for bacterial contamination and algal loss. The high nitrate and phosphate concentration in wastewater decreases the harvesting time and the need for growth supplements. During the use of this resource, the removal of organic contaminants has been shown in wastewater samples, improving yet another aspect of water treatment and reducing the overall cost of the process [31].

Microalgal growth using wastewater reduces the requirement for fresh water, therefore reducing the price of the nutrients. It is also said that some microalgae do not require any sort of pesticide or fertilizer to grow, another improvement for the environmental aspect of the production [41].

11.3 Conventional scale-up method: success and failures

Based on Tercero et al. [42], to produce 10 Mgal year^{-1} of biodiesel, an area of 19.5 km^2 is needed. Their calculations were based on a generic alga that can accumulate up to 40% of lipids with a productivity of 39.2 g m^2 day^{-1}. They also performed an economic profitability analysis and to achieve a 10% rate of return on the investment, the selling price should be equal to $18.35 per gallon, which is well above the fossil diesel fuel. They conclude that the main cost of production is related to the temperature control system. In an another CO_2 to fuel life-cycle assessment (LCA) study [43], they missed the opportunity to add in their sensitivity analysis: the area needed for such a facility and the CO_2 emissions by the maintenance supplies, such as cars and trucks.

There is a reason for this missing data. Microalgae production today is performed in (i) small-scale reactors (from 100 to 1,000 m^3) when comparing it with other technologies, such as conventional fermenters already used in the food and energy industry, or at (ii) small microalgae farms (from 1 to 10 ha) when comparing it with large crops, such as those producing wheat or corn (Figure 11.2). To enlarge the volume and surface of microalgae-related processes, significant efforts to develop

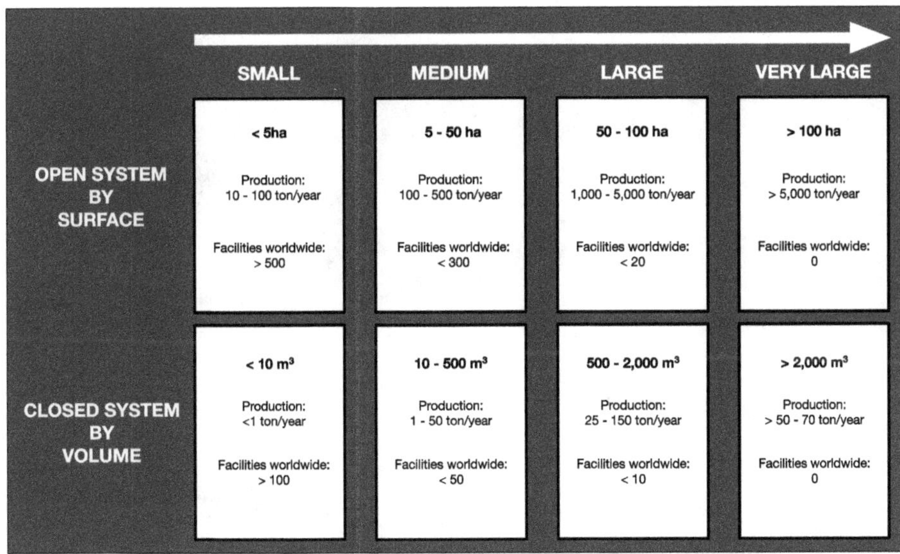

Figure 11.2: Comparison of industrial production facilities as a function of technology and size (modified from [14]).

new technologies are needed [14]. Moreover, the technologies already being used were optimized over lengthy periods of fails being large.

The main reasons for these fails were the extrapolation of data from laboratory or small pilot scale to large volumes, much higher than 10–20 times as rational criteria. It should be noted that at laboratory scale it is possible to provide all the requirements of microalgae cells, such as light, temperature, pH, O_2/CO_2 and nutrients quite easily although at really high cost. However, to achieve successful industrial processes, these requirements must be provided at a rational cost, with the maximal production cost of microalgae biomass ranging from \$100 to \$1 kg^{-1} when considering high or low value applications.

The main reasons why large microalgae facilities failed are (i) the inadequate extrapolation of large facilities by repeating non-optimally designed units, (ii) the inadequate extrapolation of the light path (tube diameter or water depth), or (iii) the excessive scale-up performed on the basis of over optimistic expectations. Figure 11.3 shows some examples of these failures. The development of large facilities by repeating small single units already evaluated at pilot scale usually fails because the reactors don't achieve a minimum volume to be profitable. Additionally the excessive number of small reactors makes it difficult to manage the entire facility. Thus, microalgae facilities based on reactors smaller than 100 m^3 are not feasible. It is recommended to develop base units larger than 1,000 m^3 to achieve successful industrial scale processes [14].

Figure 11.3: Examples of PBR scale-up failures: (a) excessive number of reactors too small to be practical; (b) inadequate tube diameter and configuration of tubular PBRs; and (c) too optimistic design of raceway reactors based on simulations (courtesy of University of Almeria, Spain).

Regarding the light path, this variable determines the light availability inside the cultures, with the excess of light (low light path) reducing the performance of the cells by photoinhibition and the deficiency of light (high light path), therefore reducing the photosynthesis rate to unsuccessful values. In general, it is assumed that light paths in the range of 0.05–0.20 m are recommended, the prior corresponding to values already recommendable for tubular reactors whereas the latter corresponding to raceway reactors. Finally, large microalgae facilities should always be composed of a repeated number of single reactors, although this fact must be carefully analyzed when designing new facilities.

Current microalgae facilities vary in size up to 10 ha and only a very few cases of 100 ha exists around the world. These huge 100 ha facilities have serious management problems, so extrapolating to larger surfaces on the basis of nonaccurate simulations is not realistic [14]. Once this scale-up problematic is solved, new problems arise, such as those related to biological systems: contamination. With the development of very large "farms," new parasites and viruses appear and the "new" biological systems thrive massively and replace the algae of interest. A major example of this issue, with large-scale reactors, is the contamination of *Haematococcus* cultures with parasites, such as *Chytrids* [44].

The major limiting factor of the photosynthesis process is the light availability [45]. It is a function of impinging light and the surface to volume ratio of the reactor. There is a large list of examples of failures related to the consideration that microalgae productivity can be maintained for multiple water depths. Raceway reactors operated at a culture depth exceeding 0.3 m are not feasible due to low light penetration, since scarce or null net biomass productivity can be measured at these conditions. The optimal water depth was found to be in the range of 0.1–0.2 m [46] and a large number of companies were already producing microalgae at these conditions. Reducing the water depth below this value would also be positive due to the reduction of the dark zone, yet still increasing the biomass productivity. However, the fluid-dynamic at these conditions become complex and culture circulation along large raceway reactors is not possible. To solve the culture circulation problem, different designs than raceway reactors are required, such as tubular or thin-layer reactors in which the culture is recirculated using pumps.

When improving the light availability, a new problem arises: the removal of oxygen and supply of CO_2 [47]. Oxygen removal is a major bottleneck in large scale reactors because the microalgae cultures become oxygen inhibited at dissolved oxygen concentrations higher than 20 mg L^{-1}. Considering that the oxygen saturation with air is 9 mg L^{-1} and the oxygen production rate can achieve values of 2 mg $L^{-1} \cdot min^{-1}$, it means that in only 5 min damaging dissolved oxygen concentrations can be achieved if oxygen is not removed. This is a significant problem in raceway reactors where the circulation time along the reactor took several minutes, with dissolved oxygen values higher than 20 mg L^{-1} being achieved for most of the daylight period, thus reducing the overall biomass productivity to 1/3 of the potentially achievable.

The supply of CO_2 is also linked to this problem. CO_2 provided to microalgae cultures is dissolved and accumulated in the culture as part of the bicarbonate buffer [32]. Since a one-to-one molar ratio of CO_2 consumption per O_2 produced by photosynthesis exists, it means that CO_2 demand values of 2.75 mg $L^{-1} \cdot min^{-1}$ already exist in microalgae cultures. Since the total inorganic carbon already present in microalgae cultures is in the range of 100 mg L^{-1} it means that the reserve of carbon in the liquid is enough to maintain the growth of microalgae for longer periods without a reduction or carbon limitation for up to 20 min. However, the consumption of CO_2 has a large impact on the variation of pH with inadequate pH dramatically reducing the performance of microalgae cells. To solve this problem optimized reactors must be designed consisting of (i) sections capable of providing CO_2 and removing O_2 in accordance with the requirements of microalgae cells and (ii) sections capable of providing the adequate light to perform photosynthesis at a maximal rate.

After solving light and O_2/CO_2-related issues, the remaining ones are related to temperature [24]. The control of temperature is a major issue in large scale reactors. It should be noted that the location where the reactors are installed completely determines the temperature to which the cultures will be exposed. In the case of a pilot or small facility it would be possible to implement control strategies for temperature, such as water spraying or internal heat exchangers through recirculating water. However, at large scale this is impossible due to the high volumes of water required or the related huge energy consumption.

It should be noted that in large reactors the mean temperature of microalgae cultures approaches the mean daily ambient temperature. However, daily variation is quite different with large peaks of temperature in microalgae cultures at noon due to the absorption of heat by radiation. Thus, the temperature of the microalgae cultures can be up to 5 °C higher than ambient temperature at noon. It means that strains not tolerating temperatures higher than 30 °C die immediately when they are produced in temperate climates from spring to summer because 30 °C are usually achieved at these locations. Low temperature is not so critical because it does not provoke the death of the cells, although microalgae performance is reduced by half when decreasing the temperature 10 °C with respect to the optimal value.

11.4 Iterative scale-up method

Scaling refers to the necessary spreading and adaptation of any developing technology into successful implementations. It answers the need of more: because of the market need for a successful product or technology, more input capacity, more output productivity, and more process efficiency are requested. Quite often, scale-up comes down to a basic challenge: how to spread a performing solution from a level where it is completely assessed and mastered to the next level where uncertainties can hardly be assessed or mitigated.

11.4.1 Computational strategies for PBR iterative scale-up

Despite the parameters and difficulties described above that are inherited from life science, PBRs are currently intended for industrial use [48]. However, there is no standard or procedure for scaling up from laboratory to industrial scale, hence the need for the iterative scale-up methodology. In addition to laboratory experiments, it is necessary to use modeling tools and complex mathematical analyses that will help, with this approach, to save time and money for a transition to the industrial stage [49].

We propose approaching the described complexity by a systemic compartmentalized computational methodology as recommended in the literature [50–52]. Keeping in mind that the system generates novel properties resulting from the physics, phases and constituents interaction, we will manage, as much as is physically relevant, to subdivide the system into interacting and participating subsystems.

This will allow a proper scale-up-related risk anticipation, a relevant cost and life cycle cost estimation and the optimization of subunit yields. As this subdivision might be too exhaustive, let us provide some guidelines on the major physical subsystems. It is essential to remember on the indispensable use of appropriate computer simulation tools for each subsystem, but also an integrative, collaborative and well-suited software platform, since it is also necessary to always combine modeling with experimental validation.

11.4.2 Biological subsystem

The first subsystem consists of the physiological study of microalgae cells, since each species will have specific requirements and tolerances toward the operational parameters. However, it is necessary to have the mathematical model that explains the algal growth, but also the accumulation of components in the cells (lipids, saccharides, etc.) as a function of temperature, pH, light intensity and distribution, availability of CO_2, and the composition of the culture medium. The complexity of this step is based on the specificity of the carbon source, the culture medium, the species, and the product that we want to obtain industrially. Since there is no single criterion that will predict the PBR performance, the modification of one of these parameters necessarily requires the re-evaluation of the model and a laboratory phase to determine the optimal parameters. This step is very time-consuming because it requires time to grow algae, test parameters and analyze samples to validate the mathematical models.

11.4.3 Hydrodynamics subsystem

After the determination of the optimal biological parameters for growth and yield, the next subsystem is the hydrodynamic simulation of the fluid in the PBR. Indeed, this step will help to design the geometry, the gas velocity insertion and spargers disposition of the PBR to achieve the favorable conditions of algae growth. The fluid movement inside the PBR will be responsible for mixing, CO_2 solubility, gas holdup, mass transfer, and light availability. In fact, to achieve optimal cycles of photosynthesis, that is, that alternate between light and dark cycles, computational fluid dynamics (CFD) will help in the simulation of the microalgae movement close to the reactor walls, where they can be under light and then move them back to the darker region of the PBR. The simulation of light penetration in PBRs is however a complex and little studied phenomenon. Mathematical models are also designed for uniform media, whereas in this case we have bubbles of different shapes and traveling speeds that will reflect and scatter the light inside the PBR. To simulate the penetration of light in these PBRs, it is necessary to simulate a complex medium and this requires a very complex mathematical model which implies an expensive computational effort to obtain a realistic simulation.

11.4.4 Complete photobioreactor model

When the results of the individual subsystems are completed and validated at the laboratory level, the combination of all subsystems can be considered the final step in a modeling process. Thus, the complete mathematical model of the PBR system is obtained. However, in order to validate this complex model, it is necessary to move away from the laboratory stage to a larger scale installation. This fits perfectly into the iterative scale-up methodology, because the larger scale installation will make it possible to verify if the optimal growth rates and yields obtained by simulation and laboratory validation are correct. If the results diverge, the larger scale installation makes it possible to measure the parameters and correct the model or to improve the PBR, until convergent results are obtained.

Since the Iterative scale-up method is accomplished by incremental improvements leading to the final design, the subdivision of the PBR into independent subsystems, guarantees a faster more effective and error tolerant industrial unit. In fact, since each technical difficulty is reduced to its smallest component the challenge can be easily grasped and overcome, by combining understandable experimental data and simulation results.

11.4.5 Acceptance assessment for scaling-up PBRs

Projects based on PBR technology fit in a wider context motivated by legal framework and public acceptance. Where regulations remain relatively understandable and anticipatable, public interaction with the implementation of novel technologies is far less predictable. Public perception includes safety and health considerations as well as economic, social, and environmental stringency [53].

Closely associated with decontamination and remediation in the industry, many emerging technologies, with a large potential for pollutant removal, carbon mitigation or waste management, have nevertheless attracted considerable public opposition in many countries recently [54, 55]. CO_2 capture and storage, CO_2 capture and utilization, methanation and gasification are some well-known examples where technology dissemination and integrative public policies development have been slowed or even halted [56–59]. It might be taken for granted that public mistrust, at least partially, results from the project stakeholders themselves (generally polluting industries) [60], but the expanding scientific literature on public acceptance provides additional tools to understand the diversity of the players and their typical and potential interactions with the technology and its scale-up process.

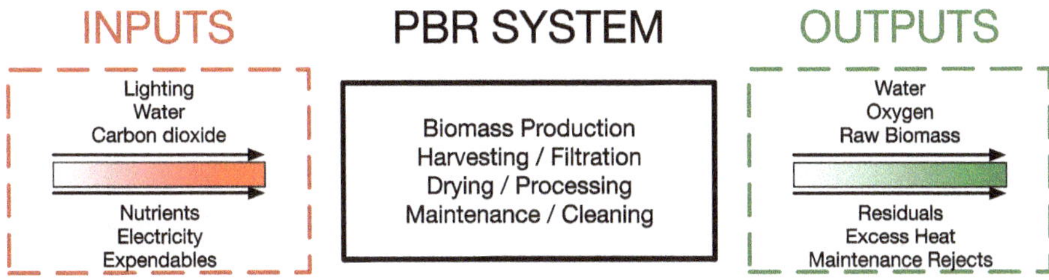

Figure 11.4: Life-cycle inventory of a typical microalgae biomass production unit.

Now, considering the PBR technology from the depollution and carbon mitigation angle, it is true that there being room for inorganic carbon conversion to organic biomass, PBRs have an important climate benefit through CO_2 sequestration from the atmosphere or centralized sources of emission such as flue gas, exhaust gas, or biogas, directly or indirectly via gas bottles (with industrial CO_2 originated from steam reforming). Moreover, the use of marginal land, preventing food cultures or environmental competition and the potential use of wastewater as a nutrient source add to the advantages of the technology.

However, as shown in previous sections, this positive impact is reduced, or even flipped, by indirect emissions resulting from the construction or the operation of the reactors and from the downstream management of the produced biomass [61–63]. This limitation would be the very essence of a potential public rejection, leading to a failure in the technology implementation, a more austere policy regarding funding and ultimately a negative driver for market success. It is therefore important to fully understand the process limitations and analyze their environmental impact during the process design and scaling. Iterative scale-up gives the means for decision-making at the different levels of maturity, preventing limitations perception from being ignored, and contributing to a better identification of the benefits as they can be perceived by the public and the decision makers. An iterative implementation of life cycle analysis, as a holistic approach, during the industrial scaling of the solutions can be a decisive way to counteract the public-related sources of scale-up failures.

LCA is a systemic methodology based on the integration of solutions (products, services, and technologies) for the complete life cycle in order to assess their impact on the environment. LCA study takes into account all the steps, starting from feedstock extraction, manufacture and processing, utilization and finally recycling or elimination, embedding logistics in every step [64, 65]. It cannot be emphasized enough that LCA should not be limited to literature data extrapolation or lab scale and small pilot preliminary data processing, but should bridge the knowledge gap iteratively moving back and forth from one maturity level to another. This will notably allow a comprehensive assessment of the sensitivity to design choices, seasonality, and site specificity.

When addressing the technology LCA, the engineer must, at every iterative step of the design and scaling process, respond to the following questions:

1. What is the ultimate goal of the assessment and what is its scope?
2. What is the exhaustive inventory of the different units of the PBR, including the downstream management?
3. What are the impacts of the process requirement, the bio-sourced product and of the generated waste?
4. What is the interpretation of these results and their perception by the public and decision makers?

First, an exhaustive assessment of the implemented solution has to be performed from cradle-to-gate. Here, the system boundaries will include all the environmental impacts due to material and energy supply of the biomass production process (Figure 11.2). If additional expendables are used (fertilizers, chemicals, etc.), they have to be included with respect to the required raw materials, production process and post-utilization processing. In decontamination or a CO_2 removal setting specifically, it will indicate the undesired compound (metal, CO_2, etc.) removal efficiency. Then, the gate-to-gate perimeter is enlarged to incorporate subsequent utilization of the bio-sourced product, the sequestered CO_2 application, indicating if CO_2 is re-emitted into the atmosphere or permanently removed. Also, substitution for fossil or polluting products by algae based products are accounted for in this wider assessment. We refer here to a cradle-to-grave study.

Plant construction and operation requirements, such as land servicing and infrastructure, process units and auxiliaries and production chain materials are assessed taking into account their management as a final waste after a potential recycling. The primary energy demand for algal cultivation in the PBR and for biomass processing must then be assessed considering the unit scaling. The electricity consumption in PBRs is mainly caused by pumping or compressed air production to ensure mixing and potentially artificial lighting. Studies in the literature report a net CO_2 negative footprint for open pond and airlift PBRs, even with current sub-optimal levels of productivity. But when adding the harvesting and biorefinery downstream processes, the system often becomes CO_2 emitting, even with a high renewable ratio in the energy mix input. It is then mandatory to keep in mind, at every step of the scaling process, to enhance biomass productivity and specific energy consumption and on a wider scale identify the main hotspots that may increase the environmental impact of the technology.

This information is provided by the life-cycle inventory, where all mass and energy flows are assembled and processed for all system inputs and outputs. It is being completed for the quantification of the product functions to serve as a basis for comparison. Interpreting environmental impacts is not always intuitive. Results are often expressed with a certain level of systemic complexity not directly corresponding to their manifestation and its perception. To facilitate the information extraction and reading, standardized categories are recommended for their relevance, transparency, applicability, ease of disseminating information and robustness.

11.5 Conclusion

Biological processes, specifically algae-based processes, present complexities as a result of the multiple interactions between system constituting elements and their environment.

Yet, due to monoexpertise team involvement, or the often siloed input of industry, engineering and research spheres, this complexity is recurrently addressed from a narrow area of expertise, whereas the design of PBRs will depend on several, intrinsically linked factors and considerations. This multiplicity can only be taken into account by the association and creation of multidisciplinary consortia. Otherwise, a single-angled approach or vision would ultimately lead to a linear setback in the scale-up approach. As we have seen, this approach is doomed to a failure, since a field-specific team will consider any parameter outside its field of expertise as "black magic," often minimized or, even worse, discarded. A multidisciplinary team can divide the problem complexity into subsystems according to the respective know how of the different horizons and build-up in a synergistic co-constructing slant, which is in line with the iterative scale-up methodology.

Ecological industry tools are also mandatory to support a wider mainstream adoption of PBR technology in the industry. From the sustainability perspective, raw biomass production, even for bioremediation applications, remains potentially mitigated and environmentally inefficient. Therefore, PBRs value chain has to be considered as a whole and a more inclusive process, co-valorizing and substitutive oriented, shall guide the technology scaling intent. Indeed, high economical and environmental added value applications, combining profitability, remediation and in many cases enhancing the fossil to a renewable shift, will constitute the growth driver of this industry.

Iterative scale-up is, from our perspective, a valuable asset, providing decisive insight in a phase where understanding and development is aggregating feedback accumulation, alimenting the mathematical modeling transposition of the biology and its complexity and aiming for an imminent impact of both economic and environmental benefits of the PBR technology.

References

[1] D. Özçimen, B. Nan, A. T. Koçer and M. Vehapi. Bioeconomic assessment of microalgal production. Microalgal Biotechnology, 195, 2018.

[2] G. L. Bhalamurugan, O. Valerie and L. Mark, Valuable bioproducts obtained from microalgal biomass and their commercial applications: a review, Environmental Engineering Research, 23(3), 229–241, 2018.

[3] F. G. Acién Fernández, A. Reis, R. H. Wijffels, M. Barbosa, V. Verdelho and B. Llamas, The role of microalgae in the bioeconomy, New Biotechnology, 61, 99–107, 2021.

[4] D. Han, Y. Li and Q. Hu, Biology and commercial aspects of haematococcus pluvialis, Handbook of Microalgal Culture: Applied Phycology and Biotechnology, 2, 388–405, 2013.

[5] P. Z. Margalith, Production of ketocarotenoids by microalgae, Applied Microbiology and Biotechnology, 51(4), 431–438, 1999.

[6] A. J. Vizcano, A. Rodiles, G. López, M. I. Sáez, M. Herrera, I. Hachero, T. F. Martnez, M. C. Cerón-Garca and F. J. Alarcón, Growth performance, body composition and digestive functionality of senegalese sole (solea senegalensis kaup, 1858) juveniles fed diets including microalgae freeze-dried biomass, Fish Physiology and Biochemistry, 44(2), 661–677, 2018.

[7] N. Abdel-Raouf, A. A. Al-Homaidan and I. B. M. Ibraheem, Agricultural importance of algae, African Journal of Biotechnology, 11(54), 11648–11658, 2012.

[8] Y. Lu and J. Xu., Phytohormones in microalgae: a new opportunity for microalgal biotechnology?, Trends in Plant Science, 20(5), 273–282, 2015.

[9] P. Pedroni, J. Davison, H. Beckert, P. Bergman and J. Benemann. International network for biofixation of co2 and greenhouse gas abatement with microalgae. In Greenhouse Gas Control Technologies-6th International Conference, 1863–1866. Elsevier, 2003.

[10] D. A. G. Guzmán, E. P. Olmos, S. Blanco, G. Acién, P. A. G. Encina, S. B. Rodrguez and R. M. Torre, et al. Evaluation of the dynamics of microalgae population structure and process performance during piggery wastewater treatment in algalbacterial photobioreactors. 2017.

[11] B. Wang, Y. Li, N. Wu and C. Q. Lan, Co2 bio-mitigation using microalgae, Applied Microbiology and Biotechnology, 79(5), 707–718, 2008.
[12] T. Mekonnen, P. Mussone, H. Khalil and D. Bressler, Progress in bio-based plastics and plasticizing modifications, Journal of Materials Chemistry A, 1(43), 13379–13398, 2013.
[13] M. G. Bastos Lima, Toward multipurpose agriculture: food, fuels, flex crops and prospects for a bioeconomy, Global Environmental Politics, 18(2), 143–150, 2018.
[14] F. G. Acién, E. Molina, J. M. Fernández-Sevilla, M. Barbosa, L. Gouveia, C. Sepúlveda, J. Bazaes and Z. Arbib, Economics of Microalgae Production, in Microalgae-based biofuels and bioproducts, Elsevier, 485–503, 2017.
[15] A. Richmond et al. Biological principles of mass cultivation. Handbook of microalgal culture: Biotechnology and applied phycology, 125–177, 2004.
[16] J. Masojdek, M. Koblzek and G. Torzillo. Photosynthesis in microalgae. Handbook of microalgal culture: biotechnology and applied phycology, 20, 2004.
[17] J. S. Burlew et al. Algal culture from laboratory to pilot plant. Algal culture from laboratory to pilot plant., 1953.
[18] M. Janssen, L. De Bresser, T. Baijens, J. Tramper, L. R. Mur, J. F. H. Snel and R. H. Wijffels, Scale-up aspects of photobioreactors: effects of mixing-induced light/dark cycles, Journal of Applied Phycology, 12(3), 225–237, 2000.
[19] E. Sforza, D. Simionato, G. M. Giacometti, A. Bertucco and T. Morosinotto, Adjusted light and dark cycles can optimize photosynthetic efficiency in algae growing in photobioreactors, PloS One, 7(6), e38975, 2012.
[20] E. Molina Grima, F. Garca Camacho, J. A. Sanchez Perez, J. M. Fernández Sevilla, F. G. Acién Fernández and A. C. Gomez, A mathematical model of microalgal growth in light-limited chemostat culture, Journal of Chemical Technology & Biotechnology: International Research in Process, Environmental and Clean Technology, 61(2), 167–173, 1994.
[21] J. U. Grobbelaar, Physiological and technological considerations for optimising mass algal cultures, Journal of Applied Phycology, 12(3), 201–206, 2000.
[22] Q. Bechet, A. Shilton, O. B. Fringer, R. Munoz and B. Guieysse, Mechanistic modeling of broth temperature in outdoor photobioreactors, Environmental Science & Technology, 44(6), 2197–2203, 2010.
[23] O. Bernard and B. Rémond, Validation of a simple model accounting for light and temperature effect on microalgal growth, Bioresource Technology, 123, 520–527, 2012.
[24] M. Ras, J.-P. Steyer and O. Bernard, Temperature effect on microalgae: a crucial factor for outdoor production, Reviews in Environmental Science and Bio/Technology, 12(2), 153–164, 2013.
[25] Q. Hu, et al., Environmental effects on cell composition, Handbook of Microalgal Culture: Biotechnology and Applied Phycology, 1, 83–93, 2004.
[26] M. A. C. L. De Oliveira, M. P. C. Monteiro, P. G. Robbs and S. G. F. Leite, Growth and chemical composition of spirulina maxima and spirulina platensis biomass at different temperatures, Aquaculture International, 7(4), 261–275, 1999.
[27] I. Rafiqul, A. Hassan, G. Sulebele, C. Orosco and P. Roustaian. Influence of temperature on growth and biochemical composition of spirulina platensis and s. fusiformis. 2003.
[28] S. J. Rowland, S. T. Belt, E. J. Wraige, G. Massé, C. Roussakis and J.-M. Robert, Effects of temperature on polyunsaturation in cytostatic lipids of haslea ostrearia, Phytochemistry, 56 (6), 597–602, 2001.
[29] Y. Lee, et al., Algal Nutrition: Heterotrophic Carbon Nutrition, in Handbook of microalgal culture: biotechnology and applied phycology, 116–124, 2004.
[30] S. P. Singh and P. Singh, Effect of co2 concentration on algal growth: a review, Renewable and Sustainable Energy Reviews, 38, 172–179, 2014.

[31] T. L. Da Silva and A. Reis, Scale-up Problems for the Large Scale Production of Algae, in Algal biorefinery: an integrated approach, Springer, 125–149, 2015.
[32] I. De Godos, J. L. Mendoza, F. G. Acién, E. Molina, C. J. Banks, S. Heaven and F. Rogalla, Evaluation of carbon dioxide mass transfer in raceway reactors for microalgae culture using flue gases, Bioresource Technology, 153, 307–314, 2014.
[33] J. C. Merchuk and X. Wu, Modeling of photobioreactors: application to bubble column simulation, Journal of Applied Phycology, 15(2), 163–169, 2003.
[34] Y. Azov, Effect of ph on inorganic carbon uptake in algal cultures, Applied and Environmental Microbiology, 43(6), 1300–1306, 1982.
[35] T. Duarte-Santos, J. L. Mendoza-Martn, F. G. Acién Fernández, E. Molina, J. A. Vieira-Costa and S. Heaven, Optimization of carbon dioxide supply in raceway reactors: influence of carbon dioxide molar fraction and gas flow rate, Bioresource Technology, 212, 72–81, 2016.
[36] C. Bergstrom, C. McKeel and S. Patel. Effects of ph on algal abundance: A model of bay harbor, michigan. 2007.
[37] P. M. Visser, B. W. Ibelings, M. Bormans and J. Huisman, Artificial mixing to control cyanobacterial blooms: a review, Aquatic Ecology, 50(3), 423–441, 2016.
[38] J. T. Lehman, Understanding the role of induced mixing for management of nuisance algal blooms in an urbanized reservoir, Lake and Reservoir Management, 30(1), 63–71, 2014.
[39] G. Padmaperuma, R. V. Kapoore, D. J. Gilmour and S. Vaidyanathan, Microbial consortia: a critical look at microalgae co-cultures for enhanced biomanufacturing, Critical Reviews in Biotechnology, 38(5), 690–703, 2018.
[40] C. Wang and C. Q. Lan, Effects of shear stress on microalgae–a review, Biotechnology Advances, 36(4), 986–1002, 2018.
[41] I. Rawat, R. Ranjith Kumar, T. Mutanda and F. Bux, Biodiesel from microalgae: a critical evaluation from laboratory to large scale production, Applied Energy, 103, 444–467, 2013.
[42] Elia Armandina Ramos Tercero, Giacomo Domenicali and Alberto Bertucco. Autotrophic production of biodiesel from microalgae: an updated process and economic analysis, Energy, 76, 807–815, 2014.
[43] J. L. Manganaro and A. Lawal, Co2 life-cycle assessment of the production of algae-based liquid fuel compared to crude oil to diesel, Energy & Fuels, 30(4), 3130–3139, 2016.
[44] J. E. Longcore, S. Qin, D. Rabern Simmons and T. Y. James, Quaeritorhiza haematococci is a new species of parasitic chytrid of the commercially grown alga, haematococcus pluvialis, Mycologia, 112(3), 606–615, 2020.
[45] W. Blanken, P. Richard Postma, L. De Winter, R. H. Wijffels and M. Janssen, Predicting microalgae growth, Algal Research, 14, 28–38, 2016.
[46] J. L. Mendoza, M. R. Granados, I. De Godos, F. G. Acién, E. Molina, C. Banks and S. Heaven, Fluid-dynamic characterization of real-scale raceway reactors for microalgae production, Biomass and Bioenergy, 54, 267–275, 2013.
[47] J. L. Mendoza, M. R. Granados, I. De Godos, F. G. Acién, E. Molina, S. Heaven and C. J. Banks, Oxygen transfer and evolution in microalgal culture in open raceways, Bioresource Technology, 137, 188–195, 2013.
[48] Uk-hyeon Yeo, In-bok Lee, Il-hwan Seo and Rack-woo Kim, Identification of the key structural parameters for the design of a large-scale pbr, Biosystems Engineering, 171, 165–178, 2018.
[49] J. U. Grobbelaar, Factors governing algal growth in photobioreactors: the “open” versus “closed” debate, Journal of Applied Phycology, 21(5), 489–492, 2009.
[50] A. D. Kroumov, A. N. Módenes, D. E. G. Trigueros, F. R. Espinoza-Quiñones, C. E. Borba, F. B. Scheufele and C. L. Hinterholz, A systems approach for co2 fixation from flue gas by microalgae – theory review, Process Biochemistry, 51(11), 1817–1832, 2016.

[51] C. L. Hinterholz, A. R. Schuelter, A. N. Módenes, D. E. G. Trigueros, C. Borba, F. Espinoza-Quiñones and A. Kroumov, Microalgae flat plate-photobioreactor (fp-pbr) system development: computational tools to improve experimental results, Acta Microbiologica Bulgarica, 33, 119–124, 2017.
[52] F. B. Scheufele, C. L. Hinterholz, M. M. Zaharieva, H. M. Najdenski, A. N. Módenes, D. E. G. Trigueros, C. E. Borba, F. R. Espinoza-Quiñones and A. D. Kroumov, Complex mathematical analysis of photobioreactor system, Engineering in Life Sciences, 19(12), 844–859, 2019.
[53] K. Itaoka, A. Saito and M. Akai, Public Acceptance of Co2 Capture and Storage Technology: A Survey of Public Opinion to Explore Influential Factors, in Greenhouse gas control technologies 7, Elsevier, 1011–1019, 2005.
[54] K. Arning, J. O.-V. Heek, A. Linzenich, A. Kätelhön, A. Sternberg, A. Bardow and M. Ziefle, Same or different? insights on public perception and acceptance of carbon capture and storage or utilization in germany, Energy Policy, 125, 235–249, 2019.
[55] K. Arning, J. O.-V. Heek, A. Sternberg, A. Bardow and M. Ziefle, Risk-benefit perceptions and public acceptance of carbon capture and utilization, Environmental Innovation and Societal Transitions, 35, 292–308, 2020.
[56] W. Kaminsky. Possibilities and limits of pyrolysis. In Makromolekulare Chemie. Macromolecular Symposia, volume 57, 145–160. Wiley Online Library, 1992.
[57] C. D. Le. Gasification of biomass: An investigation of key challenges to advance acceptance of the technology. PhD thesis, University of Bath, 2012.
[58] N. Hauke, Die grüne revolution an der tankstelle? die relevanz politischer narrative am beispiel der einführung des biokraftstoffes e10, in Politische narrative, Springer, 173–197, 2014.
[59] M. Thema, F. Bauer and M. Sterner, Power-to-gas: electrolysis and methanation status review, Renewable and Sustainable Energy Reviews, 112, 775–787, 2019.
[60] B. W. Terwel, F. Harinck, N. Ellemers and D. D. L. Daamen, Going beyond the properties of co2 capture and storage (ccs) technology: how trust in stakeholders affects public acceptance of ccs, International Journal of Greenhouse Gas Control, 5(2), 181–188, 2011.
[61] A. G. Silva, R. Carter, F. L. M. Merss, D. O. Corrêa, J. V. C. Vargas, A. B. Mariano, J. C. Ordonez and M. D. Scherer, Life cycle assessment of biomass production in microalgae compact photobioreactors, Gcb Bioenergy, 7(2), 184–194, 2015.
[62] P. Pérez-López, J. H. De Vree, G. Feijoo, R. Bosma, M. J. Barbosa, M. T. Moreira, R. H. Wijffels, A. J. B. Van Boxtel and D. M. M. Kleinegris, Comparative life cycle assessment of real pilot reactors for microalgae cultivation in different seasons, Applied Energy, 205, 1151–1164, 2017.
[63] T. Sarat Chandra, M. Maneesh Kumar, S. Mukherji, V. S. Chauhan, R. Sarada and S. N. Mudliar, Comparative life cycle assessment of microalgae-mediated co 2 capture in open raceway pond and airlift photobioreactor system, Clean Technologies and Environmental Policy, 20(10), 2357–2364, 2018.
[64] International Organization for Standardization. Environmental Management: Life Cycle Assessment; Principles and Framework, volume 14044. ISO, 2006.
[65] International Organization for Standardization. Environmental management: life cycle assessment; requirements and guidelines. ISO Geneva, Switzerland, 2006.

Rahmat Sotudeh-Gharebagh, Jamal Chaouki, Fatemeh Vatankhah, Jaber Shabanian, Mohammad Latifi

12 Discussion and concluding remarks

Abstract: This chapter summarizes the concepts and basics, enabling tools and techniques, and promising applications of the iterative scale-up method for six different processes. Three open cases are also defined herein. They provide an opportunity for readers, engineers, practitioners, and scientists to practice the implementation of the iterative scale-up method and gain expertise before applying it to their specific cases. We believe that with the practical information reported in the previous chapters of this book and the power of information technology, significant time and money can be saved if a company skips piloting before the commercial design. This is in broad agreement with the new normal where companies show less interest in the application of the iterative scale-up method to commercialize their processes as their resources become scarcer. It would also allow new case studies based on this method to emerge with the help of universities, companies, and engineering firms. The new scale-up method will be viewed as the new blood in process innovation, design, and scale-up in favor of decreasing costs and preparation time to commercialization. We hope that this scale-up method will be integrated into design courses to train a new generation of graduates for its successful commercial implementation.

Keywords: conventional scale-up, iterative scale-up, case study, design, training

Acknowledgments: The authors are grateful to Dr. Bahman Yari and Dr. Rouzbeh Jafari for valuable discussions concerning the MDA process and Dr. Milad Aghabararnejad concerning CLC process.

Rahmat Sotudeh-Gharebagh, School of Chemical Engineering, College of Engineering, University of Tehran, P.O. Box 11155-4563, Tehran, Iran, e-mail: sotudeh@ut.ac.ir
Jamal Chaouki, Department of Chemical Engineering, Polytechnique Montréal, 2500 Chemin de Polytechnique, Montréal, Québec, H3T 1J4, Canada, e-mail: jamal.chaouki@polymtl.ca
Fatemeh Vatankhah, Department of Chemical Engineering, Polytechnique Montréal, 2500 Chemin de Polytechnique, Montréal, Québec, H3T 1J4, Canada, e-mail: fatemeh.vatankhah@polymtl.ca
Jaber Shabanian, Process Engineering Advanced Research Lab, Department of Chemical Engineering, Polytechnique Montréal, C.P. 6079, Station Center-Ville, Montréal, Québec, H3C 3A7, Canada, e-mail: jaber.shabanian@polymtl.ca
Mohammad Latifi, Department of Chemical Engineering, Polytechnique Montréal, 2500 Chemin de Polytechnique, Montréal, Québec, H3T 1J4, Canada, e-mail: mohammad.latifi@polymtl.ca

https://doi.org/10.1515/9783110713985-012

12.1 Introduction

The key difference between conventional and iterative scale-up methods is related to how plant designers can take advantage of pilot plants in the commercialization of a process as shown in Figure 12.1.

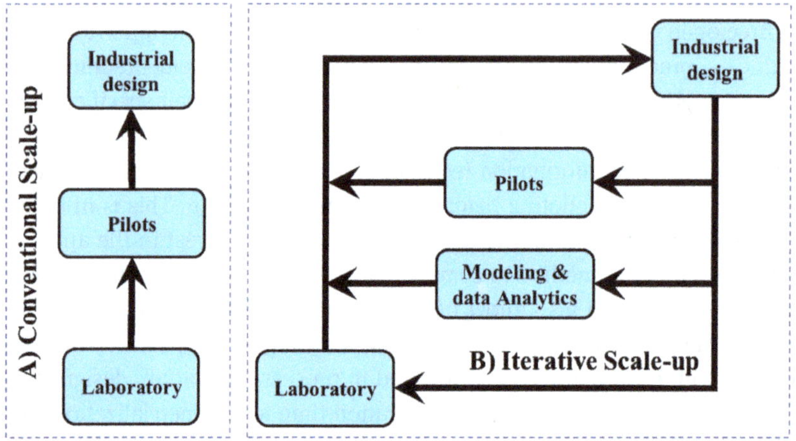

Figure 12.1: Schematic representations of conventional and iterative scale-up methods.

In the iterative scale-up method with four major components, piloting, which is central to the conventional scale-up method, is potentially (but not necessarily) removed from the scale-up path. Chemical and biological process industries avoid operating uncertain pilot plants due to their minimal benefits, operational concerns, and rather significant expenses. This leads to a direct pre-design of a commercial plant based on the data available in literature or that obtained from experimental trials on laboratory, bench, and small pilot scale plants, as well as process modeling and simulation. The novel scale-up method, however, helps remedy the shortcomings of the conventional scale-up method and fills the knowledge gap created in the wake of the pilot plant removal. In the following sections, the key features and shortcomings of the conventional scale-up method and the promising future of the iterative scale-up are briefly discussed.

12.1.1 Conventional scale-up method: challenges and opportunities

In Chapter 1, prepared by Sotudeh-Gharebagh and Chaouki [1], it was concluded that although the conventional scale-up method can deal with gas and liquid processing, the particulate processes face serious challenges. The uncertain sizes of pilot plants are perceived as increasing costs and seemingly slowing the scale-up process before industrial design as they increase not only costs but time investment. Although a pilot plant's value is often underappreciated until a failure occurs, going forward companies will choose their processes to pilot more selectively as resources are stretched thin. As evident in the literature, the scale-up of solid processing units is more challenging due to the tediousness of characterizing and fully understanding them. Irregular process interactions occur with various internals within unit operations, equipment, and instrumentation. There is also no clear procedure for verifying if suppliers or third-party design firms meet regulatory specifications. Mathematical modeling, on the other hand, has high potential and could be applied to many problems alongside with other tools. Presently, however, scale-up practices are failing to fill the gap between scale-up and start-up and only provide incremental progress in some areas. The path forward based on process simulation, down-scaled pilot plants, re-visiting laboratory data and the progress a design knowledge repository for the identification and minimization of risks to an acceptable level is clear and scale-up efforts on many fronts should continue.

12.1.2 Iterative scale-up method, concept, and basics

Chapter 2, prepared by Chaouki and Sotudeh-Gharebagh [2], introduced the basics and concepts behind the iterative scale-up method. Figure 12.1 illustrated how the conventional scale-up begins with the laboratory unit, progresses to the pilot plant, and finishes with the industrial design, while in the iterative scale-up method, we move directly from a laboratory-scale to commercial design. At this stage, a preliminary design uncovers all potential technical and economic problems that are then solved as accurately as possible using various tools and techniques through iteration. Therefore, to improve the key characteristics of the design, it may be necessary depending on the conditions to use new data generated from either small-scale tests, modeling, or data analytics. It is only at this stage that a pilot should be considered, and the decision made regarding its size, subset, operation, and way to generate useful data since the objective of this step is to reveal any ambiguities. Each iteration generates increased knowledge to identify and mitigate more risks to an acceptable level so that the commercial design improves. Risk identification is carried out several times during the scale-up for project extrapolation. The final solution can be obtained through progressive iterations until satisfactory conditions are met. The above practice and paradigm shift would not only bridge the gap between the conventional scale-up time and start-up time of projects but would reduce the design, implementation, and operational costs in the process deployment.

12.2 Tools and techniques

Proper application of enabling tools and techniques is central to the iterative scale-up method. By integrating these tools into the scale-up, the commercial design is progressively improved during the iteration to minimize technical, economic, and environmental uncertainties. The summary of key points developed in Chapters 3–5 is concluded in this section.

12.2.1 Process extrapolation by simulation

In Chapter 3, prepared by Zarghami et al. [3], it was shown that the process simulation can provide valuable information and minimize the scaling costs by decreasing the size of the pilot or even changing it to an e-pilot. The chapter explored the advantages and limitations of e-pilots in an iterative scale-up method and introduced the role and importance of simulation in process and equipment scales. Process simulators, used at the preliminary design stage, predict the behavior of the chemical process. Unit simulators (like CFD[1] codes), on the other hand, are mostly developed for use in the detailed design. The flow of information between process and unit simulations within a multiscale modeling approach is then explored. To overcome the CFD limits and shortcomings, the compartmental modeling (CM) and hybrid CFD–CM models based on hydrodynamic similarity are introduced. These models have simplified and/or reduced models that are still capable of simulating higher unit scales with good precision and acceptable computation time. The CFD e-pilot can only be applied to the unit scales depending on availability of high computational resources. Reliable e-pilots do, however, require practical verification and validation methods. Also, the uncertainty quantification methods are essential for quantifying the various sources of uncertainty within the e-pilot.

12.2.2 Transition from e-pilot to full commercial scale

In Chapter 4, prepared by Bashiri et al. [4], the conventional scale-up method for the design and scale-up of multiphase mixing systems is discussed, highlighting challenges and risks. As an alternative to this method, applications of the iterative scale-up method in the form of e-pilot are presented to provide a short and reliable path to large-scale technology implementation. This chapter introduced both a methodology for developing hybrid multiscale models, which is an efficient design tool for iterative scale-up, and multiscale gas/liquid flow models to predict the performance

1 Computational Fluid Dynamics (CFD)

of multiphase mixing systems. The models are used for the iterative scale-up of industrial processes, such as the production of specialty and bulk chemicals in stirred tank reactors, and CO_2 capture and acid gas removal using a rotating packed bed in which the mass transfer between the gas and liquid phase is vital for the success of large-scale operations.

12.2.3 Life-cycle assessment and technology scale-up

Chapter 5, developed by Majeau-Bettez [5], addressed the environmental performance of processes under development. To jointly manage the various types of environmental impacts and mitigate the corresponding risks, a systematic approach is necessary. A life-cycle assessment (LCA) provides this system perspective and informs technology choices although certain obstacles have limited the usefulness of conventional LCAs in process design. To provide effective guidance, LCAs can benefit from the iterative, uncertainty-focused, model-based approach to process development that is presented throughout this book. Exploring different strategies, this chapter examined how to make the best use of available data for environmental assessment from lab scale to commercialization and even mass deployment. These strategies range from qualitative early technology screening to detailed modeling and analyses. Furthering iterative refinement, this chapter argued that reducing uncertainty as a guiding principle to focus LCA efforts and a closer integration of detailed engineering, process simulation, and environmental system modeling deserves a more central role. This integration would put new requirements not only on engineers' best practices, but also on software and database development efforts.

12.3 Complete case studies

The main challenge with the iterative scale-up is related to confidentiality regarding the release of data from existing practices. Due to a lack of reliable information on the attempted conventional scale-ups and their link with industrial failures, process designers still believe that pilot plants should be in place, even though dealing with the uncertainties of the common pilot sizes is not possible. It will take time to overcome these cultural barriers since a complete acceptance of iterative scale-up and collaborative efforts are needed for this purpose. Companies may be more selective in the future on choosing processes to pilot as resources become scarcer [6]. However, if alternative and less challenging solutions are proposed, it could attract their attention. At the same time, the enabling technologies also have their own limitations in dealing with uncertainties as detailed in Chapter 3. These existing barriers may partly slow the development of the iterative scale-up, but since there is a strong desire to take advantage of rapidly growing digitization tools in the process industry, the future will be much more promising. In the following case studies, the application enabling technologies and tools are shown to be easing the scale-up process.

12.3.1 Case study I: *n*-Butane partial oxidation to maleic anhydride – VPP manufacture

The case study in Chapter 6 was developed by Patience [7] and showed the significant uncertainties the DuPont's research and development (R&D) team faced throughout the design and manufacture of the vanadium pyrophosphate (VPP) catalyst. The scale-up from experimental to commercial design was rigorous and the uncertainties in the process were successfully addressed. Attrition resistance was the original focus and although the catalyst did not fall apart in the pilot plant, it was still recommended to assume high attrition rates in commercial economic calculations. Unfortunately, the final commercial design was severely impacted and it viability was compromised. The original calcination activation protocol called for a temperature hold of 18 h at 460 °C, which the commercial team rejected because it required stainless steel and would double the reactor cost. The research team overcame this constraint by carrying out several hundred tests in a pressurized fluidized bed and developed a procedure for a better catalyst. The thermal and chemical stabilities of the catalyst were excellent. The VPP was operated at a pressure of 350 kPa (absolute) and 400 °C with 9% n-C_4H_{10} and 3–4% O_2 at the exit. The commercial facility was operated for almost a decade primarily with the same catalyst from the start-up period. Demonstrating considerable ingenuity, the R&D team developed a way to recycle catalyst further reducing operating costs.

12.3.2 Case study II: *n*-Butane partial oxidation to maleic anhydride – commercial design

In Chapter 7, a case study developed by Patience and Chaouki [8], the next step in building single-train reactors to produce specialty chemicals was DuPont's commercialized process. Coupling the partial oxidation of butane with the hydrogenation of maleic acid to tetrahydrofuran in an intensified reactive distillation unit was certainly more economical than the technology used at the time. Applying an iterative scale-up method, they went from laboratory scale to commercial design and then building a pilot plant and cold flow facilities to resolve uncertainties posed by the scale-up. The plant was operated for almost 10 years before being shutdown due to design flaws, market pressures, and a change in corporate vision. The plant required more modifications to its configuration and the significant financial investment could only be justified if management believed that the market would continue to grow and that they had a low-cost process. The DuPont case underscores the importance of identifying and addressing all technical uncertainties during the commercial design. Failure to recognize and/or assess these uncertainties due to a lack of experience or time can be catastrophic. It also demonstrates the importance of leadership to provide the technical team with sufficient time to resolve the problems encountered during the piloting and commercial start-up.

12.3.3 Case study III: Methanol to olefins

In Chapter 8, which is a case study developed by Shabanian et al. [9], it was shown how the production of methanol to olefins (MTO) was industrially implemented over the past decade. Among major technologies, the DMTO[2] technology is dominant. Research on how this technology was successfully developed and scaled up its reactor, forms the core of this chapter. DICP successfully scaled up its DMTO fluidized bed reactor after carrying out experiments at four different scales from laboratory to pilot, demonstration and commercial. After more than 25 years, the fundamental research on the reactor selection and scale-up produced the world's first MTO commercial plant in 2010. The proper selection of the reactor and its design were key challenges in the successful development of many processes, particularly those with catalytic reactions. Using the iterative scale-up method, they demonstrated how to select the most appropriate reactor and its critical operating conditions within a tighter timeframe. To implement the new scale-up method, they adopted critical background information and experimental data at laboratory scale, and a decision-making procedure for the selection of a proper gas–solid catalytic reactor based on weight hourly space velocity. To minimize the risks of the new method, they identified its uncertainties and presented mitigation plans to address them by supplementary laboratory- and pilot-scale results. Directions on how to further improve the profitability of available technology with the help of process simulation were also discussed.

12.3.4 Case study IV: Hydropotash from potassium feldspar

In Chapter 9, which is a case study developed by Latifi et al. [10], the feasibility of the production of high temperature potash (HTP)[3] from potassium feldspar was investigated using the iterative scale-up method. Lab-scale data revealed that HTP process operates at a temperature much lower than that in similar high temperature processes. On the other hand, applying the iterative scale-up method discovered that the scale-up of the process is not economically viable at present due not only to

2 Dalian Institute of Chemical Physics

3 High Temperature Potash

the cost of the HTP process, but also to the limitation caused by the market. After economically evaluating the process that included in situ regeneration of calcium nitrate, a potential pathway to commercial scale using a scalable process opens up. The iterative scale-up method extracted results demonstrating that it is still too early to investigate the process operation at a pilot-scale, especially due to the estimated cost still being above the criteria. For in situ regeneration of the calcium nitrate unit, the process was only based on the flowsheet developed by extracting information from literature and simulation data. Thus, a new campaign of lab-scale tests will be required to verify the feasibility of calcium nitrate regeneration after a systematic kinetic study. That is, based on the iterative scale-up method, the scale-up of the HTP process might become feasible if the fertilizer was in such great demand that it caused a dramatic spike in the price or if an economic regeneration of calcium nitrate with a cost-effective raw material, like calcium carbonate, was used.

12.3.5 Case study V: Lactide production process development

In the case study in Chapter 10, developed by Yazdanpanah and Chaouki [11], the iterative scale-up methodology was applied to develop, evaluate, and benchmark an industrial process. A simple, basic flow diagram marked the beginning of the process development. The technology was improved step by step until it covered all the necessary stages from feed treatment to product purification, including heat integration. Taking the overall performance into consideration, the optimal local selection was determined after proposing and analyzing various options for each step. To design different sections, experimental tests and modeling were used. Heat and process integration and intensification are recommended to identify potential improvements. Finally, the novel technology was economically evaluated to compare it with existing and alternative processes and identify areas where further improvement is required in the next steps. A significant advantage of the iterative method is that it carries out the industrial design in less time. Four patents have been filed for the industrial-scale process design. A series of steps were followed in an iterative method, including the definition of design basis, the development of a preliminary process flow diagram, the detailed study of process elements, process simulation, equipment design, and CAPEX[4] and OPEX[5] estimation. These steps are all interdependent and require integrated continuous improvement during development. Finally, a steady-state pilot design is proposed for the next development stage.

4 Capital Expenditures
5 Operating Expenditures

12.3.6 Case study VI: Microalgae-based technologies

In the case study in Chapter 11, developed by El Bahraoui et al. [12], it was shown that among emerging technologies for CO_2 capturing, biological mitigation by microalgae proves to be a promising alternative to physical methods. Microalgae uses light and inorganic compounds to produce complex organic molecules through photosynthesis. The process converts light into chemical energy in the microorganism cells leading to natural conversion of CO_2 to glucose and oxygen. As microalgae grow, the carbohydrate, protein, and lipid cell content tend to be higher and the harvest can be processed into biofuels. Processing microalgae into biofuels has the potential to mitigate global warming effects and alleviate the energy crisis. In an industrial application, the main challenge for massive microalgae production is light limitation due to substantial light losses and heat generation inside the bioreactor. Using traditional scaling, the chapter shows that there have been many failures: industrial bioreactors are in 2D and produce very little compared to the areas occupied by these processes. With an iterative scaling, to go from the laboratory scale to the industrial scale, it is necessary to carry out a very complex modeling and simulation, which must bring together multidisciplinary teams and carry out a LCA beforehand. In fact, several microalgae processes emit more CO_2 than they consume. This type of scaling makes it possible to better address the industrial problem and correct the failures before their implementation.

12.4 Open case studies

12.4.1 Case study N-I: Chemical looping combustion process

12.4.1.1 Problem statement

The chemical looping combustion (CLC) process is a leading future technology in CO_2 capture. This process consists of two reactors, that is, air and fuel reactors, which operate at moderately high temperatures, ~800 °C. The selection of an outperforming redox and cost-effective metallic particles is a critical step in the CLC process since these particles are circulated between reactors serving as oxygen carriers (OCs). The metallic particles, for example, Ni, Cu, Cd, Mn, and Fe, and minerals, which have a high potential to be reduced and oxidized periodically, are adopted as OCs. As illustrated in Figure 12.2, in the air reactor, the metallic particles react with oxygen from air to form metal oxides (MeOs), which are then transported to the fuel reactor to provide the required oxygen for the combustion of fuels. The reduced metallic particles are then recycled to the air reactor for regeneration, that is, re-oxidation. With this compact configuration, oxygen is indirectly supplied to the combustion reaction to avoid the direct contact of fuel with the air. This promising mode of contact and the advantages of the operation at rather high temperatures allow the formation of fuel and thermal NO_x to be minimized. As a result, the exhaust of the fuel reactor would mostly consist of steam (H_2O) and carbon dioxide (CO_2) and would be free of nitrogen, oxygen, and NO_x. Steam can be conveniently separated from the exhaust stream by condensation and a relatively pure stream of CO_2 can be obtained, which can be further captured for sequestration or utilization in various industrial applications, for example, CO_2 to the chemical, or food and beverage industries.

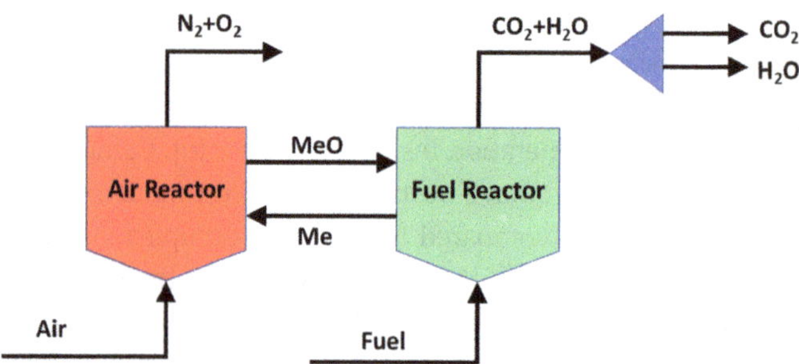

Figure 12.2: Schematic representation of a chemical looping combustion unit.

Most experimental trials have been either unsuccessful or ended in pilot plants with capacities of 10–200 kW as shown in Table 12.1. No commercial CLC unit has been constructed due to reaction and hydrodynamic uncertainties and only conceptual designs involving 100–1000 MW were proposed in the literature [13, 14]. Lyngfelt et al. [13] fully reviewed 11,000 h of CLC operations highlighting the current technology status and future perspectives. Adánez and Abad [15] reviewed the status and future research needs on CLC processes for industrial and power plant applications. They have also mentioned that the CLC process with gaseous fuels was demonstrated mainly at small scale units ranging from 10 to 120 kW and that it is time for a scale-up.

12.4.1.2 Challenges with conventional scale-up method

The main challenges with the conventional scale-up of CLC units are related to the considerable budget and time needed to construct a properly sized pilot plant suitable to drive correlations for successful commercial scale-up. Hydrodynamic and reaction uncertainties, which may lead to serious failures, would make managerial decisions very tedious and costly. There have been enormous studies conducted at laboratory and pilot scales dedicated to CLC units and selected works are reported in Table 12.1. However, the commercialization of CLC units has not been very successful

Table 12.1: Selected CLC units at pilot scales with their key characteristics.

Firm	Fuel	Oxygen carrier	Temperature (°C)	Power (kW)	Conversion (%)	Remark
Chalmers Univ. [16]	Methane	Nickel-based	950	10	95	100 h operation One batch of particles Need for longer operation
Chalmers Univ. [13]	NG,[6] SG,[7] PC,[8] LHC,[9] Coal, Biomass etc	ores or waste NiO, Fe_2O_3, CuO Mn_3O_4, CaO	Variable	0.3–250	Variable	11,000 h operation (total) operated in 46 units
Chalmers Univ. [17]	Coal	Ilmenite[10]	850	10	50–79	22 h operation 140 h circulation of OC OC price ≪ NI-based
Chalmers Univ. [18]	Petroleum coke	Ilmenite	950–990	10	66–78	11 h operation improved FR cyclone low OC market price
Vienna Univ. [14]	Natural gas $CH_4 + CO + H_2$	Nickel based	950	120	–	Good gas–solid contact air to fuel ratio = 1.2 Design and erecting
Instituto de Carboquımica [19]	Methane	Copper based	800	10	100	200 h operation OC to fuel ratio > 1.5 No agglomeration
Southeast Univ. [20]	Biomass	Iron based	750–950	10	66–50	30 h operation One batch of OC Surface sintering of OC
DOE[11] [21]	Natural gas	Nickel–iron based	800–960	40–60	70–90	11 h operation Attrition resistance OC material cost R_{OC}[12]=1 60–300 kg h^{-1}

6 Natural Gas
7 Syngas
8 Petroleum Coke
9 Lower HydroCarbons
10 Natural iron titanium oxide
11 US Department of Energy (DOE)
12 Oxygen carrier circulation rate

yet due to major scale-up risks, which include technical and economical obstacles leading to many added costs or failures [13]. Some of these technical uncertainties are related to, for example, the complexity of a CLC operation dealing with two reactors requiring different residence times, solid circulation rate, proper controlling of the redox rates of OC, attrition and agglomerations, heat integration, and operation at rather elevated pressures and temperatures. This is the reason why the conventional scale-up is rather costly, tedious, and time-consuming.

12.4.1.3 Promises with iterative scale-up method

The iterative scale-up method can be adopted to scale-up commercial CLC units and the relevant key challenges can be properly addressed through an iterative procedure either by re-evaluating the literature, carrying out new experiments in the laboratory on small scales and using cold models, as needed. In addition, the application of e-piloting through modeling and simulation tools detailed in Chapter 3 would ensure minimizing the uncertainties to an acceptable level at minimal cost. Some studies that have also been reported on CLC pilots showed the importance of steady state and transient modeling, and CAPEX and OPEX analysis [22–24]. These and similar studies demonstrate how the iterative scale-up can help ease the commercialization of CLC processes due to its iterative nature. More efforts in this direction are being planned in the future, which will significantly decrease the time and budget requirements and overcome the current scale-up barriers.

12.4.1.4 Key issues to be addressed

The iterative scale-up will not only define new uncertainties, but will also, through the estimation of CAPEX and OPEX, quantify the order of magnitude of uncertain key parameters, for example, the minimum capacity of the oxygen transported by the OC, which helps determine the OC residence times in the fuel and air reactors, as well as its circulation rate between reactors, the minimum allocated attrition related to the loss of solids and the adsorption and desorption kinetics. However, the key challenges for the successful commercialization of CLC processes are as follows:

a. To develop and test advanced and less expensive OCs, which are the core of CLC units [15]
b. To deal with chemical and mechanical properties, attrition rate and agglomeration of OCs
c. To deal with oxidization and regeneration rates of MeOs on a commercial scale
d. To optimize oxygen adsorption/desorption rates and OCs to fuel ratio
e. To manage heat integration between reactors
f. To test various feeds, including gaseous, liquid, and solids feeds

g. To correlate reaction rates with solids and gas residence times and optimal conversions
h. To carry life cycle analysis (LCA) to effectively quantify its efficiency in terms of CO_2 separation

Among these, the most tedious challenge is the choice of efficient and less expensive OCs. Although many OCs have been tested in several laboratory and pilot units as shown in Table 12.1, they do not adequately meet the process requirements as the CAPEX and OPEX of the CLC process using them are high. OCs play a vital role in the performance of the CLC process in terms of the net energy, product and by-product generation, fuel conversion, and air requirement. OCs may show totally different behaviors at various unit scales.

12.4.2 Case study N-II: Methane dehydroaromatization

12.4.2.1 Problem statement

Although methane, that is, the main component of natural gas, is a valuable resource for both energy and chemical production, it is one of the principal sources of greenhouse gas emissions. In some cases, methane is preferably flared and converted to carbon dioxide, which has a lower negative impact on the environment. The methane dehydroaromatization (MDA) process applies a heterogeneous catalytic reaction to directly convert methane into highly aromatic, in-demand hydrocarbons along with hydrogen, the most environmentally friendly energy carrier. Owing to a recent drop in the price of natural gas, where methane acts as its main component, the MDA process has proved to be a potential alternative for the production of aromatics and minimization of industrial dependency on crude oil [25]. However, this reaction is thermodynamically limited, that is, it has a maximum achievable single-pass equilibrium conversion and requires optimum operating conditions, a solid catalyst, and a reactor for the selective conversion of methane to the desired products. The MDA process has not been commercialized yet, due mainly to the unavailability of an effective reactor design on a commercial scale.

Critical elements ought to be comprehensively investigated for the successful commercialization of a chemical process based on a catalytic gas–solid reaction. They include thermodynamics, catalyst, reaction mechanism and kinetics, and reactor design. The molybdenum (Mo) zeolite supported catalyst of Mo-HZSM-5 demonstrated the best catalytic performance for the MDA reaction [26]. On the one hand, the highly endothermic MDA reaction should be achieved at elevated temperatures, 700–800 °C, owing to thermodynamic requirements. On the other hand, the inevitable formation of undesired coke at high temperature yields catalyst deactivation and lower selectivity toward desired products. Hence, an effectively designed reactor should help continuously regenerate catalyst at high regeneration and MDA reaction temperatures. A dual fluidized bed reactor design is currently the largest MDA unit that annually produces 35 tons of benzene and 5 tons of naphthalene at 800 °C and 0.3 MPa [27].

12.4.2.2 Challenges with conventional scale-up method

An effectively designed catalytic gas–solid reactor for the MDA process should satisfy certain requirements. The most critical ones include the following: it should (i) properly operate on a commercial scale and provide (ii) the highest possible gas–solid contact, (iii) online removal of produced hydrogen to ensure the highest methane conversion and selectivity toward desired products, and (iv) uniform temperature distribution to avoid excessive coke formation and inadequate methane conversion in hot and cold zones of the reactor, respectively. In addition to the MDA reactor, a proper catalyst regenerator operating under optimum conditions is required to preserve catalyst-active sites. A low carbon footprint is also necessary for the whole process, including energy supplies and regeneration agent, to design a carbon neutral MDA process.

Some reactor designs are effective only on a laboratory scale and do not work properly or at all on a commercial scale. Although packed bed reactors show acceptable performances for the MDA reaction, they still face serious challenges, such as non-uniform temperature distribution, a high-pressure gradient, and a large investment for a multiple packed bed reactor configuration to continuously regenerate catalysts [27]. Fluidized bed reactors, however, provide a better gas–solid contact, uniform temperature distribution, and better catalysts circulation for their continuous regeneration. Nevertheless, they suffer from a low single-pass conversion. Membrane reactors can enhance the methane single pass conversion by continuous hydrogen removal [28]. A large required membrane area on a commercial scale decelerates the commercialization of MDA membrane reactors. A thorough evaluation of the process on laboratory, pilot, demonstration, and commercial scales is needed to commercialize the MDA process through the conventional scale-up method. Considering the challenges discussed here, it will be a very expensive and time- and energy-consuming task to commercialize the MDA process based on a single reactor type using the conventional scale-up method and it may result in an unrecoverable process development failure. A combination of proper reactor types is preferred to satisfy all corresponding requirements.

12.4.2.3 Promises with iterative scale-up method

To adequately address the requirements of effective MDA reactor–regenerator design, we propose a combined reactor configuration comprised of a circulating fluidized bed reactor–regenerator assembly with embedded membranes and a microwave heating feature. Technical challenges in the successful scale-up of the proposed configuration include: (i) coke formation rate, which influences the coke content, aromatics selectivity, regeneration cycles, and catalyst residence time in the regenerator, (ii) residence times of catalysts and fluidizing/reacting gases in the MDA reactor and regenerator, while catalysts should stay longer in the regenerator, which leads to a different design of the regenerator compared to the MDA reactor, (iii) catalyst attrition that directly affects catalyst synthesis, and MDA reactor and regenerator designs, (iv) total membrane area required to effectively separate hydrogen during the gas residence time in the reactor; this factor varies with the permeability of the selected membrane, (v) membrane mechanical strength to be viable on a commercial scale, and (vi) the amount of required energy on commercial scale.

12.4.2.4 Key challenges to be addressed

The iterative scale-up method, compared to the conventional one, guarantees a faster and more effective and economical MDA process commercialization. This is accomplished by incremental improvements on the final commercial design by meticulously pondering the technical challenges and resolving them iteratively when employing experimental data from small scales and/or process/unit modeling/simulation results. The proposed MDA reactor-regenerator configuration can address all design requirements to successfully commercialize the MDA process. It also orients the research activities toward the application of successfully combined reactors and green and effective process electrification (PE), which is in line with the fourth industrial revolution. Following the iterative scale-up method, the proposed reactor-regeneration configuration needs supplementary experiments on a laboratory scale to delineate process/design unknowns prior to the commercial scale design. Experimental data from the laboratory/pilot scale helps calculate/determine the most critical factors, namely the optimal inlet feed to catalyst ratio, catalyst attrition, weight hourly space velocities and gas and catalyst residence times in the MDA reactor and regenerator, required membrane area, as well as the number, spatial distribution, and power of magnetrons (microwave generators) around the MDA reactor together with their operational frequencies. In addition, modeling and simulation of the new reactor design allow multiple virtual experiments to be performed on the target assembly and ensure a more confident reactor scale-up before/without spending a lot of time, energy, and money at the pilot and demonstration scales.

12.4.3 Case study N-III: Process electrification

12.4.3.1 Problem statement

Process Electrification (PE) is one of the most promising transition paths to low-carbon-footprint manufacturing of fuels, chemicals, and materials. If electricity is produced from a renewable energy resource, such as biomass, wind, or solar resources, the process could be carbon neutral. Various unconventional heating technologies, such as microwave heating, induction heating, plasma, and ultrasound, have demonstrated promising results in terms of efficient heating and simultaneous enhancement of conversion and selectivity in chemical reactions [29–32]. The latter improvements can result in the generation of fewer by-products and residual streams, which consequently require less complex downstream purification processes that are often energy intensive. The temperature difference in multiphase/multicomponent systems, resulting from different radiation susceptibilities of materials, is an advantageous feature in PE. For instance, in gas–solid catalytic reactions with microwave heating, while the catalyst particles are microwave receptors, their temperatures are locally higher than the bulk temperature. Consequently, the gaseous products desorbed from the surface of the catalysts are quickly quenched as they enter the bulk medium of the reactor, thus preventing undesired reactions in the gas phase. Induction and microwave heating are non-invasive and inside-out heating methods that are supplied by electricity. They provide noticeable advantages [31, 32] including but not limited to selective heating, a 60–80% decrease in energy consumption [33–35], and high heat transfer rates [29, 36]. The transformation efficiency of microwaves to thermal energy has been reported between 70% and 95% [37]. There are still numerous research and development gaps to overcome before attaining an electrified process on a commercial scale. For the case of microwave heating-assisted processes, microwave receptors and catalysts are both solid particles in a heterogeneous solid-fluid reactor, wherein there should be a proper coupling/integration between them to avoid parasitic reactions and enhance the heating performance.

12.4.3.2 Key issues to be addressed

The most critical challenges include the reaction kinetics, temperature measurements in multiphase/component systems, process scale-up, leakage and loss of electromagnetic field, and safety. For instance, there is not a global agreement on the impact of microwave heating on kinetic parameters, such as activation energy and reaction constant [32]. Process scalability is one of the main issues in the scale-up of electrified processes, that is, either by microwave heating, induction heating, ultrasound or otherwise. Process scalability, for instance, in microwave heating is governed by several parameters, mostly the penetration limits of the applied waves and the exposed materials. In addition, if the microwave irradiations are not absorbed by the material, they could initiate sparks on the mechanical parts of the reactor depending on the reactor geometry or the reflected waves may damage the wave generator.

12.4.3.3 Challenges with conventional scale-up method

The presence of an electric, magnetic, or electromagnetic field inside a multiphase reactor implies complex interactions among materials, and hydrodynamic and electromagnetic forces. These interactions must be investigated to identify the impact of each force on reactor performance. In addition, these interactions depend on the type of electrification mechanism, that is, microwave, induction, ultrasound, and on the physicochemical properties of materials, such as chemical composition, crystallography, mineralogy, purity, heat capacity, electrical conductivity, particle size distribution. Lab-scale investigations are mainly focused on heating a small volume of sample to determine reaction kinetics, whereas the importance of electromagnetic phenomena and their sole interaction with materials are usually undermined. The mechanical design of the reactor, power supply, and distribution of electromagnetic waves at a larger scale are certainly different than in a lab-scale rig since the reactor must be able to operate steadily, safely, and effectively under desired conditions. The construction of a pilot-scale electrified reactor based on lab-scale data, and without employing a prior delineated simulation, could result in manufacturing a couple of different prototypes, while losing a significant amount of available capital investment. Hence, with respect to aforementioned unknowns, adopting a conventional scale-up method for an electrified process will cause a substantial financial and technical failure.

12.4.3.4 Promises with iterative scale-up method

Some critical questions in the scale-up of an electrified process are how different is it going to be from a similar process that employs a conventional heating mechanism and how can it guarantee a uniform and effective distribution of the electromagnetic field to the material in a large-scale reactor. These questions could be answered by employing the iterative scale-up method. An e-pilot can be envisioned through a systematic campaign of simulation and modeling activities. Maxwell's electromagnetic equations should be solved simultaneously with Navier–Stokes' fluid mechanic equations. This integrated approach should consider various operating conditions inside multiphase reactors, physicochemical properties of materials, hydrodynamics, and reactor geometries. The development of the e-pilot can start with some simple assumptions and gradually become more complex through a series of iterations. Consequently, the impact of electromagnetic and hydrodynamic phenomena on reactor performance can be identified by conducting a series of sensitivity analyses. For instance, in the case of microwave heating-assisted reactors, following the uniform exposure of the multiphase bulk to the waves by a properly designed magnetron, wave guides and the mixing pattern of the multiphase bulk can be analyzed. Subsequently, a minimum viable prototype (MVP) can be physically manufactured to validate the simulation and modeling data. Experimental data and modeling data can then be circulated between the MVP and e-pilot until an optimized design of the MVP is achieved. The optimized MVP must safely operate with zero wave leakage, while providing a reliable temperature control. Currently, there is almost no scaled-up microwave heating-assisted process, even at a pilot-scale and especially at temperatures as high as in a pyrolysis reactor. More innovation is still required to enhance the adequate distribution of microwaves within the heating target, while decreasing the initial cost of microwave generators. The iterative method is deemed to be the right path for the scale-up of electrified processes due to its cost effectiveness, accuracy, and speed.

12.5 Concluding remarks

In the future companies will have little interest in the conventional scale-up method for their process development as both financial resources become scarcer and time-to-market is reduced. Therefore, more opportunities will be created for process exploitation through the iterative scale-up method as conceptualized with the basics and case studies reported throughout this book. This new method will shorten the time span between the lab scale and commercial design using advanced enabling tools and techniques instead of uncertain direct piloting. It will decrease the design and implementation time, CAPEX and OPEX in the process deployment chain. However, to widely illustrate the effectiveness of the iterative scale-up method, it can be implemented in successfully developed processes depending on the availability of real plant data to show how it can save time, energy, and expenses for similar or even better results. We believe that with the accessibility of high-performance computing infrastructure for multiscale problems and the availability of massive industrial data from the process industry, this innovative scale-up method will create more added value in the chemical, mineral, and biochemical process industry. This is what the fourth industrial revolution needs to empower companies with smarter ways of doing business, which will improve productivity, safety, and profitability. However, there are some cultural and organizational barriers and bottlenecks to overcome within companies in order to slightly change their design routine and decrease costs, for example, piloting, releasing more data to verify and validate the iterative scale-up method. At this stage, greater confidence and powerful functionalities will be achieved enabling companies to update their design practices and engineering schools to adapt their design teaching to this emerging method. Finally, we strongly believe that the future of the iterative scale-up method is very bright and will have a huge impact on the success of the chemical industry.

References

[1] R. Sotudeh-Gharebagh and J. Chaouki, Chapter 1: Conventional Scale-up: Challenges and Opportunities, in Scale-up processes: iterative methods for the chemical and biochemical industries, J. Chaouki and R. Sotudeh-Gharebagh, Eds., Berlin: Walter de Gruyter GmbH, 2021. doi: https://doi.org/10.1515/9783110713985-001.

[2] J. Chaouki and R. Sotudeh-Gharebagh, Chapter 2: Iterative Scale-Up: Concept and Basics, in Scale-up processes: iterative methods for the chemical and biochemical industries, J. Chaouki and R. Sotudeh-Gharebagh, Eds., Berlin: Walter de Gruyter GmbH, 2021. doi: https://doi.org/10.1515/9783110713985-002.

[3] R. Zarghami, S.-G. R. B. Blais, N. Mostoufi and J. Chaouki, Chapter 3: Process Extrapolation by Simulation, in Scale-up processes: iterative methods for the chemical and biochemical industries, J. Chaouki and R. Sotudeh-Gharebagh, Eds., Berlin: Walter de Gruyter GmbH, 2021. doi: https://doi.org/10.1515/9783110713985-003.

[4] H. Bashiri, R. Rabiee, A. Shams, S. Golshanb and B. C. J. Blais, Chapter 4: Transition from e-Pilot to Full Commercial Scale, in Scale-up processes: iterative methods for the chemical and biochemical industries, J. Chaouki and R. Sotudeh-Gharebagh, Eds., Berlin: Walter de Gruyter GmbH, 2021. doi: https://doi.org/10.1515/9783110713985-004.

[5] G. Majeau-Bettez, Chapter 5: Life-Cycle Assessment and Technology Scale-Up, in Scale-up processes: iterative methods for the chemical and biochemical industries, J. Chaouki and R. Sotudeh-Gharebagh, Eds., Berlin: Walter de Gruyter GmbH, 2021. doi: https://doi.org/10.1515/9783110713985-005.

[6] D. A. Berg, Technology Evaluation, 2018. [Online]. Available: https://chemeng.queensu.ca/courses/CHEE400/TechEval.pdf. [Accessed August 2020].

[7] G. S. Patience, Chapter 6: Case Study I: n-Butane Partial Oxidation to Maleic Anhydride: VPP Manufacture, in Scale-up processes: iterative methods for the chemical and biochemical industries, J. Chaouki and R. Sotudeh-Gharebagh, Eds., Berlin: Walter de Gruyter GmbH, 2021. doi: https://doi.org/10.1515/9783110713985-006.

[8] G. S. Patience and J. Chaouki, Chapter 7: Case Study II: n-Butane Partial Oxidation to Maleic Anhydride: Commercial Design, in Scale-up processes: iterative methods for the chemical and biochemical industries, J. Chaouki and R. Sotudeh-Gharebagh, Eds., Berlin: Walter de Gruyter GmbH, 2021. doi: https://doi.org/10.1515/9783110713985-007.

[9] J. Shabanian, G. S. Patience and J. Chaouki, Chapter 8: Case Study III – Methanol to Olefins, Jaber, Gregory S. Patience, Jamal Chaouki, in Scale-up processes: iterative methods for the chemical and biochemical industries, J. Chaouki and R. Sotudeh-Gharebagh, Eds., Berlin: Walter de Gruyter GmbH, 2021. doi: https://doi.org/10.1515/9783110713985-008.

[10] M. Latifi, M. Khodabandehloo and J. Chaouki, Chapter 9: Case Study IV – Hydropotash from Potassium Feldspar, in Scale-up processes: iterative methods for the chemical and biochemical industries, J. Chaouki and R. Sotudeh-Gharebagh, Eds., Berlin: Walter de Gruyter GmbH, 2021. doi: https://doi.org/10.1515/9783110713985-009.

[11] M. Yazdanpanah and J. Chaouki, Chapter 10: Case Study V – Lactide Production Process Development, in Scale-up processes: iterative methods for the chemical and biochemical industries, J. Chaouki and R. Sotudeh-Gharebagh, Eds., Berlin: Walter de Gruyter GmbH, 2021. doi: https://doi.org/10.1515/9783110713985-010.

[12] N. El Bahraoui, S. Chidami, R. Rihani, G. Acien and J. Chaouki, Chapter 11: Case Study VI – Microalgae-Based Technologies, in Scale-up processes: iterative methods for the chemical and biochemical industries, J. Chaouki and R. Sotudeh-Gharebagh, Eds., Berlin: Walter de Gruyter GmbH, 2021. doi: https://doi.org/10.1515/9783110713985-011.

[13] A. Lyngfelt, A. Brink, Ø. Langørgen, T. Mattisson, M. Rydén and C. Linderholm, 11,000 h of chemical-looping combustion operation – Where are we and where do we want to go?, International Journal of Greenhouse Gas Control, 88, 38–56, 2019.

[14] P. Kolbitsch, T. Pröll, J. Bolhar-Nordenkampf and H. Hofbauer, Design of a chemical looping combustor using a dual circulating fluidized bed (DCFB) reactor system, Chemical Engineering & Technology, 32(3), 398–403, 2009.

[15] J. A. A. Adánez, Chemical-looping combustion: status and research needs, Proceedings of the Combustion Institute, 37, 4303–4317, 2019.

[16] A. Lyngfelt and H. Thunman, Chapter 36 construction and 100 h of operational experience of a 10-kW chemical looping combustor, Elsevier Science, 2005. [Online]. Available: http://www.entek.chalmers.se/~anly/co2/prototype%20paper%20chapter%2036.PDF. [Accessed on March 2021].

[17] N. Berguerand and A. Lyngfelt, Design and operation of a 10kWth chemical-looping combustor for solid fuels – Testing with South African coal, Fuel, 87(12), 2713–2726, 2008.

[18] N. Berguerand and A. Lyngfelt, The use of petroleum coke as fuel in a 10kWth chemical-looping combustor, International Journal of Greenhouse Gas Control, 2(2), 169–179, 2008.

[19] L. F. De Diego, F. Garcı´a-labiano, P. Gayán, J. Celaya, J. M. Palacios and J. Adánez, Operation of a 10kWth chemical-looping combustor during 200h with a CuO–Al2O3 oxygen carrier, Fuel, 86(7–8), 1036–1045, 2007.

[20] L. Shen, J. Wu, J. Xiao, Q. Song and R. Xiao, Chemical-looping combustion of biomass in a 10 kwth reactor with iron oxide as an oxygen carrier, Energy & Fuels, 23(5), 2498–2505, 2009.

[21] S. Bayham, D. Straub and J. Weber, Operation of a 50-kWth chemical looping combustion test facility under autothermal conditions, International Journal of Greenhouse Gas Control, 87, 211–220, 2019.

[22] M. Aghabararnejad, G. S. Patience and J. Chaouki, Techno-economic comparison of a 7-mwthbiomass chemical looping gasification unit with conventional systems, Chemical Engineering & Technology, 38(5), 867–878, 2015.

[23] A. Bischi, Ø. Langørgen, I. Saanum, J. Bakken, M. Seljeskog, M. Bysveen, J.-X. Morin and O. Bolland, Design study of a 150 kWth double loop circulating fluidized bed reactor system for chemical looping combustion with focus on industrial applicability and pressurization, International Journal of Greenhouse Gas Control, 5(3), 467–474, 2011.

[24] M. Lucio and L. A. Ricardez-Sandoval, Dynamic modelling and optimal control strategies for chemical-looping combustion in an industrial-scale packed bed reactor, Fuel, 262(116544), 1–19, 15 February 2020.

[25] U. Menon, M. Rahman and S. J. Khatib, A critical literature Review of the advances in methane dehydroaromatization over multifunctional metal-promoted zeolite catalysts, Applied Catalysis A: General, 608, 117870, 2020.

[26] N. Kosinov and E. J. M. Hensen, Reactivity, selectivity, and stability of zeolite-based catalysts for methane dehydroaromatization, Advanced Materials, 32(44), 2002565, 2020.

[27] Z.-G. Zhang, Process, reactor and catalyst design: Towards application of direct conversion of methane to aromatics under nonoxidative conditions, Carbon Resources Conversion, 2, 157–174, 2019.

[28] J. Xue, Y. Chen, Y. Wei, Feldhoff, A. and H. C. J. Wang, Gas to liquids: natural gas conversion to aromatic fuels and chemicals in a hydrogen-permeable ceramic hollow fiber membrane reactor, ACS Catalysis, 6(4), 2448–2451, 2016.

[29] M. Latifi, F. Berruti and C. Briens, A novel fluidized and induction heated microreactor for catalyst testing, AIChE Journal, 60(9), 3107–3122, 2014.

[30] M. Latifi, F. Berruti and C. Briens, Non-catalytic and catalytic steam reforming of a bio-oil model compound in a novel "Jiggle Bed" Reactor, Fuel, 129, 278–291, 2014.

[31] M. Latifi and J. Chaouki, A novel induction heating fluidized bed reactor: its design and applications in high temperature screening tests with solid feedstocks and prediction of defluidization state, AIChE Journal, 61(5), 1507–1523, 2015.

[32] J. Chaouki, S. Farag, M. Attia and J. Doucet, The development of industrial (thermal) processes in the context of sustainability: The case for microwave heating, The Canadian Journal of Chemical Engineering, 98(4), 832–847, 2020.

[33] M. Gupta and E. Leong, Microwaves and metals, John Wiley & Sons, 2007.

[34] A. Aguilar-Reynosa, A. Romaní, R. Ma. Rodríguez-Jasso, C. N. Aguilar, G. Garrote and H. A. Ruiz, Microwave heating processing as alternative of pretreatment in second-generation biorefinery: an overview, Energy Conversion and Management, 136, 50–65, 2017.

[35] A. Amini, M. Latifi and J. Chaouki, Electrification of materials processing via microwave irradiation: a review of mechanism and applications, Renewable and Sustainable Energy Reviews Journal, p. Submitted, 2020.

[36] S. Samih, M. Latifi, S. Farag, P. Leclerc and J. Chaouki, From complex feedstocks to new processes: the role of the newly developed micro-reactors, Chemical Engineering and Processing -Process Intensification, 131, 92–105, 2018.

[37] K. Kashimura, M. Sato, M. Hotta, D. Kumar Agrawal, K. Nagata, M. Hayashi, T. Mitani and N. Shinohara, Iron production from Fe3O4 and graphite by applying 915MHz microwaves, Materials Science and Engineering: A, 556, 977–979, 2012.

Index

https://doi.org/10.1515/9783110713985-013

www.ingramcontent.com/pod-product-compliance
Lightning Source LLC
Chambersburg PA
CBHW080911220326
41598CB00034B/5542

9783110713930